Analyzing Data Through Probabilistic Modeling in Statistics

Dariusz Jacek Jakóbczak
Koszalin University of Technology, Poland

IGI Global
PUBLISHER of TIMELY KNOWLEDGE

A volume in the Advances in Data Mining and Database Management (ADMDM) Book Series

Published in the United States of America by
IGI Global
Engineering Science Reference (an imprint of IGI Global)
701 E. Chocolate Avenue
Hershey PA, USA 17033
Tel: 717-533-8845
Fax: 717-533-8661
E-mail: cust@igi-global.com
Web site: http://www.igi-global.com

Library of Congress Cataloging-in-Publication Data

Names: Jakóbczak, Dariusz Jacek, 1965- editor.
Title: Analyzing data through probabilistic modeling in statistics / Dariusz Jacek Jakóbczak, editor.
Description: Hershey, PA : Engineering Science Reference, an imprint of IGI Global, [2021] | Includes bibliographical references and index. | Summary: "This book addresses different aspects of probabilistic modeling, stochastic methods, probabilistic distributions, data analysis, optimization methods, and probabilistic methods in risk analysis"-- Provided by publisher.
Identifiers: LCCN 2020006877 (print) | LCCN 2020006878 (ebook) | ISBN 9781799847069 (hardcover) | ISBN 9781799854937 (paperback) | ISBN 9781799847076 (ebook)
Subjects: LCSH: Social sciences--Statistical methods. | Probabilities.
Classification: LCC HA29 .A5826 2021 (print) | LCC HA29 (ebook) | DDC 001.4/22--dc23
LC record available at https://lccn.loc.gov/2020006877
LC ebook record available at https://lccn.loc.gov/2020006878

This book is published in the IGI Global book series Advances in Data Mining and Database Management (ADMDM) (ISSN: 2327-1981; eISSN: 2327-199X)

British Cataloguing in Publication Data
A Cataloguing in Publication record for this book is available from the British Library.

For electronic access to this publication, please contact: eresources@igi-global.com.

Advances in Data Mining and Database Management (ADMDM) Book Series

David Taniar
Monash University, Australia

ISSN:2327-1981
EISSN:2327-199X

Mission

With the large amounts of information available to organizations in today's digital world, there is a need for continual research surrounding emerging methods and tools for collecting, analyzing, and storing data.

The **Advances in Data Mining & Database Management (ADMDM)** series aims to bring together research in information retrieval, data analysis, data warehousing, and related areas in order to become an ideal resource for those working and studying in these fields. IT professionals, software engineers, academicians and upper-level students will find titles within the ADMDM book series particularly useful for staying up-to-date on emerging research, theories, and applications in the fields of data mining and database management.

Coverage

- Decision Support Systems
- Neural Networks
- Profiling Practices
- Enterprise Systems
- Educational Data Mining
- Association Rule Learning
- Data Quality
- Predictive Analysis
- Database Testing
- Quantitative Structure–Activity Relationship

IGI Global is currently accepting manuscripts for publication within this series. To submit a proposal for a volume in this series, please contact our Acquisition Editors at Acquisitions@igi-global.com or visit: http://www.igi-global.com/publish/.

The Advances in Data Mining and Database Management (ADMDM) Book Series (ISSN 2327-1981) is published by IGI Global, 701 E. Chocolate Avenue, Hershey, PA 17033-1240, USA, www.igi-global.com. This series is composed of titles available for purchase individually; each title is edited to be contextually exclusive from any other title within the series. For pricing and ordering information please visit http://www.igi-global.com/book-series/advances-data-mining-database-management/37146. Postmaster: Send all address changes to above address.

Titles in this Series

For a list of additional titles in this series, please visit: www.igi-global.com/book-series
http://www.igi-global.com/book-series/advances-data-mining-database-management/37146

Challenges and Applications of Data Analytics in Social Prspectives
V. Sathiyamoorthi (Sona College of Technology, India) and Atilla Elci (Hasan Kalyoncu University, Turkey)
Engineering Science Reference • ©2021 • 324pp • H/C (ISBN: 9781799825661) • US $245.00

Multidisciplinary Functions of Blockchain Technology in AI and IoT Applications
Niaz Chowdhury (The Open University, Milton Keynes, UK) and Ganesh Chandra Deka (Ministry of Skill Development and Entrepreneurship, New Delhi, India)
Engineering Science Reference • ©2021 • 255pp • H/C (ISBN: 9781799858768) • US $245.00

Handbook of Research on Engineering, Business, and Healthcare Applications of Data Science and Analytics
Bhushan Patil (Independent Researcher, India) and Manisha Vohra (Independent Researcher, India)
Engineering Science Reference • ©2021 • 583pp • H/C (ISBN: 9781799830535) • US $345.00

Advanced Deep Learning Applications in Big Data Analytics
Hadj Ahmed Bouarara (Tahar Moulay University of Saida, Algeria)
Engineering Science Reference • ©2021 • 351pp • H/C (ISBN: 9781799827917) • US $245.00

Opportunities and Challenges for Blockchain Technology in Autonomous Vehicles
Amit Kumar Tyagi (Research Division of Advanced Data Science, Vellore Institute of Technolgy, Chennai, India)
Gillala Rekha (K. L. University, India) and N. Sreenath (Pondicherry Engineering College, India)
Engineering Science Reference • ©2021 • 316pp • H/C (ISBN: 9781799832959) • US $245.00

Cross-Industry Use of Blockchain Technology and Opportunities for the Future
Idongesit Williams (Aalborg University, Denmark)
Engineering Science Reference • ©2020 • 228pp • H/C (ISBN: 9781799836322) • US $225.00

Applications and Developments in Semantic Process Mining
Kingsley Okoye (University of East London, UK)
Engineering Science Reference • ©2020 • 248pp • H/C (ISBN: 9781799826682) • US $195.00

Handling Priority Inversion in Time-Constrained Distributed Databases
Udai Shanker (Madan Mohan Malaviya University of Technology, India) and Sarvesh Pandey (Madan Mohan Malaviya University of Technology, India)
Engineering Science Reference • ©2020 • 338pp • H/C (ISBN: 9781799824916) • US $225.00

IGI Global
PUBLISHER of TIMELY KNOWLEDGE

701 East Chocolate Avenue, Hershey, PA 17033, USA
Tel: 717-533-8845 x100 • Fax: 717-533-8661
E-Mail: cust@igi-global.com • www.igi-global.com

Editorial Advisory Board

Table of Contents

Detailed Table of Contents

The present study was conducted to determine the reasons affecting poverty in Turkey and specify the level of significance of these reasons to explain poverty. The data which was analyzed in the study were retrieved from the Household Budget Research which was published by the Turkish Statistical Institute. In the study, logit models have been established by taking into account the demographic and socioeconomic indicators and the characteristics of the household. For each of these models, ROC curves were drawn, and the best model was found with the help of the areas under the curve. The results showed that the model which was developed based on the variables related to housing and the model according to consumption expenditure were determined to be more significant to explain poverty. Also, the rental value of the house, the floor type of the house and household's head educational level were found to be the most significant determinants of poverty according to the analysis of the results.

Object recognition is one of the topics of artificial intelligence, computer vision, image processing, and machine vision. The classical problem in these areas of computer science is that of determining object via characteristic features. An important feature of the object is its contour. Accurate reconstruction of contour points leads to possibility to compare the unknown object with models of specified objects. The key information about the object is the set of contour points which are treated as interpolation nodes. Classical interpolations (Lagrange or Newton polynomials) are useless for precise reconstruction of the contour. The chapter is dealing with proposed method of contour reconstruction via curves interpolation. First stage consists in computing the contour points of the object to be recognized. Then one can compare models of known objects, given by the sets of contour points, with coordinates of interpolated points of

In this chapter, the authors take a closer look into the economic relation with cybercrime and an analytics method to combat that. At first, they examine whether the increase in the unemployment rate among youths is the prime cause of the growth of cybercrime or not. They proposed a model with the help of the Phillips curve and Okun's law to get hold of the assumptions. A brief discussion of the impact of cybercrime in economic growth is also presented in this paper. Crime pattern detection and the impact of bitcoin in the current digital currency market have also been discussed. They have proposed an analytic method to combat the crime using the concept of game theory. They have tested the vulnerability of the cloud datacenter using game theory where two players will play the game in non-cooperative strategy in the Nash equilibrium state. Through the rational state decisions of the players and implementation MSWA algorithm, they have simulated the results through which they can check the dysfunctionality probabilities of the datacenters.

In this chapter, the authors take a walkthrough in BCI technology. At first, they took a closer look into the kind of waves that are being generated by our brain (i.e., the EEG and ECoG waves). In the next section, they have discussed about patients affected by CLIS and ALS-CLIS and how they can be treated or be benefitted using BCI technology. Visually evoked potential-based BCI technology has also been thoroughly discussed in this chapter. The application of machine learning and deep learning in this field are also being discussed with the need for feature engineering in this paradigm also been said. In the final section, they have done a thorough literature survey on various research-related to this field with proposed methodology and results.

The most talked about disease of our era, cancer, has taken many lives, and most of them are due to late prognosis. Statistical data shows around 10 million people lose their lives per year due to cancer globally. With every passing year, the malignant cancer cells are evolving at a rapid pace. The cancer cells are mutating with time, and it's becoming much more dangerous than before. In the chapter, the authors propose a DCGAN-based neural net architecture that will generate synthetic blood cancer cell images from fed data. The images, which will be generated, don't exist but can be formed in the near future due to constant mutation of the virus. Afterwards, the synthetic image is passes through a CNN net architecture which will predict the output class of the synthetic image. The novelty in this chapter is that it will generate some cancer cell images that can be generated after mutation, and it will predict the class of the image, whether it's malignant or benign through the proposed CNN architecture.

The huge amount of data burst which occurred with the arrival of economic access to the internet led to the rise of market of cloud computing which stores this data. And obtaining results from these data led to the growth of the "big data" industry which analyses this humongous amount of data and retrieve conclusion using various algorithms. Hadoop as a big data platform certainly uses map-reduce framework to give an analysis report of big data. The term "big data" can be defined as modern technique to store, capture, and manage data which are in the scale of petabytes or larger sized dataset with high-velocity and various structures. To address this massive growth of data or big data requires a huge computing space to ensure fruitful results through processing of data, and cloud computing is that technology that can perform huge-scale and computation which are very complex in nature. Cloud analytics does enable organizations to perform better business intelligence, data warehouse operation, and online analytical processing (OLAP).

This research investigates the comparison of inventory management planning in Verhult's demand and exponentially increasing demand. The working process is different in both the cases coupling the parameters and points out the constraints for the optimal total cost in both the cases. This analysis shows that rate of deterioration and percentage of reworkable items is considered as decision variable in both (1) exponentially increasing demand and (2) Verhult's demand. While comparing, the total cost in Verhult's demand pattern is more profitable production process. A substantial numerical example is considered to investigate the effect of change in the total cost in both the demand function. A sensitivity analysis is developed to study the effect of changes in total cost.

The white noise process, the Ornstein-Uhlenbeck process, and coloured noise process are salient noise processes to model the effect of random perturbations. In this chapter, the statistical properties, the master's equations for the Brownian noise process, coloured noise process, and the OU process are summarized. The results associated with the white noise process would be derived as the special cases of the Brownian and the OU noise processes. This chapter also formalizes stochastic differential rules for the Brownian motion and the OU process-driven vector stochastic differential systems in detail. Moreover, the master equations, especially for the coloured noise-driven stochastic differential system as well as the OU noise process-driven, are recast in the operator form involving the drift and modified diffusion operators involving an additional correction term to the standard diffusion operator. The results summarized in this chapter will be useful for modelling a random walk in stochastic systems.

The universality of the Kalman filtering can be found in the control theory. The Kalman filter has found its applications in sophisticated autonomous systems and smart products, which are attributed to its realization in a single complex chip. In this chapter, considering the Kalman filter from the perspective of conditional characteristic function evolution and Itô calculus, three Kalman filtering theorems and their formal proof are developed. Most notably, this chapter reveals the following: (1) Kalman filtering equations are a consequence of the 'evolution of conditional characteristic function' for the linear stochastic differential system coupled with the linear discrete measurement system. (2) The Kalman filtering is a consequence of the 'stochastic evolution of conditional characteristic function' for the linear stochastic differential system coupled with the linear continuous measurement system. (3) The structure of the Kalman filter remains invariant under two popular stochastic interpretations, the Itô vs Stratonovich.

The capability to elaborate a reliable estimate at completion for a project since the early stage of project execution is the prerequisite in order to provide an effective project control. The non-repetitive and uncertain nature of projects and the involvement of multiple stakeholders increase project complexity and raise the need to exploit all the available knowledge sources in order to improve the forecasting process. Therefore, drawing on a set of case studies, this chapter proposes a Bayesian approach to support the elaboration of the estimate at completion in those industrial fields where projects are denoted by a high level of uncertainty and complexity. The Bayesian approach allows the authors to integrate experts' opinions, data records related to past projects, and data related to the current performance of the ongoing project. Data from past projects are selected through a similarity analysis. The proposed approach shows a higher accuracy in comparison with the traditional formulas typical of the earned value management (EVM) methodology.

Preface

The Advances in Logistics, Operations, and Management Science (ALOMS) Book Series provides a collection of reference publications on the current trends, applications, theories, and practices in the management science field. Providing relevant and current research, this series and its individual publications would be useful for academics, researchers, scholars, and practitioners interested in improving decision making models and business functions.

Probabilistic modeling represents a subject arising in many branches of mathematics, economics and computer science. Such modeling connects pure mathematics with applied sciences. Statistics similarly is situated on the border between pure mathematics and applied sciences. So when probabilistic modeling meets statistics, it is very interesting occasion. Our life and work are impossible without planning, time-tabling, scheduling, decision making, optimization, simulation, data analysis, risk analysis and process modeling. Thus, it is a part of management science or decision science.

This book looks to discuss and address the difficulties and challenges that occur during the process of planning or decision making. The editors have found the chapters that address different aspects of probabilistic modeling, stochastic methods, probabilistic distributions, data analysis, optimization methods, probabilistic methods in risk analysis, and related topics. Additionally, the book explores the impact of such probabilistic modeling with other approaches.

This comprehensive and timely publication aims to be an essential reference source, building on the available literature in the field of statistics, probabilistic modeling, operational research, planning and scheduling, data extrapolation in decision making, probabilistic interpolation and extrapolation in simulation, stochastic processes, and decision analysis. It is hoped that this text will provide the resources necessary for economics and management sciences, also for mathematics and computer sciences.

Decision makers, academicians, researchers, advanced-level students, technology developers, and government officials will find this text useful in furthering their research exposure to pertinent topics in operations research and assisting in furthering their own research efforts in this field.

Book topics include the following:

- Probabilistic Modeling
- Statistics
- Operations research
- Stochastic Methods
- Probabilistic Methods in Planning
- Decision Making
- Data Analysis

- Optimization Methods
- Probabilistic Methods in Risk Analysis
- Probabilistic Interpolation and Extrapolation
- Process Modeling
- Data Simulation
- Decision Analysis
- Stochastic Processes
- Probabilistic Optimization
- Data Mining
- Mathematical Modeling
- Probabilistic Models in Scheduling
- Time-Tabling
- Data Extrapolation in Planning and Decision Making

The book is divided into three sections and 12 chapters:

Section 1: Probabilistic Modeling in Statistics;

Section 2: Dual Approach of Data Analytics and Machine Learning Modelling in Real Case Scenarios;

Section 3: Case Studies From Business and Industry.

Section 1 consists of three chapters about mathematical modeling in analyzing risk. First chapter is presented by Z. Çiçek and H. Demirgil: "Determination of Poverty Indicators Using Roc Curves in Turkey." The present study was conducted to determine the reasons affecting poverty in Turkey and specify the level of significance of these reasons to explain poverty. The data which was analyzed in the study were retrieved from the Household Budget Research which was published by the Turkish Statistical Institute. In the study, Logit models have been established by taking into account the demographic and socioeconomic indicators and the characteristics of the household. For each of these models, ROC curves were drawn and the best model was found with the help of the areas under the curve. The results showed that the model which was developed based on the variables related to housing and the model according to consumption expenditure were determined to be more significant to explain poverty. Also, the rental value of the house, the floor type of the house and household's head educational level were found to be the most significant determinants of poverty according to the analysis of the results.

Second chapter is presented by D. J. Jakóbczak: "Data Analyzing via Probabilistic Modeling: Interpolation and Extrapolation." Object recognition is one of the topics of artificial intelligence, computer vision, image processing and machine vision. The classical problem in these areas of computer science is that of determining object via characteristic features. Important feature of the object is its contour. Accurate reconstruction of contour points leads to possibility to compare the unknown object with models of specified objects. The key information about the object is the set of contour points which are treated as interpolation nodes. Classical interpolations (Lagrange or Newton polynomials) are useless for precise reconstruction of the contour. The chapter is dealing with proposed method of contour reconstruction via curves interpolation. First stage consists in computing the contour points of the object to be recognized. Then one can compare models of known objects, given by the sets of contour points, with coordinates of interpolated points of unknown object. Contour points reconstruction and curve interpolation is possible using new method of Hurwitz - Radon Matrices.

The third chapter by D. J. Jakóbczak consists of "Decision Making and Data Analysis: Curve Modeling via Probabilistic Method." Proposed method, called Probabilistic Nodes Combination (PNC), is the method of 2D curve modeling and handwriting identification by using the set of key points. Nodes are treated as characteristic points of signature or handwriting for modeling and writer recognition. Identification of handwritten letters or symbols need modeling and the model of each individual symbol or character is built by a choice of probability distribution function and nodes combination. PNC modeling via nodes combination and parameter γ as probability distribution function enables curve parameterization and interpolation for each specific letter or symbol. Two-dimensional curve is modeled and interpolated via nodes combination and different functions as continuous probability distribution functions: polynomial, sine, cosine, tangent, cotangent, logarithm, exponent, arc sin, arc cos, arc tan, arc cot or power function.

Section 2 consists of five chapters about "Dual Approach of Data Analytics and Machine Learning Modelling in Real Case Scenarios." The first chapter of this section has been authored by A. Chatterjee, S.Roy, and T. Sinha titled as "Patient Arrival to Public OPDs: Analysis and Use of Statistical Distribution for Improving Performance Indicators in Rural Hospitals." In this chapter, the random effect model has been assessed on the data to choose the respective hypothesis. Random Variable Effect model has been opted for as the hypothesis as Fixed Effect model has been ruled out by Haussmann Test. A Bayesian Risk Analysis has also been carried out on the mortality rate of the patients on which factors it depends. An analysis on the risk factors and key indicators on which the survival rates of the patient depends have been done through Bayesian Hierchichal Modeling, from which it is observed that blood facility in hospitals with active number of presence of doctors play a key role in this. Then a Poisson distribution model is fitted to get some insights on the data. The data has been considered as cross-sectional panel data collected from government reports and open source archives. Next, the number of patients that will arrive in 2021 has been predicted along with the shortfall of hospitals has been projected and remedies have been suggested for the same. Finally, econometric analyses in the healthcare domain have been carried out in reference to improve the performance indicators.

The second chapter has been authored by S. Roy, T. Sinha, A. Chatterjee entitled as "An Econometric Overview on Growth and Impact of Online Crimeand Analytics View to Combat Them." In this chapter, analytics method has been used to combat the economic relation with cyber crime. A model has been proposed with the help of the Phillips Curve and Okun's law to get hold of the assumptions made. Crime pattern detection and the impact of Bitcoin in the current digital currency market have been discussed broadly in this chapter. The vulnerability of the cloud datacenter using the concepts of game theory has been tested. The basic idea used behind the methodology is that two players will play the game in non-cooperative strategy in the Nash equilibrium state. Through the rational state decisions of the players and implementation of MSWA algorithm, the results through which we can check the dysfunctionality probabilities of the datacenters have been successfully simulated. Through experiment it is concluded that economic condition of a country plays a pivotal role in the increment of cybercrime across the country. In addition to that, poor unemployment rate gives rise to increased crime rate and which, in fact, is increasing hand in hand with inflation.

The third chapter is presented by A. Chatterjee, A. Mandal, S. Roy, S. Sinha, A. Priya, and Y. Gupta, entitled as "A Decadal Walk on BCI Technology: A Walkthrough." This chapter enlightens us the major pillars of BCI technology. It gives a closer look into the kind of waves like EEG and ECoG waves which are generated by our brain. Also application of BCI in the fields of treatment of CLIS Patient, VEP testing as well as application of machine learning and deep learning in this field has been established here. The authors also discussed few research works which have been done on this field and how BCI can be helpful in our daily life that has been illustrated here.

The fourth chapter is presented by A. Chatterjee, S. Roy entitled as "A Fusion Based Approach to Generate and Classify Synthetic Cancer Cell Image Using DCGAN and CNN Architecture." In this paper, the authors want to focus to the fact that every 6th death is due to cancer and the malignant cancer cells are evolving at a rapid pace and cells are becoming much more dangerous after mutating with time and so it is very important to detect the cancer cells. The author proposed a DCGAN based neural net architecture which will generate synthetic blood cancer cell images from the data and predict the output class of the synthetic image. From the DCGAN image a rough estimation can be done regarding the cells produced in near future due to constant mutation. With this architecture a high accuracy of 92.32% is achieved on the validation set, and through which high probability of getting correct class in the output is established even if the synthetic image is passed. Thus the author gives an advanced model through which detection of category of future cells can be done.

The fifth and final chapter of this section is presented by D. J. Jakóbczak and A. Chatterjee entitled as "The Rise of 'Big Data' in the Field of Cloud Analytics." This chapter mainly focuses on the liaison between the convergence of Analytics field and Cloud computing field. This paper gives an insights how analytics is influencing the cloud platform with the Map Reduce Framework coming into the game along with Hadoop platform where the big data platform is being framed. Addressing the massive growth of data or big data requires a huge computing space to ensure fruitful results through processing of data, and cloud computing is that technology which can perform huge-scale and computation which are very complex in nature. Cloud Analytics does enables organizations to perform better business intelligence, data warehouse operation and Online Analytical Processing (OLAP). This paper includes characteristics, classification of big data applications with implementation through cloud computing. Moreover we will have a look into how to apply big data analytics to create accurate measures.

Section 3 consists of five chapters about case studies of economy, business and industry. Chapter 9, by Kuppulakshmi V., Sugapriya C., Kathirvel Jeganathan and Nagarajan Deivanayagampillai, is called "Analyzing EPQ Inventory Model With Comparison of Exponentially Increasing Demand and Verhult's Demand." This research investigates the comparison of inventory management planning in Verhult's demand and exponentially increasing demand. The working process is different in both the cases coupling the parameters and point out the constraints for the optimal total cost in both the cases. This analysis shows that rate of deterioration and percentage of reworkable items is considered as decision variable in both (i) Exponentially increasing demand and (ii) Verhult's demand. While comparing, the total cost in Verhult's demand pattern is more profitable production process. A substantial numerical example is considered to investigate the effect of change in the total cost in both the demand function. A sensitivity analysis is developed to study the effect of changes in total cost.

The authors of next chapter: Shaival Hemant Nagarsheth and Shambhu Nath Sharma are dealing with "Statistics of an Appealing Class of Random Processes." The white noise process, the Ornstein-Uhlenbeck process, and coloured noise process are salient noise processes to model the effect of random perturbations. In this chapter, the statistical properties, the master's equations for the Brownian noise process, coloured noise process, and the OU process are summarized. The results associated with the white noise process would be derived as the special cases of the Brownian and the OU noise processes. This chapter also formalizes stochastic differential rules for the Brownian motion and the OU process-driven vector stochastic differential systems in detail. Moreover, the master equations especially for the coloured noise-driven stochastic differential system as well as the OU noise process-driven, are recast in the operator form involving the drift and modified diffusion operators involving an additional correction term to the standard diffusion operator. The results summarized in this chapter will be useful for modelling a random walk in stochastic systems.

The eleventh chapter by Sandhya Rathore, Shambhu Nath Sharma and Shaival Hemant Nagarsheth is called "The Universality of the Kalman Filter: A Conditional Characteristic Function Perspective." The universality of the Kalman filtering can be found in the control theory. The Kalman filter has found its applications in sophisticated autonomous systems and smart products, which are attributed to its realization in a single complex chip. In this chapter, considering the Kalman filter from the perspective of conditional characteristic function evolution and Itô calculus three Kalman filtering Theorems and their formal proof are developed. Most notably, this chapter reveals the following: (i) Kalman filtering equations are a consequence of the 'evolution of conditional characteristic function' for the linear stochastic differential system coupled with the linear discrete measurement system. (ii) The Kalman filtering is a consequence of the 'stochastic evolution of conditional characteristic function' for the linear stochastic differential system coupled with the linear continuous measurement system. (iii) The structure of the Kalman filter remains invariant under two popular stochastic interpretations, the Itô vs Stratonovich.

Chapter 12 by F. Caron is known as "Project Control: A Bayesian Model." The capability to elaborate a reliable estimate at completion for a project since the early stage of project execution is the prerequisite in order to provide an effective project control. The non-repetitive and uncertain nature of projects and the involvement of multiple stakeholders increase project complexity and raise the need to exploit all the available knowledge sources in order to improve the forecasting process. Therefore, drawing on a set of case studies, this paper proposes a Bayesian approach to support the elaboration of the estimate at completion in those industrial fields where projects are denoted by a high level of uncertainty and complexity. The Bayesian approach allows to integrate experts' opinions, data records related to past projects and data related to the current performance of the ongoing project. Data from past projects are selected through a similarity analysis. The proposed approach shows a higher accuracy in comparison with the traditional formulas typical of the Earned Value Management (EVM) methodology.

The editor, the publisher and the authors hope that this book *Analyzing Data Through Probabilistic Modeling in Statistics* will be a heavy brick in the construction of the House of Science. Please read it!

This book is published by IGI Global (formerly Idea Group Inc.), publisher of the Information Science Reference (formerly Idea Group Reference), Medical Information Science Reference, Business Science Reference, and Engineering Science Reference imprints. For additional information regarding the publisher, please visit www.igi-global.com.

October 2020

Acknowledgment

The Editor wish to acknowledge the contributions of all Authors who elaborated and presented in their chapters valuable scientific and research results. I am also indebted to all reviewers for their competence, professionalism and diligence demonstrated during the review process. They have made a significant contribution to the final version of this book. Thanks are also due to all members of IGI Global team involved in the preparation of this book for their consistent support.

Dariusz Jacek Jakóbczak
Koszalin University of Technology, Poland

Section 1
Probabilistic Modeling in Statistics

Chapter 1
Determination of Poverty Indicators Using Roc Curves in Turkey

Zübeyde Çiçek
https://orcid.org/0000-0003-1914-1228
Süleyman Demirel University, Turkey

Hakan Demirgil
Suleyman Demirel University, Turkey

ABSTRACT

The present study was conducted to determine the reasons affecting poverty in Turkey and specify the level of significance of these reasons to explain poverty. The data which was analyzed in the study were retrieved from the Household Budget Research which was published by the Turkish Statistical Institute. In the study, logit models have been established by taking into account the demographic and socioeconomic indicators and the characteristics of the household. For each of these models, ROC curves were drawn, and the best model was found with the help of the areas under the curve. The results showed that the model which was developed based on the variables related to housing and the model according to consumption expenditure were determined to be more significant to explain poverty. Also, the rental value of the house, the floor type of the house and household's head educational level were found to be the most significant determinants of poverty according to the analysis of the results.

INTRODUCTION

Poverty which has been observed in almost all phases of the history of the mankind is a difficult concept to determine and evaluate as it is a multi-faceted concept. Although there is not a basic definition for poverty, it is possible to generalize it as individuals' not having the adequate amount of income to meet their basic needs. Poverty is defined as economic welfare's level being lower than under a minimum level of income which has been determined based on one individual or more in absolute terms or based

DOI: 10.4018/978-1-7998-4706-9.ch001

on a particular society's standards. Poverty has been generally defined in a narrow way in developing countries with the aim of referring to an individual's good and service consumption. The appropriate minimum level was defined in line with the prespecified "basic consumption needs" and especially nutrition (Lipton & Ravallion, 1995).

Working through poverty reduction in the society becomes more of an issue as poverty has remained for a long time. The actions to be taken to reduce poverty in the society have importance as a matter of the fact that it has negative influence on social structure. Being able to define the significance of the rising threat with the problem of poverty and figure out the process throughout it has intensified are correlated with the way the scope of poverty is measured (Bourguignon & Chakravarty, 2003). The first challenge to measure poverty is to determine the poverty line with respect to the income or consumption level. The individuals with lower income than the specified level are described as poor. (Sen, 1976). This level is generally determined according to absolute and relative poverty approach in scientific studies. Absolute poverty is defined as the required consumption level for individuals to maintain their lives in physical terms. Consequently, minimum need of the individuals is stated considering food and non-food constituents while determining the absolute poverty (TÜİK, 2008). As for relative poverty, a specific percentage of the society' average income is settled as the poverty line. The individuals or household under this level are categorized as poor (Anand, 1983). The welfare criterion which is taken into consideration is regarded with consumption expenditure as well as income level. Income or consumption expenditures might be preferred in line with the aim it is used for. Poverty calculations based on income which are also applied by Eurostat and OECD are used since they provide advantages to make comparisons between countries in methodological terms (TÜİK, 2008).

Poverty is generally considered as a situation due to insufficient income level demographic, socio-economic and housing characteristics such as the level of education, age, number of households, location of housing, characteristics of housing are factors that influence the determination of basic poverty indicators such as income and consumption. Therefore, in this study, Household Budget Survey data published by Turkish Statistical Institute in 2010, 2011 and 2012 were used. The relative poverty line was calculated by taking into account the income or consumption expenditure levels used in the measurement of poverty. The poverty variable was determined as dependent variable by identifying the poor and non-poor. This study aims to determine the important demographic, socioeconomic and residential properties affecting poverty by using logistic regression and ROC analysis methods.

First, we establish the significance of the general field of the poverty studies and then identify a place where a new contribution could be made. Unlike previous studies, additional variables have been included in the variable that captures the effect of housing characteristics (such as heating system, floor type, electronic devices, etc.) and individual characteristics in poverty. The validity of the significant variables that affect poverty and the general validity of the logistic model was evaluated by using the area under the roc curve.

LITERATURE

Just as in the whole world, poverty is a major problem that cannot be ignored in Turkey. Therefore, a significant number of studies are conducted on poverty. Some studies about poverty in Turkey and world literature: Dumanlı (1996), calculated the poverty line on the basis of the minimum number of calories required for an individual's nourishment throughout a day for the years 1987 and 1994. The significance of poverty in Turkish context was also analyzed considering different regions and years and some suggestions were made to reduce the poverty in the study mentioned above.

Wodon (1997), compared the significance of the targeted indicators to define poverty by using ROC curve in a study which was carried out in Bangladesh. ROC analysis was first used in that study for the first time to analyze poverty although it had been used in quite a lot of studies before. Education, job and settlement are the best indicators used to define poverty across the country. While education is more important than land ownership in urban areas, land ownership is more important than education in rural areas. Baulch and Minot (2002), attempted to develop poverty reduction policies by combining the data gained from Vietnamese Life Quality Study in 1998 and population census results in 1999. They compared the significance of the variables of poverty for urban and rural areas; and the whole country. Baulch (2002), modelled the indicators of poverty by analyzing the data retrieved form Vietnamese Life Quality Questionnaire in 1997 and 1998 with Probit method and analyzed the significance of these indicators by means of ROC curve. Deaton (2003), analyzed Indian National Household data in the years 1999 and 2000 by developing different Logit models for different regions and maintained that poverty ratios varied in different regions.

Kızılgöl (2008), applied the method of least squares and ordered Logit model by using Household Budget Questionnaire data in 2002 and 2005 with the aim of specifying the indicators of household poverty in terms of consumption. The results showed that the most important determinants of poverty are the educational status of the household members, the size of the household and the place where the household resides. Epo (2010) applied a binomial and polychotomous logit regression in Cameroon to investigate the causes of poverty using the ECAM II Household Consumption Surveys. In the study, the variables of education, age of household head, and proportion of working adult household members decrease the risk of household poverty. Living in rural areas increases the risk of poverty. Canbay and Selim (2010), used Household Budget Questionnaire data in 2004 to analyze determining factors of poverty in Turkey and their further analyses focused on comparison between urban and rural areas. They used Logit model in an attempt of stating the most significant indicators of poverty. As a result of the analysis, it has been observed that the most important determinants of poverty are the situation of the household head at work, workplace activity and the size of the household. Dal (2013), stated the poverty line as $ 4.3 by using the data gained from Household Budget Questionnaire in 2010 according to purchasing power parity for 2010. In this study, Logit model was developed to specify the determinants of poverty and analyses were made by means of ROC curve. It has been observed that women, age 64 and above, seasonal workers are at higher risk of being poor. Ownership of electronic equipment, toilet ownership, and ease of access to the public transportation of the house also decrease the possibility of being poor. The number of children in the household also increases the risk of poverty.

Tatlıdil and Demirağ (2014), analyzed the data retrieved from Income and Life Quality Study which was carried out in 2009 by stating significant variables to determine the poverty level of households and applying logistic regression and MARS methods. According to the results obtained, it was determined that variables such as age, marital status, education status, employment status, number of households, housing type, property status, residential area, heating system type, electronic equipment etc. are important variables in explaining poverty. Biyase and Zwane (2015), using the probit model, determine the factors that influence poverty and household welfare in South Africa. According to the estimation results, levels of education, race, dependency ratio, gender, employment status and marital status are statistically significant determinants of household welfare. Abrar Ul Haq, Ayub and Ullah (2015), carried out a study by using the data which was gained from Household Budget Questionnaires applied in rural areas in Southern Punjab and made analyses on significance of the variables affecting the rural household poverty by developing Logit models. In the study, household size, number of people per room, female / male ratio are positively associated with household poverty. Female labor participation, market access, education, gender of household head, assets and livestock has inverse relationship with poverty. Garza -Rodriguez (2015), used 2008 National Survey of Income and Expenditures of Households survey data in Mexica. It was aimed to examine the determinants or correlations of poverty using the logistic regression model. As a result of the study, the education level and age of the household head are inversely related to poverty. Being a household size, agriculture and service worker is positively associated with poverty. The gender and household location of the household head were not statistically significant.

Oluwatayo and Babalola (2020), carried out to examine asset ownership and income as determinants of household poverty in South Africa. The logistic regression model was established using the National Income Dynamics Study (NIDS) data. The results, ownership of non-monetary assets, income and household size had a positive influence on the household poverty status.

MATERIALS AND METHODS

Being able to define poverty in different contexts has considerable importance in order to specify the causes of it and come up with solutions for it since it shows varieties up to different places and times. It is essential to define poverty according to different contexts so as to state the reasons for poverty and develop solutions for it as it changes in line with different times and places.

The concepts such as absolute and relative poverty are used in quantitative studies. However, relative poverty is more likely to be used more as it is more valid in international studies. In the current study, relative poverty lines were analyzed according to the income and consumption level based on the data retrieved from Household Budget Questionnaires which were applied in the years 2010, 2011 and 2012. The study was conducted with the sample of total 29058 households by combining individual and household data set with regard to their bulletin numbers on the basis of household head which is available in Household Budget Questionnaires in the years 2010, 2011 and 2012. 16 explanatory variables were used in the study. These are: household head's gender, household head's marital status, household's educational level, household's having health insurance or not, household type, access to the transportation facilities on daily bases, the floor type of the rooms, the number of the rooms, the length of the time span during which household dwell in the house, the amount of the rent paid per month, the number of telephones, mobile phones, computers, LCD televisions and the heating system (Table 1).

Firstly, aggregated effects by means of Logit model were analyzed by using individuals' demographic information, some household characteristics and the housing characteristics. Two different models were developed according to the binary poverty indicators which were based on equivalent members' yearly income and monthly consumption expenditures. The reliability of these models was measured according to the area under the ROC curve (AUC). The comparison of the imputed values attained by using Logit model were also analyzed by means of ROC curves by putting different variables into two different categories such as personal characteristics and housing characteristics.

Finally, the significance levels of the univariate variables for poverty were analyzed in the present study. ROC curves were drawn for specific variables and their area under ROC Curve (AUC) levels were compared in terms of their significance to explain poverty. Each and every member of the household was

Table 1. Explanatory variables

Household Head's Gender	1. Male 2. Female
Household Head's Marital Status	1. He/She never married. 2. Married. 3. His/Her wife / husband died. 4. Divorced
Household Head's Age Group	1. 15-24 Ages 2. 25-35 Ages 3. 35-44 Ages 4. 45-54 Ages 5. 55-64 Ages 6. 65+ Ages
Household Head's Educational Level	1. Illiterate 2. Literate but haven't finished any schools 3. Primary and elementary school 4. Secondary school or vocational school equivalent to secondary school 5. High school or vocational school equivalent to high school 6. Higher education, Faculty, MA and PhD
Health Insurance	1. No 2. Yes
Household Type	1. Family with no children 2. Family with one adult 3. Nuclear family with one child 4. Nuclear family with two children 5. Nuclear family with three or more children 6. Extended families 7. People sharing the same house
Easy Access to Public Transportation Facilities	1. Very difficult 2. Difficult 3. Easy 4. Very easy
Floor type of the rooms	1. Concrete 2. Wooden 3. Fitted carpet 4. Ceramic
The number of the rooms	The parts of the houses which are surrounded by walls and at least 4 square meters are included in the number of the rooms (Reception rooms are also included in the total number)
The length of the time span during which household members dwell in the house	If the family have lived in a house for less than six months, it was written "00". If the family have lived in a house for a time period between 6 months and one year, it was written as "01"
Monthly Rental Value	This was determined in Turkish currency (Turkish lira) by taking the rent the people who live in the house as tenants, lodgment residents and others every month into consideration and the monthly rent for a similar house in the same neighborhood considering the conditions in the month which the questionnaire was applied.
Telephone, Mobile phone, Computer, LCD TV	0. No If yes, the numbers: 1-9
Heating System	1. Stove 2. Air conditioning 3. Central heating 4. Room heater / Combi boiler

Source: Household Budget Questionnaire Data Set Definitions, 2010-2011-2012, Ankara, TÜİK

analyzed according to poverty which was settled as an independent variable on the basis of household income and consumption expenditure.

Determination of the Poverty Line

It has considerable importance to specify the number of the members who share the household income and consumption expenditures in terms of determining poverty level. The amount of additional expenditures depends on a member's demographic information. The amount of the contribution that each person makes to the household income and their consumption levels are assumed to be different because of the number of the household members and the varieties between household members in terms of age and gender. Equivalency scales are aimed to make an accurate comparison through making each member of the household equivalent in terms of the total numbers of household members.

The Income or Consumption Expenditure for each Equivalent Household Member according to the Updated OECD Modified Equivalence Scale;

The Income or Consumption Expenditure for each Equivalent Household Member = Household Income or Consumption Expenditure / [1+(0.5*(The number of members who are aged 14 or under -1) + 0.3*(The number of members under 14))]

is calculated by applying the formula stated above (TÜİK, 2011). The values concerning the income and consumption expenditures in 2010, 2011 and 2012 were turned into real values based on consumer price index for the current study. The income and consumption expenditures which were turned into real values were divided by the updated OECD Modified Equivalence Scale with the renewed values and the income and consumption expenditures per member were calculated. The median value of the income and consumption expenditure for each member was calculated to find the relative poverty line for each of the values in accordance with the income and consumption expenditure. The poverty line was calculated by getting 50% of the median value of the income and consumption expenditures. Each member with lower income and consumption expenditure than this value was defined as poor (1) and each member with more income than this income was stated as non-poor.

ROC Curve

Receiver Characteristic Curve (ROC) analysis was designed by mathematics professors at Michigan University in 1950s to identify the impending aircrafts. It is grounded on the basic principles of Statistical Decision Theory and Signal Detection Theory (Metz, 2008). ROC analysis is used to measure the discriminative value of the diagnostic test results and assemblies which are set up with different variables (Ahmet Dirican 2001; James A. Hanley and Barbara McNeil 1982). This analysis uses accuracy besides sensitivity and specificity that are used while analyzing diagnostic tests (Metz, 1978).

ROC curve is drawn by using true positiveness (sensitivity) and false positiveness (1-specificity) values by means of getting threshold value for different values (Figure 1) (Tomak & Bek, 2009). Vertical axis represents sensitivity and horizontal axis represents 1-specificity value.

Different sensitivity and specificity ratios are calculated for each and every threshold value throughout the process of the analysis of a diagnostic test. Observation A means that the positive values in actual fact are also positive according to the results of diagnostic tests in other words it means true positive, Observation "D" is named as true negative or it represents the values which are negative both in actual fact and the results of the diagnostic tests (Table 2) (Dirican, 2001).

Figure 1. ROC Curve

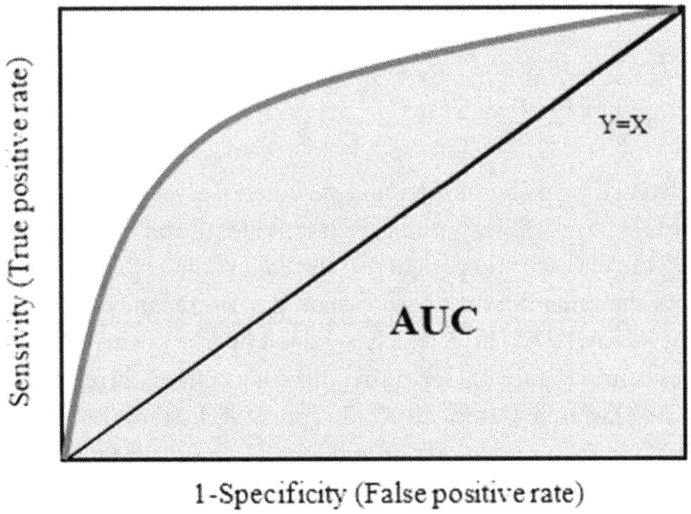

$$\text{Sensitivity(SE)} = \frac{\text{Number of True Positive Decisions}}{\text{Number of Actually Positive Cases}} = \frac{A}{A+C} \tag{1}$$

$$\text{Specificity(SP)} = \frac{\text{Number of True Negative Decisions}}{\text{Number of Actually Negative Cases}} = \frac{D}{B+D} \tag{2}$$

Sensitivity is the rate of the positive things according to the diagnostic tests into the positive things which are positive in actual fact. Specificity is the rate of the negative things according to the results of the diagnostic tests into the things which are already negative in fact. Sensitivity and specificity represent two types of accuracy for negative and positive cases. “Sensitivity” is also named as true positive rate and “1-Specificity” is also defined as false positive rate. The tests with high true positive rate and low false negative rate are preferred to be applied after the assessment of diagnostic tests (Metz, 1978; Wodon, 1997; Tomak & Bek, 2009).

Accuracy is attained by getting the rate of positive and negative decisions in fact in the number of all cases. Sensitivity is dependent upon on only positive phenomena and specificity is only based on

Table 2. Predicted cases while analyzing diagnostic tests

		Fact	
		Positive	**Negative**
Diagnostic Test Results	**Positive**	A (SE)	B (1-SP)
	Negative	C (1-SE)	D (SP)

the measurement of negative phenomena. As a matter of fact, accuracy rate is frequently used (Dirican, 2001; Lasko, Bhagwat, Zou & Ohno-Machado, 2005).

$$\text{Accuracy}=\frac{\text{Number of True Positive and Negative Decisions}}{\text{Number of All Cases}}=\frac{A+D}{A+B+C+D} \quad (3)$$

Accuracy rate is calculated by using the formula stated above. Accuracy measures the reliability of alternative diagnostic tests in terms of discrimination. It provides the facility of detecting the most reliable diagnostic test by stating the classification quality of the data gained from the test (Zweig & Campbell, 1993). ROC curve shows the reliability of a test to state the discrepancy which is analyzed for all test breakpoints in an extensive way. ROC curve which is gained by true positive rate which is drawn against the false positive rate is seen in Figure 1. Generally, there is positive correlation between true positive rate and false positive rate (Zweig & Campbell,1993; Tomak & Bek, 2009).

The diagnostic test with high true positive rate and low false positive rate is generally preferred to be used as a result of the comparison of the diagnostic tests. The most reliable diagnostic test has a curve which is the closest to the upper left corner. The reliability goes down as it gets closer to y=x curve. The curve on y=x curve is the random predictivity curve (Flach, Blockeel, Ferri, Hernández-Orallo & Stuyf, 2003). The area under the curve (AUC) is generally taken into consideration while analyzing the reliability of the ROC curve (Bradley, 1997). This shows to what extent it is reliable in terms of discriminating positive and negative concepts. This value is on the range between 0.5 and 1. As the value of the area gets closer to the extreme "1", the reliability of the ROC curve to make discrimination also gets closer to the excellent. If AUC value is 0.5, the diagnostic test is stated to have random discriminative reliability (Zweig & Campbell, 1993).

RESULTS

Poverty analysis was made by using the data from Household Budget Questionnaire in 2010, 2011 and 2012 for the present study. Logit models were developed based on the level of household's yearly income and monthly consumption expenditures. The Logit models were analyzed by means of ROC curves. The values about household and specific members were compared with AUC. Lastly, ROC curves were drawn for each and every variable which was defined in the study. Stata 12 package program was used to analyze the data used in the study.

Multivariate Models

Logit model results which were developed for poverty binary variable that was determined on the basis of the level of household's equivalent yearly income and monthly consumption expenditures are shown in Table 3 and Table 4 in sequence. The models were stated to be significant in statistical terms according to the 5% significance level ($p<0.05$).

The likelihood value of the gender was found to be 0.4009 in the model based on income. Odds ratio which was calculated according to likelihood value was attained as 1.4932 from $e^{0.4009}$. The likelihood ratio of being poor was stated to be 1.49 times higher for the households with female heads compared

Table 3. Logit model results (household's equivalent yearly income)

		Poverty according to Equivalent Members' Monthly Income	
Variables	**Categories**	**Coefficient**	**Odds**
GENDER	-	0.4009*	1.4932
AGE	25-34 Ages	-0.9633*	0.3816
	35-44 Ages	-0.7874*	0.4550
	45-54 Ages	-1.0556*	0.3480
	55-64 Ages	-1.3836*	0.2507
	65+ Ages	-1.2521*	0.2859
HEALTH INSURANCE	-	-0.5925*	0.5529
EDUCATION	Literate but not finished any schools	-0.3837*	0.6814
	Elementary and Primary school	-0.9894*	0.3718
	Secondary School or Vocational School equivalent to Secondary School	-1.2733*	0.2799
	High School or Vocational School equivalent to High School	-1.4972*	0.2237
	Higher Education, Faculty, MA and PhD	-2.6079*	0.0737
MARITAL STATUS	Married	0.3733***	1.4525
	His/her wife/husband died.	-0.5866**	0.5562
	Divorced	-0.2107	0.8100
HOUSEHOLD TYPE	Family with One Adult	0.6524*	1.9201
	Nuclear Family with One Child	0.4546*	1.5756
	Nuclear family with Two Children	1.1228*	3.0733
	Nuclear Family with Three or More Children	2.0509*	7.7747
	Extended family	1.5999*	4.9524
	People Sharing the Same House	0.7344*	2.0843
Rental Value	-	-0.0071*	0.9930
The length of the time span in which family members dwell in the house	-	-0.0065*	0.9935
The Number of the Rooms	-	-0.1395*	0.8698
Floor Type	Wooden	-0.5787*	0.5606
	Fitted Carpet	-0.5209**	0.5940
	Ceramic	-0.2215**	0.8013
Heating System	Air Conditioning	-0.5806**	0.5596
	Central Heating	-0.5011*	0.6059
	Room Heater-Combi Boller	-0.2960**	0.7438
Difficulty of Access to the Transportation Facilities	-	-0.0689**	0.9334
Telephone	-	-0.6366*	0.5291
Mobile Phone	-	-0.2182*	0.8039
Computer	-	-0.4018*	0.6691
LCD TV	-	-0.3245*	0.7229
Fixed	-	3.0213*	20.5172
	Number of obs = 29058 Pseudo R2 = 0.3657 Log likelihood = -7645.1839 Prob > chi2 = 0.0000 LR chi2(47) = 8814.18		

***, ** and * represent significance at 10% 5% and 1% levels respectively.

Table 4. Logit model results (monthly consumption expenditures)

		Poverty according to Equivalent Members' Monthly Income	
Variables	**Categories**	**Coefficient**	**Odds**
GENDER	-	0.6781*	1.9701
AGE	25-34 Ages	-0.1104	0.8955
	35-44 Ages	0.1522	1.1644
	45-54 Ages	-0.0062	0.9938
	55-64 Ages	-0.1612	0.8512
	65+ Ages	0.2046	1.2271
HEALTH INSURANCE	-	-0.3823*	0.6823
EDUCATION	Literate but not finished any Schools	-0.3096**	0.7338
	Primary and Elementary School	-0.7698*	0.4631
	Secondary School or Vocational School equivalent to Secondary School	-1.0546*	0.3483
	High School or Vocational School equivalent to High School	-1.2132*	0.2972
	Higher Education, Faculty, MA and PhD	-1.9731*	0.1390
MARITAL STATUS	Married	0.1627	1.1767
	His/Her wife/husband died.	-0.9610*	0.3825
	Divorced	-0.2423	0.7849
HOUSEHOLD TYPE	Family with one adult	0.5981*	1.8187
	Nuclear family with one child	0.4193*	1.5208
	Nuclear family with two children	1.0418*	2.8342
	Nuclear family with three or more children	1.8311*	6.2409
	Extended family	1.7048*	5.5001
	People sharing the same house	0.5274**	1.6945
Rental Value	-	-0.0087*	0.9914
The Length of Time Span during which Household Members Dwell in the House	-	-0.0031***	0.9969
The Number of Rooms	-	-0.0900*	0.9139
Floor Type	Wooden	-0.4566*	0.6335
	Fitted Carpet	-0.3987**	0.6712
	Ceramic	-0.2160**	0.8057
Heating System	Air Conditioning	-0.5306***	0.5883
	Central Heating	-0.4477**	0.6391
	Room Heater-Combi Boiler	-0.4170**	0.6590
Difficulty of Access to the Transportation Facilities	-	-0.0434	0.9576
Telephone	-	-0.7790*	0.4589
Mobile Phone	-	-0.3182*	0.7275
Computer	-	-0.5433*	0.5808
LCD TV	-	-0.3563*	0.7003
Fixed	-	1.8527*	6.3770
	Number of obs = 29058 Pseudo R2 = 0.3809 Log likelihood = -6971.3535 Prob > chi2 = 0.0000 LR chi2(47) = 8578.83		

***, ** and * represent significance at 10% 5% and 1% levels respectively.

to the households with male heads. The likelihood value was attained as 0.6781 in the model based on consumption and odds ratio was 1.9701. The risk of poverty was found to be higher for the households with female heads regarding the both models. In the studies in the literature, it was emphasized that women who are household heads are at a higher risk of being poor than men (Dansuk, 1997; Dal, 2013; Biyase & Zwane, 2015).

The age group of 15-24 was defined as reference category for age variable. All subcategories of age with the significance level of 1% were found to have significant effect on poverty. It was stated that the subcategories of age were not statistically significant for the poverty according to consumption. The risk of poverty was found to be 2.6206 (1/0.3816) times higher for households with the heads who were in 15-24 age group compared to the ones with the heads who were 25-34 according to the model which was developed for income. 71.41 value was attained by formularizing 0.2859 which was the odds value of the samples over 65 as (1-0.2859) *100. In line with this, the likelihood of poverty was found to be 71.41% times lower compared to the samples who were in the 15-24 age group. The present study showed a negative correlation between the household head's age and the likelihood of poverty. In other words, the likelihood ratio of poverty decreased significantly as the household's head got older. In the studies in the literature, there is an important relationship between age and poverty (Epo, 2010; Dal 2013; Abrar ul haq, Ayup & Ullah, 2015; Garza-Rodriguez, 2015).

The variable health insurance was found to be significant with the significance level of 1% in terms of both models. The likelihood ratio of poverty for the samples without health insurance in comparison to the samples with health insurance were calculated as 1.8087 (1/0.5529) and 1.4656 (1/0.6823) in terms of income and consumption respectively. Health insurance was defined as a significant indicator of poverty on the basis of the results mentioned above.

Educational level was found to be another significant variable for poverty. The subcategories in both models were stated to be significant compared to the reference level of the illiterate samples. The results showed that coefficient decreased as educational level increased and that means the level of poverty decreased. While the likelihood ratio of households with heads who graduated from primary school was found to be significantly lower with the rate of %31.86 ((1-0.6814) *100), the likelihood of poverty for households with heads with higher education was lower with the rate of %92.63 ((1-0.0737) *100) according to the reference level of the illiterate samples in the model which was developed according to income level. The same rates with the criteria and samples mentioned above were %26.62 ((1-0.7338) *100) and %86.1 ((1 0.1390) *100) in sequence in the model developed considering consumption level. It was found out that there was a significant decrease in the likelihood of ratio as the educational level increased. The studies in the literature are also supported, and it is emphasized that the probability of poverty decreases as the level of education increases (Epo, 2010; Tatlıdil & Demirağ, 2014; Garza-Rodriguez, 2015; Biyase & Zwane, 2017).

The samples who never got married were defined as the reference level for the variable "marital status" for both models. The coefficient value for the samples whose husbands or wives died was determined to be significant in statistical terms compared to the reference levels which were specified for both models. The odds ratio was divided by 1 and reference levels were changed as "their husbands or wives died" category with the aim of making the interpretation phase more practical. The likelihood ratio of poverty for a sample who never got married were attained as 1.7979 (1/0.5562) and 2.6144 (1/0.3825) compared to a sample whose husband or wife died for the models which were developed based on income and consumption in order. It was determined that the likelihood ratio of poverty for samples who never got married was found to be higher when compared to the samples whose husbands or wives died. This result coincides with the findings obtained by Tatlıdil and Demirağ (2014) in their study.

As for the variable namely household type, households without children was defined as the reference level. The coefficient predictive values for all subcategories were found to be significant in statistical terms with 1% significance level for both models. The rates are stated respectively for the models developed based on income and consumption level. The likelihood ratios of the households with one adult compared to the households without children were 1.9201 and 1.8187; the likelihood ratios of poverty for the nuclear families with one child were determined as 1.5756 and 1.5208; these ratios for the nuclear families with two children were found to be 3.0733 and 2.8342; the likelihood ratios of the poverty for the nuclear families with three or more children were 7.7747 and 6.2409; the likelihood ratios of poverty for the extended families were 4.9524 and 5.5001 and the likelihood ratios of poverty for the samples sharing the same house were 2.0843 and 1.6945 times higher. The conclusion that as the number of the children increased in the household, the likelihood ratios of poverty increased according to the both models might be drawn from the set of the results stated above. This result supports the results in the literature (Abrar, Ayub & Ullah, 2015; Garza-Rodriguez, 2015; Oluwatayo & Babalola, 2020).

The characteristics of the house in which the household members dwell was determined as a significant factor affecting poverty according to the results of the present study. The rental value was calculated by combining relative and standard rental values. In other words, this variable was developed considering the rental value of the house in which the household members dwelled even though they did not pay a rent. The likelihood ratio of poverty was stated to decrease as the rental value increased. One-unit increase in the rental value decreased the likelihood ratio approximately by 0.7% ((1-0.9930) *100) on the model based on income. As for the model based on consumption, one-unit increase in the rental value gave a rise to a decrease by about 0.86% ((1-0.9914) *100 in the likelihood of poverty.

Negative correlation was determined between the length of the time span during which household members dwell in the house and poverty. This demonstrated that as the length of the time span increased, the likelihood ratio of poverty decreased. One-year increase in the length of the time span was found to affect poverty 0.9935 times. The same ratio on the model based on consumption expenditures was calculated as 0.9969.

A significant decrease in the likelihood ratio of poverty was observed as the number of the rooms increased. For the model based on income, one-unit increase in the number of the rooms leaded to a decrease in the likelihood ratio of poverty by the rate of 13.02% ((1-0.8698) *100. This rate was found to be 8.61% ((1-0.9139) *100) for the model based on consumption expenditures.

The floor type of the rooms in the house which the household dwell in was integrated into the model as the variable so called “floor type”. Another variable which was assumed to be significant in terms of poverty was floor type. Concrete floor was defined as the reference level. The likelihood ratio of poverty for the households who lived in a house with wooden floor was found to be lower by 43.94% ((1-0.5606) *100 on the model based on income. The likelihood ratio of poverty for the households who lived in a house with fitted carpet was determined to be lower by 40.6% ((1-0.5940) *100) on the model based on income. The likelihood ratio of poverty for the households who lived in a house with ceramic floor was shown to be lower by 19.87% ((1-0.8013) *100) on the model based on income.

The likelihood ratio of poverty for the households who lived in a house with wooden floor was found to be lower by 36.65% ((1-0.6335) * 100 on the model which was developed based on the independent variable consumption expenditure. The likelihood ratio of poverty for the households who lived in a house with a fitted carpet was lower according to the rate 32.88% ((1-0.6712) *100) on the model which was developed based on the independent variable consumption expenditure. The likelihood ratio of poverty for the households who lived in houses with ceramic floor was lower in line with the rate 19.43% ((1-0.8057) *100) on the model which was developed based on the independent variable consumption expenditure. The likelihood ratio of poverty for the households living in a house with concrete floor was shown to be higher on both models.

The reference level for the type of heating was defined as “stove” The likelihood ratio of poverty for the households with air conditioning was found to be significantly lower by the rate 44.04% ((1-0.5596) *100) compared to the ones with stoves on the model which was developed based on the dependent variable “income”. The likelihood ratio of poverty was found to be lower for the households with central heating by 39.41% ((1-0.6059)*100) compared to the ones with stoves. The likelihood ratio of poverty for the households with room heater and combi boiler was found to be lower by 29.52% ((1-0.7048)*100). As for the model which was developed in line with the independent variable “consumption expenditures”, the likelihood ratio of poverty for households with air conditioning was lower by %36.65% ((1-0.6335) *100) in a comparison with the ones with stoves. The likelihood ratio of poverty for the households with central heating was lower by 32.88% ((1-0.6712)*100) compared to the ones with stoves. Finally, the likelihood ratio of poverty for the households with room heater and combi boiler was determined to be lower by the rate 19.43% ((1-0.8057)*100) compared to the ones with stoves. The likelihood of poverty for the households with stoves was found to be higher on both models developed on the basis of two dependent variables.

The state of having easy access to public transportation facilities in accordance with the location of the house was integrated into the model as “difficulty of access to the transportation facilities”. Negative correlation was found between poverty and difficulty of access to the transportation facilities. In other words, as the access to the transportation facilities became easier, the likelihood ratio of poverty decreased. The coefficient which was gained was significant in statistical terms on the model which was developed for income. On the other hand, the coefficient was not shown to be significant on the model which was developed on the basis of consumption expenditures. The results showed that one-unit increase in the access to the transportation facilities caused the likelihood ratio of poverty to decrease by 6.66% ((1-0.9334)*100).

The increase in the number of the mobiles, mobile phones, computer and LCD televisions in the house was also found to decrease the likelihood ratio of poverty. One-unit increase in the number of the telephones decreased the likelihood ratio of poverty by 47.09% ((1-0.5291)*100) on the model which was developed according to income. One-unit increase in the number of mobile phones decreased the likelihood ratio of poverty by the rate 19.61% ((1-0.8039)*100) on the model which was developed according to income. One-unit increase in the number of computers decreased the likelihood ratio of poverty by the rate 33.09% ((1-0.6691)*100 on the model which was developed according to income. This rate for the number of computers was 33.09% ((1-0.6691)*100) for the number of the computers in the house. This rate was found to be as 27.71% ((1-0.7229)*100) for the number of LCD televisions. One-unit increase in the number of telephones was found to give a rise to a decline in the likelihood ratio of poverty by 54.11% ((1-0.4589)*100) on the model which was developed according to consumption expenditures. This rate was found to be 27.25% ((1-0.7275)*100) for the variable the number of mobile phones. For the number of computers, one-unit increase was found to lead to a decrease by 41.92% ((1-0.5808)*100. Finally, one unit increase in the number of LCD televisions was found to decrease the likelihood ratio of poverty by the rate 29.97% ((1-0.7003)*100). These findings confirmed the conclusions of other studies (Dal 2013; Tatlıdil and Demirağ 2014).

The classification table of the model which was developed on the basis of income is shown in Table 5. The sensitivity ratio showing to what extent the test is reliable to determine the true positives was found to be 42.56% from 1799/4227. This demonstrated that this test predicted 43% of the people who were poor in actual fact. The specificity ratio indicating to what extent the text is reliable to determine the true negatives was calculated as 96.53% from 23969/24831. That is to say, it predicted 97% of the people who were not actually poor. True classification ratio of the model was attained as 88.68% from (1799+23969)/29058. That means the model predicted the state of poverty 88.68% of the households which were investigated in the present study accurately.

Table 5. The classification table of the model which was developed on the basis of income

		Actual Fact		
		Poor (1)	**Non-poor (0)**	**Total**
The results of the diagnostic test	**Poor (1)**	1799 **(TP)**	862 **(TP)**	2661
	Non-poor (0)	2428 **(FN)**	23969 **(FN)**	26397
	Total	4227	24831	29058
Sensitivity = 42.56% Specificity= 96.53% True Classification = 88.68%				

Classification table is generally used to determine the goodness of fit of the model. Nevertheless, more reliable results are gained when the analysis is made for classifications. Otherwise, it is not likely to gain reliable results (Bircan, 2004). Another method which is used for measurement of the goodness of fit is the ROC Curve model which is attained by comparing the predictive value and observed value of each and every observation. The goodness of fit for the model is analyzed by means of interpreting AUC in this method. AUC is 0.8968 for the poverty Logit model which was developed considering each equivalent member's yearly income (Figure 2). This maintains it predicted the poor and non-poor accurately by the rate of 90%.

The classification of the model based on consumption expenditures are shown in Table 6. The sensitivity rate which is used to determine to what extent the rest is sensitive to determine true positives was calculated as 40.97% from 1555/3795. The specificity rate showing how accurate the test is in terms of determining the true negatives was found to be 96.77% on the basis of 24447/25263. True classification rate of the model was 89.48% based on (1555+24447)/29058. This model was found to predict the poverty of 89.48% of the households which were investigated within the scope of the current study.

AUC for the Logit model which was developed according to the yearly income per equivalent member was attained as 0.9061 (Figure 3.). It is highly possible to state that this model was significantly discriminative as this value was in the range: $0.9 \leq AUC \leq 1$. It was found to predict the poor and non-poor accurately by the rate 91%.

Personal and Housing Characteristics

The indicators included in the definition of poverty are investigated under two categories such as economic and non-economic. Gross domestic product, unemployment, gini coefficient, real prices consti-

Figure 2. ROC curve of the logit model which was developed on income

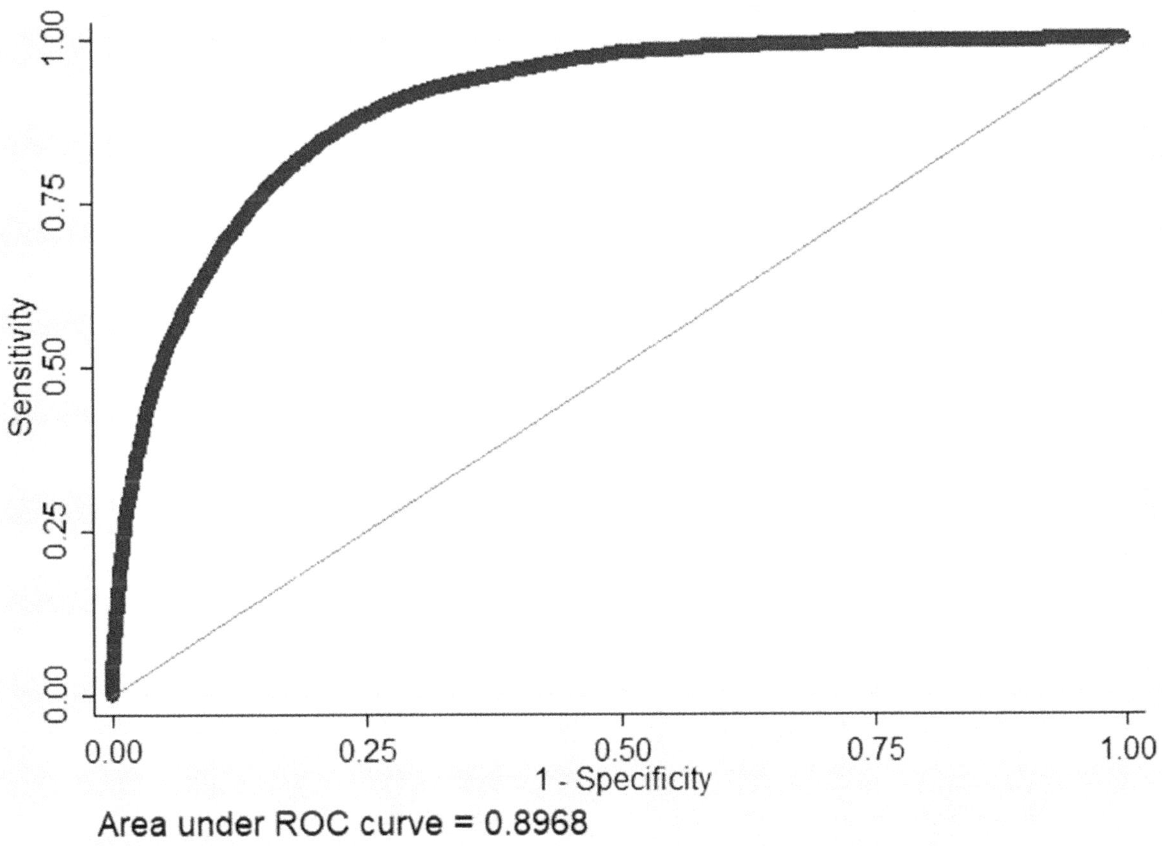

tute the economic indicators and literacy rate, educational level, household infrastructure are stated as non-economic indicators. (Sumner, 2007). ROC curves of the models which were developed in terms of income and consumption expenditures in the distinction of personal and housing characteristics were drawn and their comparisons were made in the present study. The variables such as gender, marital status, age, educational level, health insurance which are household's head characteristics were analyzed in terms of their significance for poverty. As for the model which was developed for the housing

Table 6. The classification table of the logit model which was developed on consumption expenditures

		Actual Fact		
		Poor (1)	**Non-poor**	**Total**
The result of the diagnostic test	**Poor (1)**	1555 (TP)	816 (FP)	2371
	Non-poor (0)	2240 (FN)	24447 (TN)	26687
	Total	3795	25263	29058
Sensitivity = % 40.97 Specificity = %96.77 True Classification = %89.48				

Figure 3. The ROC Curve of the logit model which was developed on consumption expenditures

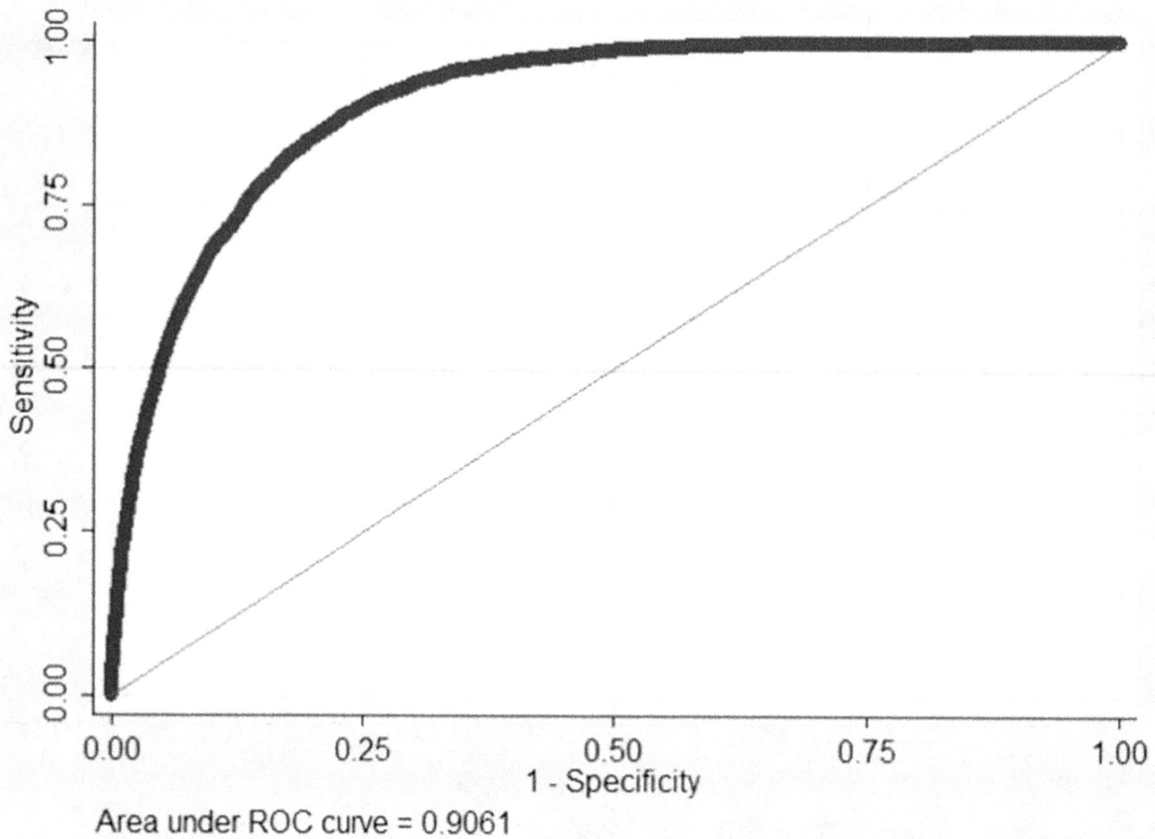

characteristics, access to the transportation facilities, floor type, the number of the rooms, rental value, heating system were investigated. Logit models were developed for each of the models and AUC values of these models are shown in Table 7.

The ROC curves which were drawn according to income level is shown in Figure 4. The AUC value of the model investigation household head's characteristics was found to be 0.771 and this value was attained as 0.8476 in the model investigating housing characteristics. Both of these models were found to provide reliable results to discriminate between poor and non-poor households. Housing characteristics were found to be more significant to determine poverty in comparison with personal characteristics. In other words, the model including housing characteristics was determined as the most significant model to affect poverty.

Table 7. AUC for two different models

Poverty Line	Household Head's Characteristics		Housing Characteristics
Income Level	0.771		0.8476
Consumption Expenditures	0.7629		0.8677

Figure 4. The ROC Curves of the models developed on personal and housing characteristics according to income level

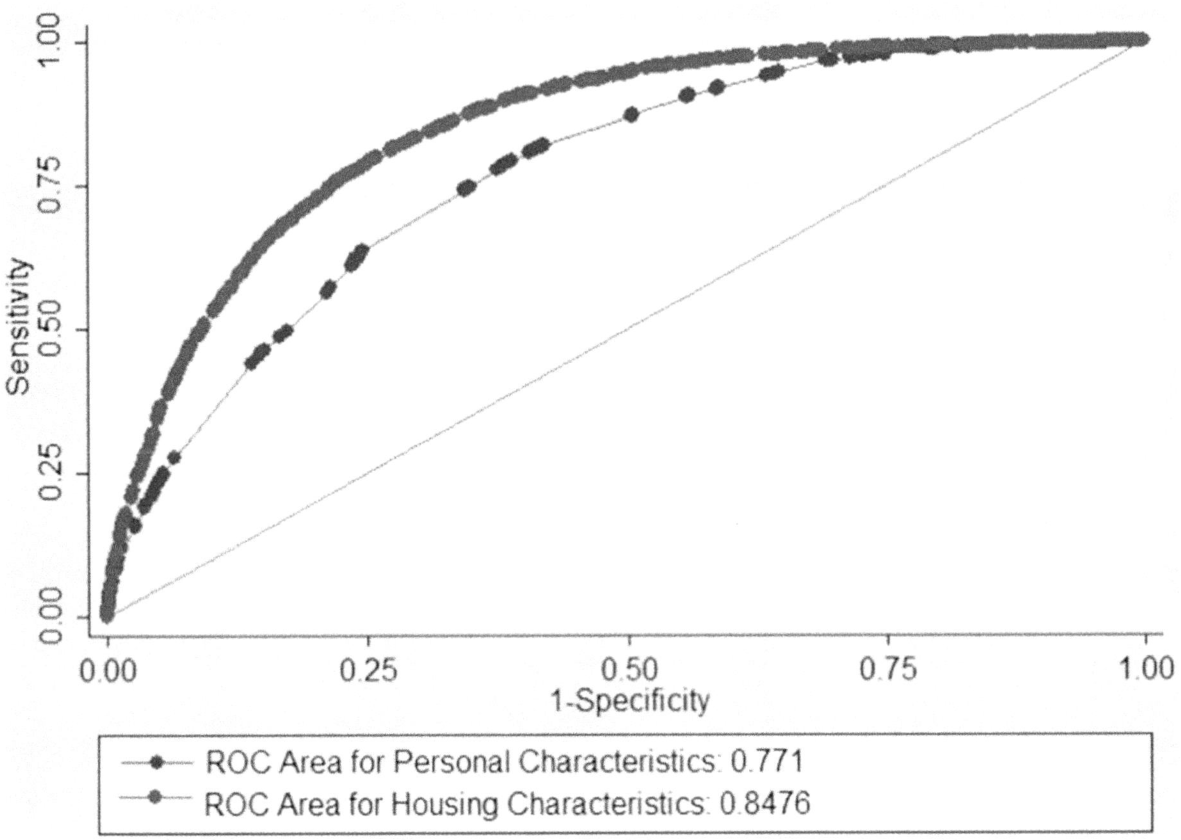

ROC curves which were drawn according to consumption expenditures are shown in Figure 5. AUC value of the model investigating household head's personal characteristics was found to be 0.7629. As for the ROC value of the model which was developed on household head's personal characteristics was determined as 0.8677. Housing characteristics were shown to be more significant in the model that was developed on consumption expenditures.

Personal Poverty Indicators

A ROC curve shows to what extent a diagnostic test discriminates between two cases accurately. The likelihood ratio of being poor for a person who is actually poor (sensitivity) is shown on the vertical axis and the likelihood ratio of being non-poor for a person who isn't actually poor (specificity) is shown on the horizontal axis (Minot & Baulch, 2002). The results were interpreted based on Area Under Curve Method (AUC) which is used to measure the reliability of ROC curves for this study. In line with this, the significance of the variable which is being investigated increases as the AUC value gets closer to +1 end on the continuum for the determination of poverty and the variable is stated to have a random effect on the determination of poverty.

Figure 5. The ROC Curves of the models developed on consumption expenditures and housing characteristics

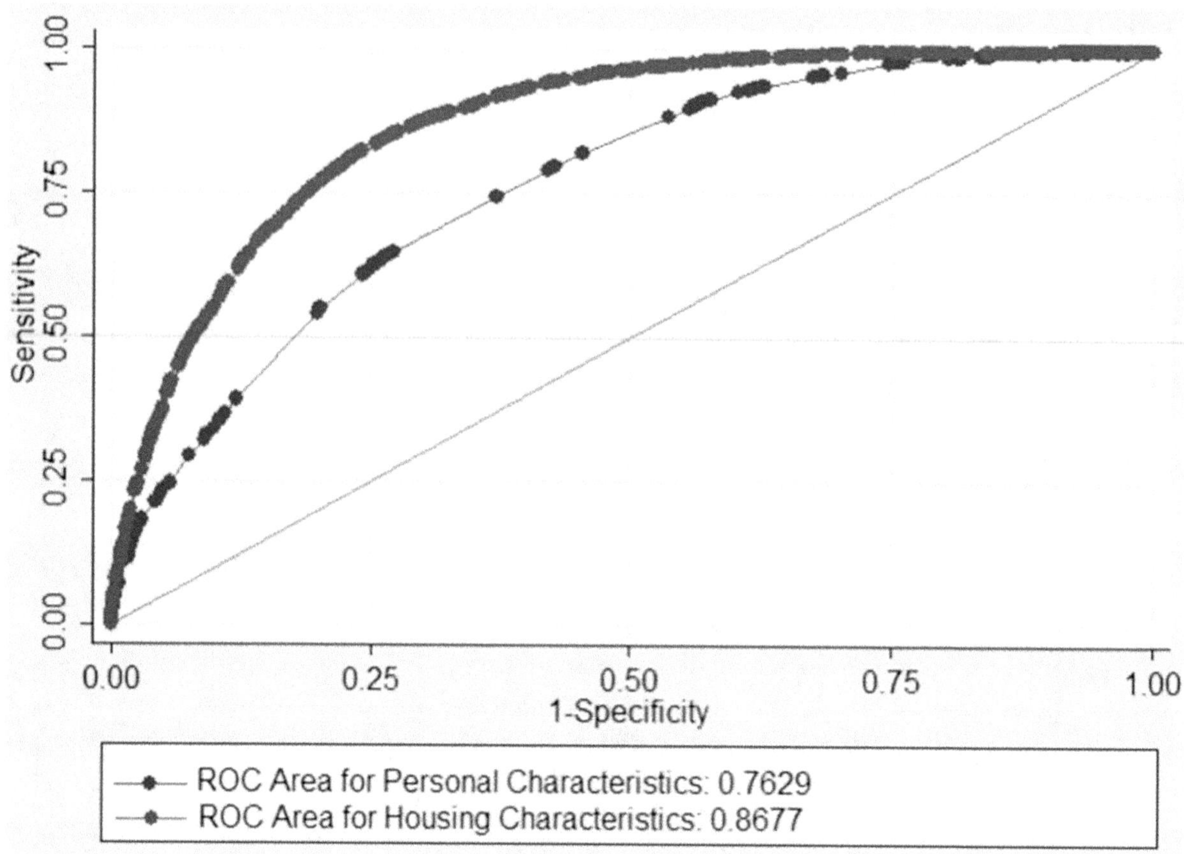

Baulch drew the ROC curve for floor types of the houses and stated subcategories for each part within the scope of the study which he conducted on the poverty in Vietnamese context in 2002. Floor types were put into four different categories such as concrete, wooden, fitted carpet and ceramic in the Household Budget Questionnaires which were applied in Turkey. The ROC curves for the poverty variable based on floor types and income level are shown in Figure 6. The results showed that the highest ratio of poverty was observed in the houses with concrete floor. As long as all the houses with concrete floor which was equivalent to 23.4% of total population is indicated as poor, the rate of poverty was shown to increase significantly. AUC value was found as 0.6875.

The state of poverty is determined considering household's income and consumption expenditures in poverty analyses. The most significant variables are specified based on these variables.

The ROC curves of each variables were drawn according to income level and consumption expenditures with the aim of determining the most significant single effects for poverty for the present study. Their AUC values are shown with their order of magnitude in Figure 6.

Rental value of the house, educational level, heating system and floor type were shown as the most significant variables to determine poverty. Minot and Baulch maintained floor type as the most significant variable in the study they conducted in 2002. Household head's educational level was also stated

Figure 6. The ROC Curve for floor types

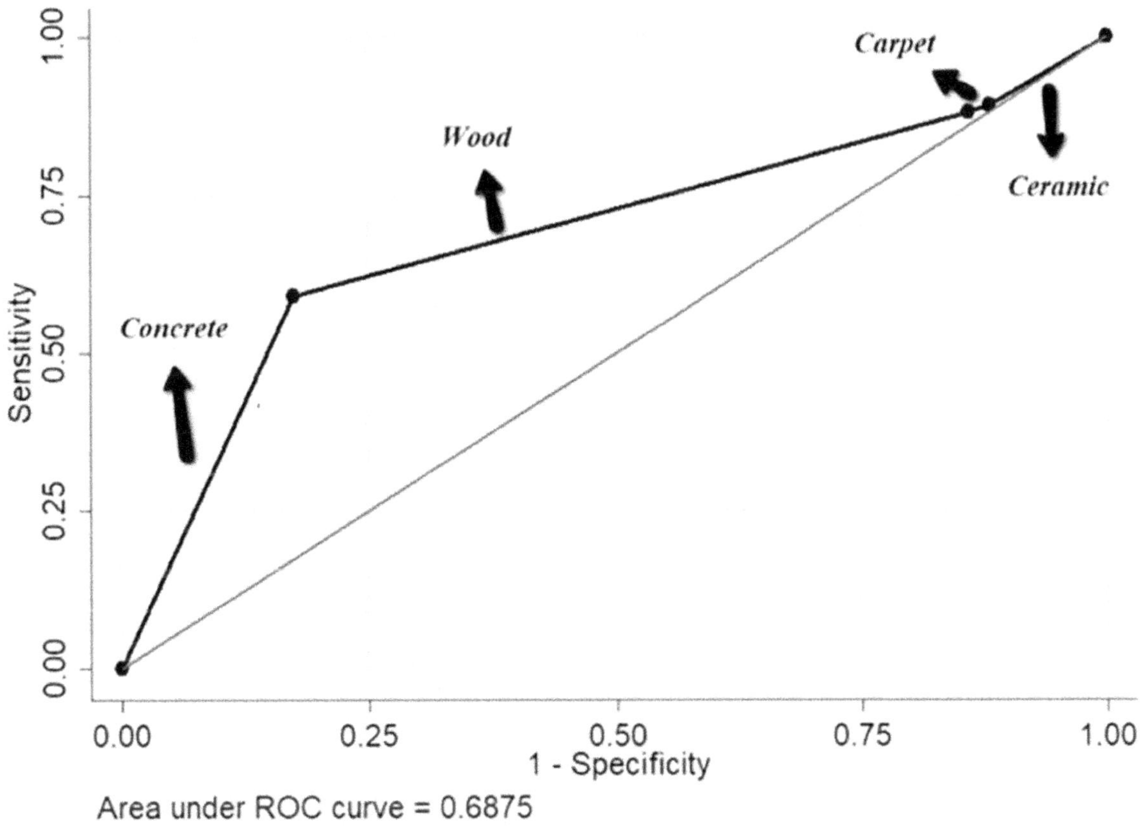

as a significant variable for poverty in the present study. Wodon also reported the same interpretation for educational level as the result of the study which he carried out in 1997. Furthermore, educational level was stated to be one of the most significant determinants of poverty in the study which was carried out by Minot and Baulch in 2002.

Possession of computers, mobile phones and LCD TVs is another significant variable to determine the poverty. The results of the present study were found to be consistent with the statement mentioned above according to the detailed analyses which were made based on AUC values. Difficulty of access to the transportation in a relationship with the location of the house and the number of the rooms in the house were also reported as significant variables for poverty. The AUC value of the ROC curve according to consumption expenditures was found to be higher than the AUC value of the ROC curve based on income as for the length of the time during which the household members dwell in the house. Household head's age, gender, marital status and his/her having health insurance were not found to be significant in terms of determining poverty (Table 8).

Table 8. AUC values of the single effects of income and consumption expenditures

	AUC			
	Income Level		Consumption Expenditures	
Gender	0.506	15	0.522	13
Age	0.530	14	0.521	14
Educational Level	0.709	2	0.726	2
Marital Status	0.503	16	0.516	15
Health Insurance	0.546	12	0.536	12
Household Type	0.651	8	0.630	9
Rental Value	0.824	1	0.851	1
The Length of the Time Span during which Household Members Dwell in the House	0.541	13	0.582	11
The Number of the Rooms	0.637	9	0.636	8
Heating System	0.695	3	0.701	3
Difficulty of Access to the Transportation Facilities	0.653	7	0.667	5
Floor Type	0.688	4	0.700	4
LCD TV	0.584	11	0.587	10
Telephone	0.670	6	0.664	6
Mobile phone	0.606	10	0.640	7
Computer	0.684	5	0.700	4

CONCLUSION

While the gross domestic product was $5482 in Turkey in 2000, it increased to $ 10150 in 2016. In addition to that, gini coefficient was 0.44 was in 2002 and it became 0,397 in 2015. It is of prime importance to determine the parameters identifying the poverty under the changing circumstances in terms of income and distribution of it. Establishing the determinants of poverty and measurement instruments will make a considerable contribution to develop solutions for it, take required actions to prevent it, develop projects and policies against it.

The present study was conducted with the aim of developing the most representative poverty model analyzing the data retrieved from Household Budget Questionnaires in the years 2010,2011 and 2012 by means of using ROC analysis and developing models to specify the indicators of poverty. The determination of poverty line was made according to two different bases such as household consumption expenditures and income level. The model which was developed in line with consumption expenditures was detected to be more significant than the model which was formed on income by a narrow margin after comparing these two models relying on the Logit models and ROC curves. The model in relation with the income level predicted the poor and non-poor household accurately by 90% and the model according to consumption expenditures predicted the poor and non-poor households accurately by the rate of %91.

As a result of the analysis, it has been seen that one of the most important determinants of poverty is education. The most important factor to be considered in the struggle against poverty is to increase the education level of the society. Especially higher education graduates have been found to have a very low risk of poverty, and in this respect, an important effect can be achieved in poverty reduction through policies that support education and especially support university education. In addition, it has been determined that women who have a head of household have a higher risk of poverty than men. For this, it is important to work on increasing women's employment and education levels. Increasing household size has been determined to significantly increase the risk of poverty, and especially families with 3 or more children are most likely to be poor. Accordingly, family planning plays an important role in the fight against poverty.

Housing characteristics were also demonstrated to be more significant indicators of poverty compared to household head's personal characteristics in line with the results of the current study. On the other hand, the decrease in the rental value and the length of the time during which household members live in the house and the number of the rooms was maintained to result in an increase in the risk of poverty. Another parameter having importance in terms of determining the poverty was reported as the floor type of the rooms. The likelihood ratio of poverty was found to be higher for the houses with concrete floors in both models. This study also showed that another significant variable determining poverty is heating system. The poorest households in all samples were stated to be the ones with stoves. This study also suggested that if the difficulty of access to the public transportation facilities increased, the likelihood ratio of poverty increased. Additionally, the likelihood ratio of poverty was stated to decrease when the number of the electrical devices in the house increased. Rental value, educational level, heating system and floor type were suggested to be significant indicators according to the poverty level according to both variables considering the single effects. This is very important for the correct determination of the households that will receive social assistance. Thus, even if poverty does not disappear completely, the poverty can be reduced by implementing anti-poverty policies and social projects. Although all these findings were attained in a limited time and scope, they were proven to be representative of the Turkish context and they could also be utilized in order to increase the efficacy of the more complicated actional policies to reduce poverty.

REFERENCES

Abrar Ul Haq, M., Ayub, K., & Ullah, M. I. (2015). Micro-level Determinants of Rural Poverty in Pakistan. *International Journal of Scientific and Research Publications*, *5*(2), 1–4.

Anand, S. (1983). *Inequality and Poverty in Malaysia: Measurement and Decomposition*. Oxford University Press.

Baulch, B. (2002). P*overty monitoring and targeting using ROC curves: Examples from Vietnam. IDS Working Paper*, *161*(August).

Bircan, H. (2004). Lojistik Regresyon Analizi : T ı p Verileri Üzerine Bir Uygulama. *Kocaeli Ünv Sosyal Bilimler Ens Derg.*, *2*, 185–208.

Biyase, M., & Zwane, T. (2018). An empirical analysis of the determinants of poverty and household welfare in South Africa. *Journal of Developing Areas*, *52*(1), 115–130. doi:10.1353/jda.2018.0008

Bourguignon, F., & Chakravarty, S. (2003). The measurement of multidimensional poverty. *The Journal of Economic Inequality*, *1*(1), 25–49. doi:10.1023/A:1023913831342

Bradley, A. P. (1997). The use of the area under the ROC curve in the evaluation of machine learning algorithms. *Pattern Recognition*, *30*(7), 1145–1159. doi:10.1016/S0031-3203(96)00142-2

Canbay, T., & Selim, S. (2010). Türkiye'de Hanehalkı Yoksulluğu. *Ege Akademik Bakış / Ege Academic Review*, *10*(2), 627–649.

Dal, S. (2013). *Türkiye'de Hanehalkı Yoksulluğunu Belirleyen Etmenler*. Gazi Üniversitesi, Sosyal Bilimler Enstitüsü.

Dansuk, E. (1997), *Türkiye'de Yoksulluğun Ölçülmesi ve Sosyo-Ekonomik Yapılarla Ölçülmesi*, (Published Specialization Thesis). DPT, Sosyal Sektörler ve Koordinasyon Genel Müdürlüğü Ücretler ve Gelirler Dairesi Başkanlığı, Yayın No: DPT: 2472, Ankara.

Deaton, A. (2003). Regional poverty estimates for India, 1999-2000. Research Program in Development Studies, Princeton University.

Dirican, A. (2001). Tani Testi Performanslarının Değerlendirilmesi Ve Kıyaslanması. *Cerrahpaşa Tıp Dergisi*, *32*(1), 25–30.

Dumanlı, R. (1996). *Yoksulluk ve Türkiye'deki Boyutları*. DPT-Sosyal Sektörler ve Koordinasyon Genel Müdürlüğü.

Epo, B. N. (2010). *Determinants of Poverty in Cameroon: A Binomial and Polychotomous Logit Analysis*. SSRN.

Flach, P., Blockeel, H., Ferri, C., Hernández-Orallo, J., & Struyf, J. (2003). Decision support for data mining. In Data Mining and Decision Support. Springer US. doi:10.1007/978-1-4615-0286-9_7

Garza-Rodríguez, J. (2015). The determinants of Poverty in the Mexican States of the US-Mexico Border. *Estudios Fronterizos (México)*, *17*(33), 141–167. doi:10.21670/ref.2016.33.a06

Hanley, A. J., & McNeil, J. B. (1982). The Meaning and Use of the Area under a Receiver Operating Characteristic (ROC) Curve. *Radiology*, *143*(1), 29–36. doi:10.1148/radiology.143.1.7063747 PMID:7063747

Kızılgöl, Ö. (2008). Yoksulluğun Belirleyicileri: Türkiye Örneği. *Finans Politik & Ekonomik Yorumlar*, *45*(526).

Lasko, T. A., Bhagwat, J. G., Zou, K. H., & Ohno-Machado, L. (2005). The use of receiver operating characteristic curves in biomedical informatics. *Journal of Biomedical Informatics*, *38*(5), 404–415. doi:10.1016/j.jbi.2005.02.008 PMID:16198999

Lipton, M., & Ravallion, M. (1995). Poverty and policy. Poverty and Policy. Handbook of Development Economics, 3, 2551–2657.

Metz, C. E. (1978). Basic principles of ROC analysis. *Seminars in Nuclear Medicine*, *8*(4), 283–298. doi:10.1016/S0001-2998(78)80014-2 PMID:112681

Metz, C. E. (2008). ROC analysis in medical imaging: A tutorial review of the literature. *Radiological Physics and Technology*, *1*(1), 2–12. doi:10.100712194-007-0002-1 PMID:20821157

Minot, N., & Baulch, B. (2002). *The spatial distribution of poverty in Vietnam and the potential for targeting*. Policy Research Working Paper; 2829.

Oluwatayo, I. B., & Babalola, M. A. (2020). Asset Ownership and Income as Drivers of Household Poverty In South Africa. *Journal of Developing Areas*, *54*(3). Advance online publication. doi:10.1353/jda.2020.0028

Sen, A. (1976). Poverty: An Ordinal Approach to Measurement. *Econometrica*, *44*(2), 219–231. doi:10.2307/1912718

Sumner, A. (2007). Meanin versus Measurement: Why Do Economic Indicators of Poverty Still Predominate? *Development in Practice*, *17*(1), 4–13. doi:10.1080/09614520601092485

Tatlıdil, H., & Demirağ, İ. (2014). Türkiye'de Yoksulluğun Sosyo-Ekonomik ve Demografik Değişkenlerle İlişkilerinin Lojistik Regresyon ve Mars Yöntemleri Kullanılarak İncelenmesi. *TISK Academy/TISK Akademi*, *9*(17).

Tomak, L., & Bek, Y. (2009). Journal of Experimental and Clinical Medicine Deneysel ve Klinik Tıp Dergisi İşlem karakteristik eğrisi analizi ve eğri altında kalan alanların karşılaştırılması. *On Dokuz Mayıs Üniversitesi. Deneysel ve Klinik Tıp Dergisi*, *27*(2), 58–65.

TÜİK. (2008). *Tüketim Harcamaları, Yoksulluk ve Gelir Dağılımı: Sorularla Resmi İstatistikler Dizisi – 6. Türkiye İstatistik Kurumu Matbaası.*

TÜİK. (2011). *Gelir ve Yaşam Koşulları Araştırması (Kesit) 2011 Mikro Veri Seti.* Author.

Wodon, Q. T. (1997). Targeting the poor using ROC curves. *World Development*, *25*(12), 2083–2092. doi:10.1016/S0305-750X(97)00108-3

Zweig, M. H., & Campbell, G. (1993). Receiver-operating characteristic (ROC) plots: A fundamental evaluation tool in clinical medicine. *Clinical Chemistry*, *39*(4), 561–577. doi:10.1093/clinchem/39.4.561 PMID:8472349

KEY TERMS AND DEFINITIONS

AUC: Area under the Roc Curve is the entire two-dimensional area under the ROC curve.

Equivalent Household Member: It is a value calculated in proportion to the size of each household and the age of the household members. It allows us to compare the income and consumption expenditures values of households.

Household Budget Survey: It is a survey of micro data on the socio-economic status of the household, the composition of the household, the employment status of the individuals and the income they earn, and the consumption expenditures based on the subgroups of the household.

OECD Modified Equivalence Scale: It is a weight of 1 for the first adult, 0.5 for a household member older than 14 or 14 years old, and 0.3 for a household member younger than 13.

Poverty: The condition of not having sufficient material property or income for a person's needs.

Poverty Line: A value of half the median income or consumption expenditures of equivalent household member.

Roc Curve: Receiver operating characteristic curve is a graph that shows the performance of a classification model at all classification thresholds.

Chapter 2
Data Analyzing via Probabilistic Modeling:
Interpolation and Extrapolation

Dariusz Jacek Jakóbczak
https://orcid.org/0000-0002-0297-6598
Koszalin University of Technology, Poland

ABSTRACT

Object recognition is one of the topics of artificial intelligence, computer vision, image processing, and machine vision. The classical problem in these areas of computer science is that of determining object via characteristic features. An important feature of the object is its contour. Accurate reconstruction of contour points leads to possibility to compare the unknown object with models of specified objects. The key information about the object is the set of contour points which are treated as interpolation nodes. Classical interpolations (Lagrange or Newton polynomials) are useless for precise reconstruction of the contour. The chapter is dealing with proposed method of contour reconstruction via curves interpolation. First stage consists in computing the contour points of the object to be recognized. Then one can compare models of known objects, given by the sets of contour points, with coordinates of interpolated points of unknown object. Contour points reconstruction and curve interpolation are possible using a new method of Hurwitz-Radon matrices.

INTRODUCTION

Method of Hurwitz - Radon Matrices (MHR), invented by the author, can be applied in reconstruction and interpolation of curves in the plane. The method is based on a family of Hurwitz - Radon (HR) matrices. The matrices are skew - symmetric and possess columns composed of orthogonal vectors. The operator of Hurwitz - Radon (OHR), built from these matrices, is described. Author explains how to create the orthogonal and discrete OHR and how to use it in a process of curve interpolation and two-dimensional data modeling. Proposed method needs suitable choice of nodes, i.e. points of the 2D curve to be interpolated or extrapolated: nodes should be settled at each extremum (minimum or maximum)

DOI: 10.4018/978-1-7998-4706-9.ch002

of one coordinate and at least one point between two successive local extrema, and nodes should be monotonic in one of coordinates (for example equidistance). Created from the family of N - 1 HR matrices and completed with the identical matrix, system of matrices is orthogonal only for vector spaces of dimensions $N = 1, 2, 4$ or 8. Orthogonality of columns and rows is very important and significant for stability and high precision of calculations. MHR method is modeling the curve point by point without using any formula of function. Main features of MHR method are: accuracy of curve reconstruction depending on number of nodes and method of choosing nodes, interpolation of L points of the curve is connected with the computational cost of rank $O(L)$, MHR interpolation is not a linear interpolation (Ullman & Basri, 1991). The problem of curve length estimation is also considered. Algorithm of MHR method and the examples of data extrapolation are described. Value anticipation is the crucial feature in risk analyzing and decision making.

BACKGROUND

The following question is important in mathematics and computer science: is it possible to find a method of curve interpolation and extrapolation in the plane without building the interpolation polynomials? This chapter aims at giving the positive answer to this question. Current methods of curve interpolation are based on classical polynomial interpolation: Newton, Lagrange or Hermite polynomials and spline curves which are piecewise polynomials (Dahlquist & Bjoerck, 1974). Classical methods are useless to interpolate the function that fails to be differentiable at one point, for example the absolute value function $f(x) = |x|$ at $x = 0$. If point (0;0) is one of the interpolation nodes, then precise polynomial interpolation of the absolute value function is impossible. Also when the graph of interpolated function differs from the shape of polynomials considerably, for example $f(x) = 1/x$, interpolation is very hard because of existing local extrema of polynomial. We cannot forget about the Runge's phenomenon: when interpolation nodes are equidistance then high-order polynomial oscillates toward the end of the interval, for example close to -1 and 1 with function $f(x) = 1/(1+25x^2)$ (Ralston, 1965). MHR method is free of these bad feature. Computational algorithm is considered and then it is important to talk about time. Complexity of calculations for one unknown point in Lagrange or Newton interpolation based on n nodes is connected with the computational cost of rank $O(n^2)$. Proposed method has lower calculation complexity.

A significant problem in risk analysis and decision making is that of appropriate data representation and extrapolation (Brachman & Levesque, 2004). Two-dimensional data can be treated as points on the curve. Classical polynomial interpolations and extrapolations (Lagrange, Newton, Hermite) are useless for data anticipation, because the stock quotations or the market prices represent discrete data and they do not preserve a shape of the polynomial. Also Richardson extrapolation has some weak sides concerning discrete data. This chapter is dealing with the method of data foreseeing and value extrapolation by using a family of Hurwitz-Radon matrices. The quotations, prices or rate of a currency, represented by curve points, consist of information which allows us to extrapolate the next data and then to make a decision (Fagin et al, 1995).

If the probabilities of possible actions are known, then some criteria are ready to be applied in decision making and analyzing risk: for example criterion of Laplace, Bayes, Wald, Hurwicz, Savage, Hodge-Lehmann (Straffin,1993) and others (Watson, 2002). But in this chapter author considers only two possibilities: to do something or not. For example to buy a share or not, to sell a currency or not. Proposed method of Hurwitz-Radon Matrices (MHR) is used in data extrapolation and then calculations for risk analyzing and decision making are described. MHR method presents new approach to extrapolation problem because it takes the interpolation nodes to create orthogonal basis as columns of matrix OHR operators. Then affine (convex) combination of such basis builds new orthogonal base for unknown coordinates of calculated points. MHR method uses two-dimensional data for knowledge representation (Markman, 1998) and for computational foundations (Sowa, 2000). Also medicine (Cierniak, 2005), industry (Jakóbczak & Kosiński, 2007) and manufacturing (Tang, 2005) are looking for the methods connected with geometry of the curves (Soussen & Mohammad-Djafari, 2004) . So suitable data representation and precise reconstruction (Latecki & Lakaemper, 1999) or extrapolation (Kozera, 2004) of the curve is a key factor in many applications of computational intelligence (Lowe, 1991 and 2001), knowledge representation and risk analysis. In comparison MHR method with Bézier curves, Hermite curves and B-curves (*B-splines*) or NURBS one unpleasant feature of these curves must be mentioned: small change of one characteristic point can make big change of whole reconstructed curve. Such an unwanted feature does not appear in MHR method. None of the methods, that are used nowadays, applies the orthogonal basis for unknown value extrapolation and foreseeing, but proposed MHR method does. Orthogonality means a very important feature for stability in calculations. Also point MHR interpolation and extrapolation via simple computations with matrices is something new in the problem of anticipation, risk analysis and decision making.

The considered problem statement looks as follow: let's assume there are given some 2D points of known data as the set of interpolation nodes. How the unknown value can be extrapolated and anticipated as the support in decision making and risk analysis, based on given characteristic (key) points?

THE METHOD OF HURWITZ - RADON MATRICES

Issues

This chapter deals with the problem of interpolation and extrapolation without computing the polynomial or any fixed function. The values of nodes are used to build the orthogonal matrix operators OHR and a linear (convex) combination of OHR operators leads to calculations of 2D curve points. Main idea of MHR method (Jakóbczak, 2006) is that the curve is interpolated or extrapolated point by point by computing the unknown coordinates of the points. The only significant factors in MHR method are: choosing the interpolation nodes and fixing the dimension of HR matrix ($N = 1, 2, 4$ or 8). Other characteristic features of function or curve, such as shape or similarity to polynomials, derivative or Runge's phenomenon, are not important in the process of MHR interpolation and extrapolation. The curve or function in MHR method is parameterized for value $\alpha \in [0;1]$ in each range of two successive interpolation nodes. Data extrapolation and foreseeing is calculated for $\alpha < 0$ or $\alpha > 1$. Estimation of the curve length with high precision is possible because of computing any number of curve points we want. Complexity of calculations for L unknown points in MHR interpolation and extrapolation, based on n nodes, is connected with the computational cost of rank $O(L)$. This is very important feature of MHR method.

The Origin of Hurwitz – Radon Matrices

Adolf Hurwitz (1859-1919) and Johann Radon (1887-1956) published separately the papers about specific class of matrices in 1923, working on the problem of quadratic forms. For example equation

$$(x_0^2 + x_1^2) \cdot (y_0^2 + y_1^2) = (z_0^2 + z_1^2)$$

is true when $z_0 = x_0y_0 - x_1y_1$, $z_1 = x_0y_1 + x_1y_0$. This result can be achieved from matrix equation:

$$\begin{bmatrix} x_0 & x_1 \\ -x_1 & x_0 \end{bmatrix} \cdot \begin{bmatrix} y_0 & y_1 \\ -y_1 & y_0 \end{bmatrix} = \begin{bmatrix} z_0 & z_1 \\ -z_1 & z_0 \end{bmatrix}.$$

Also equation

$$(x_0^2 + x_1^2 + x_2^2 + x_3^2) \cdot (y_0^2 + y_1^2 + y_2^2 + y_3^2) = z_0^2 + z_1^2 + z_2^2 + z_3^2$$

has solution:

$$z_0 = x_0y_0 - x_1y_1 - x_2y_2 - x_3y_3,$$

$$z_1 = x_0y_1 + x_1y_0 + x_2y_3 - x_3y_2,$$

$$z_2 = x_0y_2 - x_1y_3 + x_2y_0 + x_3y_1,$$

$$z_3 = x_0y_3 + x_1y_2 - x_2y_1 + x_3y_0.$$

This result can be achieved from matrix equation of dimension $N = 4$:

$$\begin{bmatrix} x_0 & x_1 & x_2 & x_3 \\ -x_1 & x_0 & -x_3 & x_2 \\ -x_2 & x_3 & x_0 & -x_1 \\ -x_3 & -x_2 & x_1 & x_0 \end{bmatrix} \cdot \begin{bmatrix} y_0 & y_1 & y_2 & y_3 \\ -y_1 & y_0 & -y_3 & y_2 \\ -y_2 & y_3 & y_0 & -y_1 \\ -y_3 & -y_2 & y_1 & y_0 \end{bmatrix} = \begin{bmatrix} z_0 & z_1 & z_2 & z_3 \\ -z_1 & z_0 & -z_3 & z_2 \\ -z_2 & z_3 & z_0 & -z_1 \\ -z_3 & -z_2 & z_1 & z_0 \end{bmatrix}.$$

Hurwitz and Radon proved that dimensions N = 1, 2, 4 and 8 are the only dimensions for these quadratic equations. When $N = 8$:

$$\left(\sum_{i=0}^{7} x_i^2\right) \cdot \left(\sum_{i=0}^{7} y_i^2\right) = \sum_{i=0}^{7} z_i^2$$

for

$$z_0 = x_0y_0 - x_1y_1 - x_2y_2 - x_3y_3 - x_4y_4 - x_5y_5 - x_6y_6 - x_7y_7,$$

$$z_1 = x_0y_1 + x_1y_0 - x_2y_3 + x_3y_2 - x_4y_5 + x_5y_4 + x_6y_7 - x_7y_6,$$

$z_2 = x_0y_2 + x_1y_3 + x_2y_0 - x_3y_1 - x_4y_6 - x_5y_7 + x_6y_4 + x_7y_{5,}$

$z_3 = x_0y_3 - x_1y_2 + x_2y_1 + x_3y_0 - x_4y_7 + x_5y_6 - x_6y_5 + x_7y_{4,}$

$z_4 = x_0y_4 + x_1y_5 + x_2y_6 + x_3y_7 + x_4y_0 - x_5y_1 - x_6y_2 - x_7y_{3,}$

$z_5 = x_0y_5 - x_1y_4 + x_2y_7 - x_3y_6 + x_4y_1 + x_5y_0 + x_6y_3 - x_7y_{2,}$

$z_6 = x_0y_6 - x_1y_7 - x_2y_4 + x_3y_5 + x_4y_2 - x_5y_3 + x_6y_0 + x_7y_{1,}$

$z_7 = x_0y_7 + x_1y_6 - x_2y_5 - x_3y_4 + x_4y_3 + x_5y_2 - x_6y_1 + x_7y_{0.}$

The matrices used to solve quadratic equations are defined: matrices A_i, $i = 1, 2, \ldots, m$ satisfying

$A_jA_k + A_kA_j = 0, A_j^2 = -I, j \neq k, j, k = 1, 2, ..., m$

are called a family of Hurwitz - Radon matrices.

A family of Hurwitz - Radon (HR) matrices has important features (Eckmann, 1999): HR matrices are skew-symmetric ($A_i^T = - A_i$) and reverse matrices are easy to find ($A_i^{-1} = - A_i$). Only for dimension $N = 1, 2, 4$ or 8 the family of HR matrices consists of N - 1 matrices. When $N = 1$ there is no matrix, just calculations with real numbers. For $N = 2$ we have one matrix:

$$A_1 = \begin{bmatrix} 0 & 1 \\ -1 & 0 \end{bmatrix}.$$

For $N = 4$ there are three HR matrices with integer entries:

$$A_1 = \begin{bmatrix} 0 & 1 & 0 & 0 \\ -1 & 0 & 0 & 0 \\ 0 & 0 & 0 & -1 \\ 0 & 0 & 1 & 0 \end{bmatrix}, A_2 = \begin{bmatrix} 0 & 0 & 1 & 0 \\ 0 & 0 & 0 & 1 \\ -1 & 0 & 0 & 0 \\ 0 & -1 & 0 & 0 \end{bmatrix}, A_3 = \begin{bmatrix} 0 & 0 & 0 & 1 \\ 0 & 0 & -1 & 0 \\ 0 & 1 & 0 & 0 \\ -1 & 0 & 0 & 0 \end{bmatrix}.$$

For $N = 8$ we have seven HR matrices with elements 0, ± 1 (Jakóbczak, 2006).

So far HR matrices are applied in electronics (Citko, Jakóbczak & Sieńko, 2005): in Space-Time Block Coding (STBC) and orthogonal design (Tarokh, Jafarkhani & Calderbank, 1999), also in signal processing (Sieńko, Citko & Wilamowski, 2002) and Hamiltonian Neural Nets (Sieńko & Citko, 2002).

The Operator of Hurwitz - Radon

Here is the beginning of proposed MHR method. Let us consider a combination of identity matrix and HR matrix of dimension $N = 2$:

$$a\begin{bmatrix}1 & 0\\0 & 1\end{bmatrix}+b\begin{bmatrix}0 & 1\\-1 & 0\end{bmatrix}=\begin{bmatrix}a & b\\-b & a\end{bmatrix}.$$

For any points $(x_1,y_1) \in \boldsymbol{R}^2$, $(x_2,y_2) \in \boldsymbol{R}^2$ matrix equation

$$\begin{bmatrix}a & b\\-b & a\end{bmatrix}\cdot\begin{bmatrix}x_1\\x_2\end{bmatrix}=\begin{bmatrix}y_1\\y_2\end{bmatrix}$$

is true with

$$a=\frac{x_1y_1+x_2y_2}{x_1^{\,2}+x_2^{\,2}},\ b=\frac{x_2y_1-x_1y_2}{x_1^{\,2}+x_2^{\,2}} \text{ and } x_1^{\,2}+x_2^{\,2}>0.$$

Reverse matrix equation

$$\begin{bmatrix}a & b\\-b & a\end{bmatrix}\cdot\begin{bmatrix}y_1\\y_2\end{bmatrix}=\begin{bmatrix}x_1\\x_2\end{bmatrix}$$

is true with

$$a=\frac{x_1y_1+x_2y_2}{y_1^{\,2}+y_2^{\,2}},\ b=\frac{-x_2y_1+x_1y_2}{y_1^{\,2}+y_2^{\,2}} \text{ and } y_1^{\,2}+y_2^{\,2}>0.$$

Also we can consider a combination of identity matrix and three HR matrices of dimension $N = 4$:

$$a\begin{bmatrix}1 & 0 & 0 & 0\\0 & 1 & 0 & 0\\0 & 0 & 1 & 0\\0 & 0 & 0 & 1\end{bmatrix}+b\begin{bmatrix}0 & 1 & 0 & 0\\-1 & 0 & 0 & 0\\0 & 0 & 0 & -1\\0 & 0 & 1 & 0\end{bmatrix}+c\begin{bmatrix}0 & 0 & 1 & 0\\0 & 0 & 0 & 1\\-1 & 0 & 0 & 0\\0 & -1 & 0 & 0\end{bmatrix}$$
$$+d\begin{bmatrix}0 & 0 & 0 & 1\\0 & 0 & -1 & 0\\0 & 1 & 0 & 0\\-1 & 0 & 0 & 0\end{bmatrix}=\begin{bmatrix}a & b & c & d\\-b & a & -d & c\\-c & d & a & -b\\-d & -c & b & a\end{bmatrix}.$$

For any points $(x_1,y_1) \in \boldsymbol{R}^2$, $(x_2,y_2) \in \boldsymbol{R}^2$, $(x_3,y_3) \in \boldsymbol{R}^2$, $(x_4,y_4) \in \boldsymbol{R}^2$ matrix equation

$$\begin{bmatrix} a & b & c & d \\ -b & a & -d & c \\ -c & d & a & -b \\ -d & -c & b & a \end{bmatrix} \cdot \begin{bmatrix} x_1 \\ x_2 \\ x_3 \\ x_4 \end{bmatrix} = \begin{bmatrix} y_1 \\ y_2 \\ y_3 \\ y_4 \end{bmatrix}$$

is satisfied with

$$a = \frac{x_1y_1 + x_2y_2 + x_3y_3 + x_4y_4}{x_1^2 + x_2^2 + x_3^2 + x_4^2},\ b = \frac{-x_1y_2 + x_2y_1 + x_3y_4 - x_4y_3}{x_1^2 + x_2^2 + x_3^2 + x_4^2},$$

$$c = \frac{-x_1y_3 - x_2y_4 + x_3y_1 + x_4y_2}{x_1^2 + x_2^2 + x_3^2 + x_4^2},\ d = \frac{-x_1y_4 + x_2y_3 - x_3y_2 + x_4y_1}{x_1^2 + x_2^2 + x_3^2 + x_4^2} \text{ and } x_1^2 + x_2^2 + x_3^2 + x_4^2 > 0.$$

Reverse matrix equation

$$\begin{bmatrix} a & b & c & d \\ -b & a & -d & c \\ -c & d & a & -b \\ -d & -c & b & a \end{bmatrix} \cdot \begin{bmatrix} y_1 \\ y_2 \\ y_3 \\ y_4 \end{bmatrix} = \begin{bmatrix} x_1 \\ x_2 \\ x_3 \\ x_4 \end{bmatrix}$$

is satisfied with

$$a = \frac{x_1y_1 + x_2y_2 + x_3y_3 + x_4y_4}{y_1^2 + y_2^2 + y_3^2 + y_4^2},\ b = \frac{x_1y_2 - x_2y_1 - x_3y_4 + x_4y_3}{y_1^2 + y_2^2 + y_3^2 + y_4^2},$$

$$c = \frac{x_1y_3 + x_2y_4 - x_3y_1 - x_4y_2}{y_1^2 + y_2^2 + y_3^2 + y_4^2},\ d = \frac{x_1y_4 - x_2y_3 + x_3y_2 - x_4y_1}{y_1^2 + y_2^2 + y_3^2 + y_4^2} \text{ and } y_1^2 + y_2^2 + y_3^2 + y_4^2 > 0.$$

A combination of identity matrix and seven HR matrices of dimension $N = 8$ looks as follows:

$$\begin{bmatrix} a_0 & a_1 & a_2 & a_3 & a_4 & a_5 & a_6 & a_7 \\ -a_1 & a_0 & a_3 & -a_2 & a_5 & -a_4 & -a_7 & a_6 \\ -a_2 & -a_3 & a_0 & a_1 & a_6 & a_7 & -a_4 & -a_5 \\ -a_3 & a_2 & -a_1 & a_0 & a_7 & -a_6 & a_5 & -a_4 \\ -a_4 & -a_5 & -a_6 & -a_7 & a_0 & a_1 & a_2 & a_3 \\ -a_5 & a_4 & -a_7 & a_6 & -a_1 & a_0 & -a_3 & a_2 \\ -a_6 & a_7 & a_4 & -a_5 & -a_2 & a_3 & a_0 & -a_1 \\ -a_7 & -a_6 & a_5 & a_4 & -a_3 & -a_2 & a_1 & a_0 \end{bmatrix}.$$

Results for matrix equations with a combination of identity matrix and seven HR matrices of dimension $N = 8$ are calculated in formulas (3), (4) and (8). Solutions of matrix equations (Sieńko, Citko & Jakóbczak, 2004) are used to build the matrix Operator of Hurwitz – Radon (OHR).

Let's assume there is given a finite set of points of the curve, called further nodes $(x_i,y_i) \in \mathbf{R}^2$ such as:

1. nodes (interpolation points) are settled at local extrema (maximum or minimum) of one of coordinates and at least one point between two successive local extrema;
2. each node (x_i,y_i) is monotonic in coordinates x_i or y_i (for example equidistance in one of coordinates).

Assume that the nodes belong to a curve in the plane. How the whole curve could be interpolated or extrapolated using this discrete set of nodes? Proposed method (Jakóbczak, 2007) is based on local, orthogonal matrix operators. The values of nodes' coordinates (x_i,y_i) are connected with HR matrices, built on N - dimensional vector space. It is important that HR matrices are skew-symmetric and only for dimensions $N = 1, 2, 4$ or 8 columns and rows of HR matrices are orthogonal (Eckmann, 1999).

If one curve is described by a set of nodes $\{(x_i,y_i), i = 1, 2, \ldots, n\}$ monotonic (for example equidistance) in coordinates x_i, then HR matrices combined with identity matrix are used to build an orthogonal and discrete Hurwitz - Radon Operator (OHR). For nodes (x_1,y_1), (x_2,y_2) OHR of dimension $N = 2$ is constructed:

$$M = \frac{1}{x_1^2 + x_2^2} \begin{bmatrix} x_1 & x_2 \\ -x_2 & x_1 \end{bmatrix} \begin{bmatrix} y_1 & -y_2 \\ y_2 & y_1 \end{bmatrix},$$

$$M = \frac{1}{x_1^2 + x_2^2} \begin{bmatrix} x_1y_1 + x_2y_2 & x_2y_1 - x_1y_2 \\ x_1y_2 - x_2y_1 & x_1y_1 + x_2y_2 \end{bmatrix}. \qquad (1)$$

For nodes (x_1,y_1), (x_2,y_2), (x_3,y_3), (x_4,y_4) monotonic (for example equidistance) in x_i OHR of dimension $N = 4$ is constructed:

$$M = \frac{1}{x_1^2 + x_2^2 + x_3^2 + x_4^2} \begin{bmatrix} u_0 & u_1 & u_2 & u_3 \\ -u_1 & u_0 & -u_3 & u_2 \\ -u_2 & u_3 & u_0 & -u_1 \\ -u_3 & -u_2 & u_1 & u_0 \end{bmatrix} \tag{2}$$

where

$u_0 = x_1y_1 + x_2y_2 + x_3y_3 + x_4y_4$, $u_1 = -x_1y_2 + x_2y_1 + x_3y_4 - x_4y_3$, $u_2 = -x_1y_3 - x_2y_4 + x_3y_1 + x_4y_2$, $u_3 = -x_1y_4 + x_2y_3 - x_3y_2 + x_4y_1$.

For nodes (x_1,y_1), (x_2,y_2), …, (x_8,y_8) monotonic in x_i OHR of dimension $N = 8$ is equal with

$$M = \frac{1}{\sum_{i=1}^{8} x_i^2} \begin{bmatrix} u_0 & u_1 & u_2 & u_3 & u_4 & u_5 & u_6 & u_7 \\ -u_1 & u_0 & u_3 & -u_2 & u_5 & -u_4 & -u_7 & u_6 \\ -u_2 & -u_3 & u_0 & u_1 & u_6 & u_7 & -u_4 & -u_5 \\ -u_3 & u_2 & -u_1 & u_0 & u_7 & -u_6 & u_5 & -u_4 \\ -u_4 & -u_5 & -u_6 & -u_7 & u_0 & u_1 & u_2 & u_3 \\ -u_5 & u_4 & -u_7 & u_6 & -u_1 & u_0 & -u_3 & u_2 \\ -u_6 & u_7 & u_4 & -u_5 & -u_2 & u_3 & u_0 & -u_1 \\ -u_7 & -u_6 & u_5 & u_4 & -u_3 & -u_2 & u_1 & u_0 \end{bmatrix} \tag{3}$$

where

$$\mathbf{u} = \begin{bmatrix} y_1 & y_2 & y_3 & y_4 & y_5 & y_6 & y_7 & y_8 \\ -y_2 & y_1 & -y_4 & y_3 & -y_6 & y_5 & y_8 & -y_7 \\ -y_3 & y_4 & y_1 & -y_2 & -y_7 & -y_8 & y_5 & y_6 \\ -y_4 & -y_3 & y_2 & y_1 & -y_8 & y_7 & -y_6 & y_5 \\ -y_5 & y_6 & y_7 & y_8 & y_1 & -y_2 & -y_3 & -y_4 \\ -y_6 & -y_5 & y_8 & -y_7 & y_2 & y_1 & y_4 & -y_3 \\ -y_7 & -y_8 & -y_5 & y_6 & y_3 & -y_4 & y_1 & y_2 \\ -y_8 & y_7 & -y_6 & -y_5 & y_4 & y_3 & -y_2 & y_1 \end{bmatrix} \cdot \begin{bmatrix} x_1 \\ x_2 \\ x_3 \\ x_4 \\ x_5 \\ x_6 \\ x_7 \\ x_8 \end{bmatrix} \tag{4}$$

We can see here that the components of the vector $\mathbf{u} - (u_0, u_1, \ldots, u_7)^T$, appearing in the matrix M in formula (3) are defined by (4) in the similar way to formula (2) but in terms of the coordinates of the above 8 nodes. Note that OHR operators (1)-(3) satisfy the condition of interpolation

$$M \cdot \mathbf{x} = \mathbf{y} \tag{5}$$

for $\mathbf{x} = (x_1, x_2 \ldots, x_N)^T \in \boldsymbol{R}^N$, $\mathbf{x} \neq \mathbf{0}$, $\mathbf{y} = (y_1, y_2 \ldots, y_N)^T \in \boldsymbol{R}^N$, $N = 1, 2, 4$ or 8.

If one curve is described by a set of nodes $\{(x_i,y_i), i = 1, 2, \ldots, n\}$ monotonic (for example equidistance) in coordinates y_i, then HR matrices combined with identity matrix are used to build an orthogonal and discrete reverse Hurwitz - Radon Operator (reverse OHR). For nodes (x_1,y_1), (x_2,y_2) reverse OHR of dimension $N = 2$ is constructed:

$$M^{-1} = \frac{1}{y_1^2 + y_2^2}\begin{bmatrix} x_1 & -x_2 \\ x_2 & x_1 \end{bmatrix}\begin{bmatrix} y_1 & y_2 \\ -y_2 & y_1 \end{bmatrix},$$

$$M^{-1} = \frac{1}{y_1^2 + y_2^2}\begin{bmatrix} x_1y_1 + x_2y_2 & -x_2y_1 + x_1y_2 \\ -x_1y_2 + x_2y_1 & x_1y_1 + x_2y_2 \end{bmatrix}. \tag{6}$$

For nodes (x_1,y_1), (x_2,y_2), (x_3,y_3), (x_4,y_4) monotonic in y_i the reverse OHR of dimension $N = 4$ is constructed with u_o, u_1, u_2 and u_3 from (2):

$$M^{-1} = \frac{1}{y_1^2 + y_2^2 + y_3^2 + y_4^2}\begin{bmatrix} u_0 & -u_1 & -u_2 & -u_3 \\ u_1 & u_0 & u_3 & -u_2 \\ u_2 & -u_3 & u_0 & u_1 \\ u_3 & u_2 & -u_1 & u_0 \end{bmatrix}. \tag{7}$$

For nodes (x_1,y_1), (x_2,y_2), …, (x_8,y_8) monotonic in y_i the reverse OHR of dimension $N = 8$ is equal with

$$M^{-1} = \frac{1}{\sum_{i=1}^{8} y_i^2}\begin{bmatrix} u_0 & -u_1 & -u_2 & -u_3 & -u_4 & -u_5 & -u_6 & -u_7 \\ u_1 & u_0 & -u_3 & u_2 & -u_5 & u_4 & u_7 & -u_6 \\ u_2 & u_3 & u_0 & -u_1 & -u_6 & -u_7 & u_4 & u_5 \\ u_3 & -u_2 & u_1 & u_0 & -u_7 & u_6 & -u_5 & u_4 \\ u_4 & u_5 & u_6 & u_7 & u_0 & -u_1 & -u_2 & -u_3 \\ u_5 & -u_4 & u_7 & -u_6 & u_1 & u_0 & u_3 & -u_2 \\ u_6 & -u_7 & -u_4 & u_5 & u_2 & -u_3 & u_0 & u_1 \\ u_7 & u_6 & -u_5 & -u_4 & u_3 & u_2 & -u_1 & u_0 \end{bmatrix}, \tag{8}$$

where the components of the vector $\mathbf{u} = (u_0, u_1, \ldots, u_7)^T$ are defined in terms of (4). Note that reverse OHR operators (6)-(8) satisfy the condition of interpolation

$$M^{-1}\cdot\mathbf{y} = \mathbf{x} \tag{9}$$

for $\mathbf{x} = (x_1,x_2\ldots,x_N)^T \in \boldsymbol{R}^N$, $\mathbf{y} = (y_1,y_2\ldots,y_N)^T \in \boldsymbol{R}^N$, $\mathbf{y} \neq \mathbf{0}$, $N = 1, 2, 4$ or 8.

Known values, for example currency rates or market prices, are represented on 2D curve by the set of nodes $(x_i,y_i) \in \boldsymbol{R}^2$ (characteristic points) as follows in proposed MHR method:

1. nodes (interpolation points) are settled at local extrema (maximum or minimum) of one of coordinates and at least one point between two successive local extrema;
2. nodes (x_i,y_i) are monotonic in coordinates x_i ($x_i < x_{i+1}$ for all i) or y_i ($y_i < y_{i+1}$);
3. one curve is represented by at least four nodes.

Condition 1 is done for the most appropriate description of a curve. The quotations or prices are real data. Condition 2 according to a graph of function means that coordinates x_i represent for example the time. Condition 3 is adequate for extrapolation, but in the case of interpolation minimal number of nodes is five.

Figure 1. Five nodes of data and a curve
Source: author, 2014

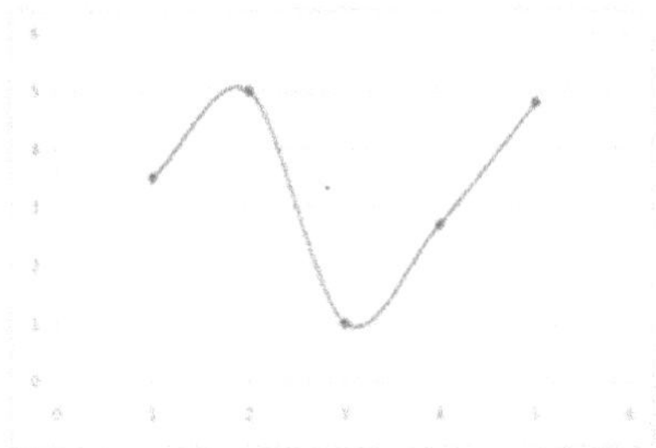

Data points are treated as interpolation nodes. How can we extrapolate continues values at time x = 5.5 for example or discrete data for next day x = 6 (Fig.1)? The anticipation of values is possible using proposed MHR method.

The Method of Hurwitz - Radon Matrices

Key question looks as follows: how can we compute coordinates of points settled between interpolation nodes or beyond the nodes? The answer is connected with MHR method for interpolation and extrapolation. On a segment of a line every number "c" situated between "a" and "b" is described by a linear (convex) combination $c = \alpha \cdot a + (1\text{-}\alpha) \cdot b$ for

$$\alpha = \frac{b-c}{b-a} \in [0;1]. \tag{10}$$

If $c < a$ then $\alpha > 1$: there is extrapolation of points situated left of nodes. If $c > b$ then $\alpha < 0$: extrapolation of points situated right of nodes.

When the nodes are monotonic in coordinates x_i, average OHR operator M_2 of dimension N = 1, 2, 4 or 8 is constructed as follows:

$$M_2 = \alpha \cdot M_0 + (1-\alpha) \cdot M_1 \quad (11)$$

with the operator M_0 built (1)-(3) by "odd" nodes $(x_1=a,y_1)$, (x_3,y_3), …, (x_{2N-1},y_{2N-1}) and M_1 built (1)-(3) by "even" nodes $(x_2=b,y_2)$, (x_4,y_4), …, (x_{2N},y_{2N}).

When the nodes are monotonic in coordinates y_i, average reverse OHR operator M_2^{-1} of dimension $N = 1, 2, 4$ or 8 is constructed as follows:

$$M_2^{-1} = \alpha \cdot M_0^{-1} + (1-\alpha) \cdot M_1^{-1} \quad (12)$$

with the reverse operator M_0^{-1} built (6)-(8) by nodes $(x_1,y_1=a)$, (x_3,y_3), …, (x_{2N-1},y_{2N-1}) and M_1^{-1} built (6)-(8) by nodes $(x_2,y_2=b)$, (x_4,y_4), …, (x_{2N},y_{2N}).

Notice that having the operator M_2 for coordinates $x_i < x_{i+1}$ it is possible to reconstruct the second coordinates of points (x,y) in terms of the vector C defined with

$$c_i = \alpha \cdot x_{2i-1} + (1-\alpha) \cdot x_{2i},\ i = 1, 2,\dots, N \quad (13)$$

as $C = [c_1, c_2,\dots, c_N]^T$. The required formula is similar to (5):

$$Y(C) = M_2 \cdot C \quad (14)$$

in which components of vector $Y(C)$ give the second coordinate of the points (x,y) corresponding to the first coordinate, given in terms of components of the vector C.

On the other hand, having the operator M_2^{-1} for coordinates $y_i < y_{i+1}$ it is possible to reconstruct the first coordinates of points (x,y) in terms of the corresponding second coordinates given by components of the new vector C defined, as previously, with

$$c_i = \alpha \cdot y_{2i-1} + (1-\alpha) \cdot y_{2i},\ i = 1, 2,\dots, N \quad (15)$$

and $C = [c_1, c_2,\dots, c_N]^T$. The final formula is similar to (9):

$$X(C) = M_2^{-1} \cdot C \quad (16)$$

in which components of the vector $X(C)$ give the first coordinate of the points (x,y) corresponding to the second coordinate, given in terms of components of the vector C.

After computing (14) or (16) for any $\alpha \in [0;1]$, we have a half of reconstructed points ($j = 1$ in algorithm 1). Now it is necessary to find second half of unknown coordinates ($j = 2$ in algorithm 1) for

$$c_i = \alpha \cdot x_{2i} + (1-\alpha) \cdot x_{2i+1},\ i = 1, 2,\dots, N \quad (17)$$

or

$$c_i = \alpha \cdot y_{2i} + (1-\alpha) \cdot y_{2i+1},\ i = 1, 2,\dots, N \quad (18)$$

depending on whether x_i (17) or y_i (18) is monotonic. There is no need to build the OHR for nodes $(x_2=a,y_2)$, (x_4,y_4), …, (x_{2N},y_{2N}) or the reverse OHR for nodes $(x_2,y_2=a)$, (x_4,y_4), …, (x_{2N},y_{2N}), because we have just found M_1 or M_1^{-1}. This operator will play the role as M_0 or M_0^{-1} in (11) or (12). New M_1 or M_1^{-1} must be computed for nodes $(x_3=b,y_3)$, …, (x_{2N-1},y_{2N-1}), (x_{2N+1},y_{2N+1}) or $(x_3,y_3=b)$, …, (x_{2N-1},y_{2N-1}), (x_{2N+1},y_{2N+1}).

As we see the minimum number of interpolation nodes $n = 2N+1 = 5$, 9 or 17 using OHR operators of dimension $N = 2$, 4 or 8 respectively. If there is more nodes than $2N+1$, the same calculations (10)-(18) have to be done for next range(s) or last range of $2N+1$ nodes. For example, if $n = 9$ then we can use OHR operators of dimension $N = 4$ or OHR operators of dimension $N = 2$ for two subsets of nodes: $\{(x_1,y_1), \ldots, (x_5,y_5)\}$ and $\{(x_5,y_5), \ldots,(x_9,y_9)\}$.

Here is the application of MHR method for function $f(x)=1/(1+25x^2)$ with five nodes equidistance in first coordinate: $x_i = -1, -0.5, 0, 0.5, 1$.

Figure 2. Twenty six interpolated points of function $f(x) = 1/(1+25x^2)$ using MHR method with 5 nodes
Source: author, 2014

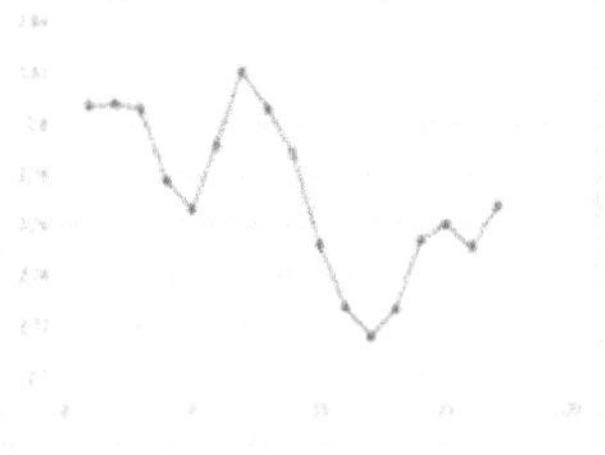

MHR interpolation of function $f(x) = 1/(1+25x^2)$ gives better result then polynomial interpolation: there is no Runge phenomenon. The same can be said for function $f(x) = 1/x$.

MHR extrapolation is valid for $\alpha < 0$ or $\alpha > 1$. In the case of continues data, parameter α is a real number. For example there are four nodes: (1;2), (1.3;5), (2;3), (2.5;6). MHR extrapolation with $\alpha = -0.01$ gives the point (2.505;6.034) and with $\alpha = -0.1$: (2.55;6.348). But the rate of a currency or the quotations are discrete data. If we assume that the rate of a currency is represented by equidistance nodes (day by day – fixed step of time $h = 1$ for coordinate x), next data or the rate on next day is extrapolated (anticipated) for $\alpha = -1$. Calculation of unknown coordinates for data points using formulas (10)-(18) is called by author the method of Hurwitz - Radon Matrices (MHR).

Algorithm and Complexity of MHR Calculations

The algorithm of points reconstruction for $2N+1 = 5$, 9 or 17 successive nodes is presented.

Algorithm 1

Let $j = 1$. *Input: Set of interpolation nodes $\{(x_i,y_i)$, $i = 1,2,\ldots,n$; $n = 5$, 9 or 17$\}$ such as:*

a) nodes are settled at local extrema (maximum or minimum) of one of coordinates and at least one point between two successive local extrema;
b) nodes (x_i,y_i) are monotonic in coordinates x_i or y_i.

Step 1: *Determine the dimension N of OHR operators:* $N = 2$ *if* $n = 5$, $N = 4$ *if* $n = 9$, $N = 8$ *if* $n = 17$.
Step 2: *If nodes are monotonic in* x_i *then build* M_0 *for nodes* $(x_1=a,y_1)$, (x_3,y_3), …, (x_{2N-1},y_{2N-1}) *and* M_1 *for nodes* $(x_2=b,y_2)$, (x_4,y_4), …, (x_{2N},y_{2N}) *from (1)-(3). If nodes are monotonic in* y_i *then build* M_0^{-1} *for nodes* $(x_1,y_1=a)$, (x_3,y_3), …,(x_{2N-1},y_{2N-1}) *and* M_1^{-1} *for nodes* $(x_2,y_2=b)$, (x_4,y_4), …, (x_{2N},y_{2N}) *from* (6)-(8).
Step 3: *Determine the number of points to be reconstructed* $K_j > 0$ *between two successive nodes, let* $k = 1$.
Step 4: *Compute* $\alpha \in [0;1]$ *from (10) for* $c_1 = c = \alpha\cdot a + (1-\alpha)\cdot b$.
Step 5: *Build* M_2 *from (11) or* M_2^{-1} *from (12).*
Step 6: *Compute vector* $C = [c_1, c_2,\ldots, c_N]^T$ *from (13) or (15).*
Step 7: *Compute unknown coordinates Y(C) from (14) or X(C) from (16).*
Step 8: *If* $k < K_j$, *set* $k = k + 1$ *and go to Step 4. Otherwise if* $j = 1$, *set* $M_0 = M_1$, $a = x_2$, $b = x_3$ *(if nodes are monotonic in* x_i*) or* $M_0^{-1} = M_1^{-1}$, $a = y_2$, $b = y_3$ *(if nodes are monotonic in* y_i*), build new* M_1 *or* M_1^{-1} *for nodes* (x_3,y_3), (x_5,y_5), …,(x_{2N+1},y_{2N+1}), *let* $j = 2$ *and go to Step 3. Otherwise, stop.*

The number of reconstructed points in algorithm 1 is $K = N(K_1+K_2)$. If there are more nodes than $2N+1 = 5$, 9 or 17, algorithm 1 has to be done for next range(s) or last range of $2N + 1$ nodes. Reconstruction of curve points using algorithm 1 is called by author the method of Hurwitz - Radon Matrices (MHR). If we have n interpolation nodes, then there is $K = L - n$ points to find using algorithm 1 and MHR method. The complexity of MHR calculations has to be considered.

Lemma 1. Let $n = 5$, 9 or 17 is the number of interpolation nodes, let MHR method (algorithm 1) is done for reconstruction of the curve consists of L points. Then MHR method is connected with the computational cost of rank $O(L)$.

Proof. Using algorithm 1 we have to reconstruct $K = L - n$ points of unknown curve. Counting the number of multiplications and divisions D in algorithm 1, for example in (20) in case $n = 5$ for each c_i at first and second half of reconstructed points, here are the results:

1) $D = 4L+7$ for $n = 5$ and $L = 2i + 5$;
2) $D = 6L+21$ for $n = 9$ and $L = 4i + 9$;
3) $D = 10L+73$ for $n = 17$ and $L = 8i + 17$; $i = 2,3,4$...

The lowest computational cost appears in MHR method with five nodes and OHR operators of dimension $N = 2$. Therefore whole set of n nodes can be divided into subsets of five nodes. Then whole curve is to be reconstructed by algorithm 1 with all subsets of five nodes: $\{(x_1,y_1), \ldots, (x_5,y_5)\}$, $\{(x_5,y_5), \ldots, (x_9,y_9)\}$,

$\{(x_9,y_9), \ldots, (x_{13},y_{13})\}$… If the last node (x_n,y_n) is indexed $n \neq 4i + 1$, then we have to use last five nodes $\{(x_{n-4},y_{n-4}), \ldots, (x_n,y_n)\}$ in algorithm 1.

The Formulas for a Single Point and Error Estimation

Now there are the formulas for computing one unknown coordinate of a single point. Assume there are given four nodes (x_1,y_1), (x_2,y_2), (x_3,y_3), (x_4,y_4) monotonic in x_i. OHR operators of dimension $N = 2$ are built (1) as follows:

$$M_0 = \frac{1}{x_1^2 + x_3^2}\begin{bmatrix} x_1y_1 + x_3y_3 & x_3y_1 - x_1y_3 \\ x_1y_3 - x_3y_1 & x_1y_1 + x_3y_3 \end{bmatrix},\ M_1 = \frac{1}{x_2^2 + x_4^2}\begin{bmatrix} x_2y_2 + x_4y_4 & x_4y_2 - x_2y_4 \\ x_2y_4 - x_4y_2 & x_2y_2 + x_4y_4 \end{bmatrix}.$$

Let first coordinate c_1 of reconstructed point is situated between x_1 and x_2:

$c_1 = \alpha \cdot x_1 + \beta \cdot x_2$ for $0 \leq \beta = 1 - \alpha \leq 1$. (19)

Compute second coordinate of reconstructed point $y(c_1)$ for $Y(C) = [y(c_1), y(c_2)]^T$ from (14):

$$\begin{bmatrix} y(c_1) \\ y(c_2) \end{bmatrix} = (\alpha \cdot M_0 + \beta \cdot M_1) \cdot \begin{bmatrix} \alpha \cdot x_1 + \beta \cdot x_2 \\ \alpha \cdot x_3 + \beta \cdot x_4 \end{bmatrix}. \tag{20}$$

After calculation (20):

$$y(c_1) = \alpha^2 \cdot y_1 + \beta^2 \cdot y_2 + \frac{\alpha \cdot \beta}{x_1^2 + x_3^2}(x_1x_2y_1 + x_2x_3y_3 + x_3x_4y_1 - x_1x_4y_3) +$$

$$+ \frac{\alpha \cdot \beta}{x_2^2 + x_4^2}(x_1x_2y_2 + x_1x_4y_4 + x_3x_4y_2 - x_2x_3y_4). \tag{21}$$

So each point of the curve $P = (c_1, y(c_1))$ settled between nodes (x_1,y_1) and (x_2,y_2) is parameterized by $P(\alpha)$ for (19), (21) and $\alpha \in [0;1]$. Similar calculations are done for nodes (x_1,y_1), (x_2,y_2), (x_3,y_3), (x_4,y_4) monotonic in y_i to compute $x(c_1)$:

$$, M_0^{-1} = \frac{1}{y_1^2 + y_3^2}\begin{bmatrix} x_1y_1 + x_3y_3 & -x_3y_1 + x_1y_3 \\ -x_1y_3 + x_3y_1 & x_1y_1 + x_3y_3 \end{bmatrix},$$

$$M_1^{-1} = \frac{1}{y_2^2 + y_4^2}\begin{bmatrix} x_2y_2 + x_4y_4 & -x_4y_2 + x_2y_4 \\ -x_2y_4 + x_4y_2 & x_2y_2 + x_4y_4 \end{bmatrix},$$

$c_1 = \alpha \cdot y_1 + \beta \cdot y_2$ for $0 \leq \beta = 1 - \alpha \leq 1$, (22)

$$\begin{bmatrix} x(c_1) \\ x(c_2) \end{bmatrix} = (\alpha \cdot M_0^{-1} + \beta \cdot M_1^{-1}) \cdot \begin{bmatrix} \alpha \cdot y_1 + \beta \cdot y_2 \\ \alpha \cdot y_3 + \beta \cdot y_4 \end{bmatrix},$$

$$x(c_1) = \alpha^2 \cdot x_1 + \beta^2 \cdot x_2 + \frac{\alpha \cdot \beta}{y_1^2 + y_3^2} r_1 + \frac{\alpha \cdot \beta}{y_2^2 + y_4^2} r_2 \tag{23}$$

for r_1 = const., r_2 = const. depending on nodes' coordinates: see formula (21). If nodes are monotonic in y_i, there is parameterization of curve points P settled between nodes (x_1,y_1) and (x_2,y_2): $P(\alpha) = (x(c_1), c_1)$ for (22), (23) and $\alpha \in [0;1]$.

If nodes (x_i,y_i) are equidistance in one coordinate, then calculation of one unknown coordinate is simpler. Let four successive nodes (x_1,y_1), (x_2,y_2), (x_3,y_3), (x_4,y_4) are equidistance in coordinate x_i and

$a = x_1$, $h/2 = x_{i+1} - x_i$ = const. Calculations in formulas (20)-(21) are done for c_1 (19):

$$y(c_1) = \alpha y_1 + \beta y_2 + \alpha\beta s \tag{24}$$

and

$$s = h\left(\frac{2ay_1 + hy_1 + hy_3}{4a^2 + 4ah + 2h^2} - \frac{2ay_2 + 2hy_2 + hy_4}{4a^2 + 8ah + 5h^2}\right). \tag{25}$$

As we can see in formulas (21) and (23)-(25), MHR interpolation is not a linear interpolation. It is possible to estimate the interpolation error of MHR method (algorithm 1) for the class of linear function f:

$$|f(c_1) - y(c_1)| = |\alpha y_1 + \beta y_2 - y(c_1)| = \alpha\beta |s|. \tag{26}$$

Notice that estimation (26) has the biggest value $0.25|s|$ for $\beta = \alpha = 0.5$, when c_1 (19) is situated in the middle between x_1 and x_2.

Having four successive nodes (x_1,y_1), (x_2,y_2), (x_3,y_3), (x_4,y_4) equidistance in coordinate x_i ($a = x_1$, $h/2 = x_{i+1} - x_i$ = const.) we can compute polynomial $W_3(x) = m_3x^3 + m_2x^2 + m_1x + t$ for these nodes and estimate the interpolation error of MHR method (algorithm 1) for the class of order three polynomials. After solving the system of equations:

$$y_1 = m_3a^3 + m_2a^2 + m_1a + t,$$

$$y_2 = m_3\left(a + \frac{h}{2}\right)^3 + m_2\left(a + \frac{h}{2}\right)^2 + m_1\left(a + \frac{h}{2}\right) + t,$$

$$y_3 = m_3(a+h)^3 + m_2(a+h)^2 + m_1(a+h) + t,$$

$$y_4 = m_3\left(a + \frac{3h}{2}\right)^3 + m_2\left(a + \frac{3h}{2}\right)^2 + m_1\left(a + \frac{3h}{2}\right) + t,$$

it is possible to compute m_3, m_2, m_1, t and the estimation for point c_1 (19):

$$|W_3(c_1) - y(c_1)| = \beta \left| \frac{1}{2} h(y_2 - y_1) \frac{12\alpha\beta a + 12a\beta^2 + 2h\beta^2 - \alpha\beta a h^2 - 12a - 2h}{6ha + 12a^2 + h^2} - \alpha \cdot s \right|. \tag{27}$$

Notice that estimations (26)-(27) are equal with zero for α = 0 or β = 0 (in nodes). For eight successive nodes (x_i, y_i), $i = 1,2,\ldots,8$ equidistance in coordinate x_i, $a = x_1$, $h/2 = x_{i+1} - x_i$ = = const., using OHR of dimension $N = 4$ in formula (14) here is a formula of second coordinate reconstruction with first coordinate c_1 (19):

So we have another parameterization (28) of the point $P(\alpha) = (c_1, y(c_1))$ for $N = 4$ and $\beta = 1 - \alpha$. Formula (28) doesn't include values y_5 and y_6: algorithm 1 with nine successive nodes (x_i, y_i), $i = 1,2,\ldots,9$ equidistance in coordinate x_i is free of using y_5 and y_6 for computing second coordinate of the point settled between first and second node.

Algorithm 1 deals with average OHR operators (11)-(12) built with two OHR. This situation leads to parameterization of reconstructed point $P(\alpha) = (c_1, y(c_1))$ or $P(\alpha) = (x(c_1), c_1)$ settled between two successive nodes, where $\alpha \in [0;1]$ is order two in formulas (21), (23)-(25) and (28). The curve or data in MHR method are parameterized for value $\alpha \in [0;1]$ in the range of two successive interpolation nodes. MHR for data extrapolation is possible with $\alpha < 0$ or $\alpha > 1$.

Risk Analysis and Decision Making via Data Extrapolation

Example 1

MHR calculations are done for true rates of euro at National Bank of Poland (NBP) from January 24th to February 14th, 2011. If last four rates are considered: (1;3.8993), (2;3.9248), (3;3.9370) and (4;3.9337), MHR extrapolation with matrices of dimension $N = 2$ gives the result (5;3.9158). So anticipated rate of euro on the day February 15th is 3.9158 (Fig.3).

Figure 3. Extrapolated rate for day 5 (February 15th) using MHR method with 4 nodes
Source: author, 2014

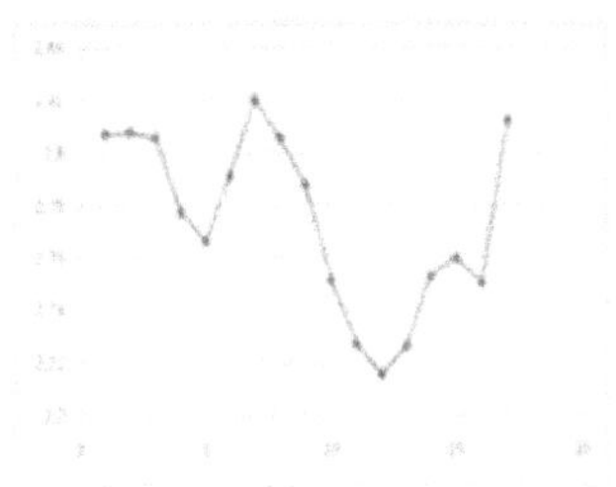

If last eight rates are considered: (1;3.9173), (2;3.9075), (3;3.8684), (4;3.8742), (5;3.8993), (6;3.9248), (7;3.9370) and (8;3.9337), MHR extrapolation with matrices of dimension $N = 4$ gives the result (9;4.0767). Anticipated rate of euro on the day February 15th is 4.0767 (Fig.4).

Figure 4. Extrapolated rate for day 9 (February 15th) using MHR method with 8 nodes
Source: author, 2014

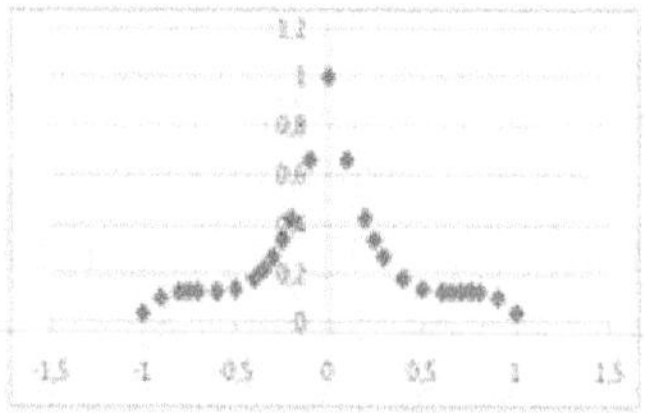

There are two extrapolated values for next day. This example gives us two anticipated rates for tomorrow: 3.9158 and 4.0767, which differs considerably. How these extrapolated values can be used in the process of decision making and analyzing risk: to buy euro or not, to sell euro or not? The proposal final anticipated rate of euro for the day February 15th (Fig.5) based on weighted mean value:

$$\frac{2 \cdot 3.9158 + 4.0767}{3} = 3.9694 \tag{29}$$

because the rate 3.9158 is calculated for $N = 2$, whereas 4.0767 is extrapolated for $N = 4$. Formula (29) takes one fact into account: dimension $N = 4$ is two times bigger than dimension $N = 2$ and the result 3.9158 has to be strengthen multiplying by two.

Figure 5. Extrapolated rate for day 9 (February 15th) using MHR method with 8 nodes and weighted mean value (29)
Source: author, 2014

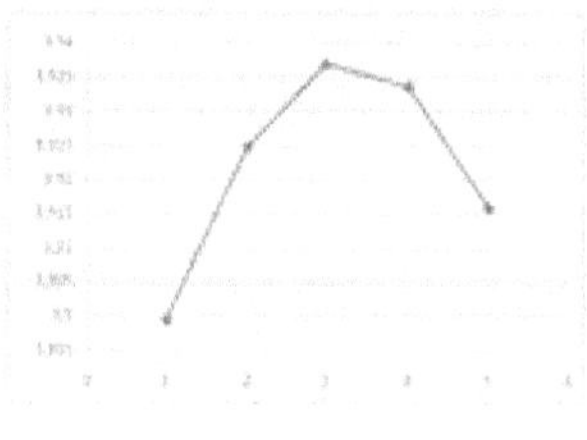

If last sixteen rates are considered, MHR extrapolation with matrices of dimension $N = 8$ has to be used. Here are the rates: (1;3.8765), (2;3.8777), (3;3.8777), (4;3.9009), (5;3.9111), (6;3.9345), (7;3.9129), (8;3.9019), (9;3.9173), (10;3.9075), (11;3.8684), (12;3.8742), (13;3.8993), (14;3.9248), (15;3.9370) and (16;3.9337). Average OHR operator M_2 and MHR calculations look as follows:

MHR extrapolation gives the result (17;3.9882). Anticipated rate of euro for the day February 15th is 3.9882 (Fig.6).

Figure 6. Extrapolated rate for day 17 (February 15th) using MHR method with 16 nodes
Source: author, 2014

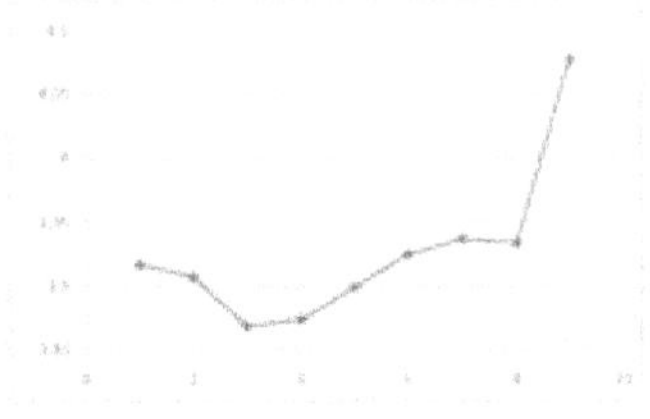

MHR extrapolation has been done for three times ($N = 2$, 4 or 8) and anticipated values are 3.9158, 4.0767 and 3.9882 respectively. The proposal final anticipated rate of euro for the day February 15th (Fig.7) based on weighted mean value:

$$\frac{4\cdot 3.9158 + 2\cdot 4.0767 + 3.9882}{7} = 3.9721 \tag{30}$$

because the rate 3.9158 is calculated with last four data points, 4.0767 is extrapolated for last eight data points and 3.9882 is computed for last sixteen data points. Formula (30) takes one fact into account: number of sixteen points is four times bigger than four and two times bigger than eigth. The result 3.9158 has to be strengthen multiplying by four and the rate 4.0767 has to be strengthen multiplying by two.

Figure 7. Extrapolated rate for day 17 (February 15th) using MHR method with 16 nodes and weighted mean value (30)
Source: author, 2014

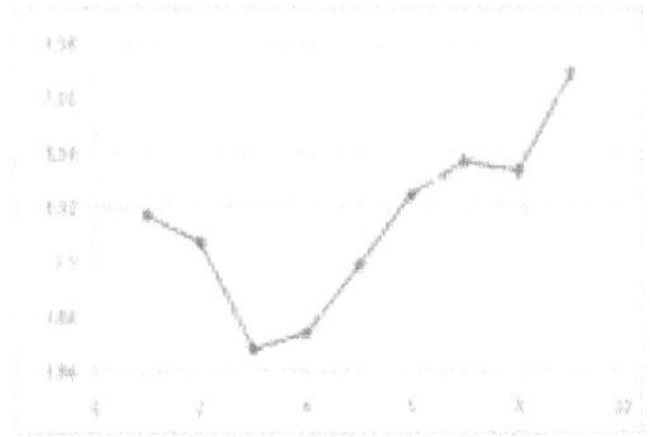

The true rate of euro for the day February 15th is 3.9398 (Fig.8).

Figure 8. The true rate of euro for day 17 (February 15th)
Source: author, 2014

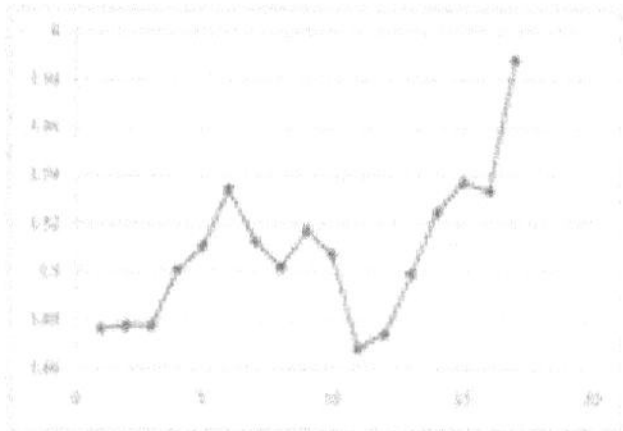

In author's opinion, values extrapolated for next day 3.9694 (13) and 3.9721 (14) are good enough to be one of the factors for making a decision of buying or selling the currency.

EXAMPLE 2

MHR calculations are done for true rates of US dollar at National Bank of Poland (NBP) from June 16th to July 8th, 2011 (Friday). If last four rates are considered: (1;2.7266), (2;2.7531), (3;2.7597) and (4;2.7505), MHR extrapolation with matrices of dimension $N = 2$ gives the result (5;2.7239):

So anticipated rate of US dollar on the day July 11th (Monday) is 2.7239.

If last eight rates are considered: (1;2.7877), (2;2.7517), (3;2.7273), (4;2.7156), (5;2.7266), (6; 2.7531), (7; 2.7597) and (8; 2.7505), MHR extrapolation with matrices of dimension $N = 4$ gives the result (9;2.8471):

Anticipated rate of US dollar on the day July 11th is 2.8471. There are two extrapolated values for next day. Example 2 gives us two anticipated rates for tomorrow: 2.7239 and 2.8471. How these extrapolated values can be used in the process of decision making: to buy dollar or not, to sell dollar or not? The proposal final anticipated rate of US dollar (Fig.9) on the day July 11th based on weighted mean value:

$$\frac{2 \cdot 2.7239 + 2.8471}{3} = 2.7650 \qquad (31)$$

because the rate 2.7239 is calculated for $N = 2$, whereas 2.8471 is extrapolated for $N = 4$.

Figure 9. Extrapolated rate for day 9 (July 11th) using MHR method with 8 nodes and weighted mean value (31)
Source: author, 2014

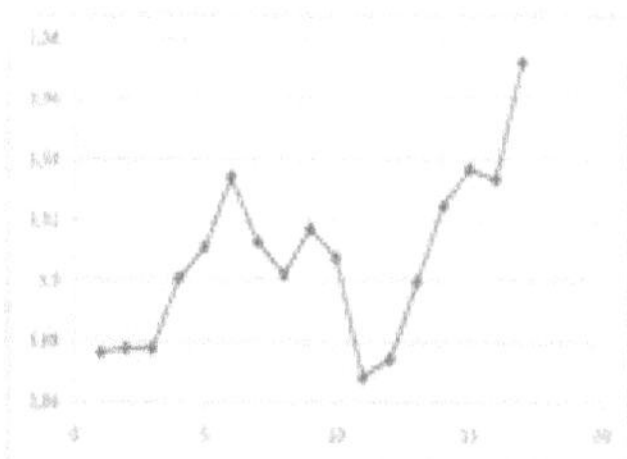

If last sixteen rates are considered, MHR extrapolation with matrices of dimension $N = 8$ has to be used. Here are the rates: (1;2.8069), (2;2.8077), (3;2.8058), (4;2.7776), (5;2.7661), (6;2.7914), (7;2.8201), (8;2.8055), (9;2.7877), (10;2.7517), (11;2.7273), (12;2.7156), (13;2.7266), (14;2.7531), (15;2.7597) and (16;2.7505). MHR extrapolation gives the result (17;2.7808). Anticipated rate of US dollar on the day July 11th is 2.7808.

MHR extrapolation has been done for three times ($N = 2$, 4 or 8) and anticipated values are 2.7239, 2.8471 and 2.7808 respectively. The proposal final anticipated rate of US dollar (Fig.10) on the day July 11th based on weighted mean value:

$$\frac{4\cdot 2.7239+2\cdot 2.8471+2.7808}{7}=2.7672 \tag{32}$$

because the rate 2.7239 is calculated with last four data points, 2.8471 is extrapolated for last eight data points and 2.7808 is computed for last sixteen data points.

Figure 10. Extrapolated rate for day 17 (July 11th) using MHR method with 16 nodes and weighted mean value (32)
Source: author, 2014

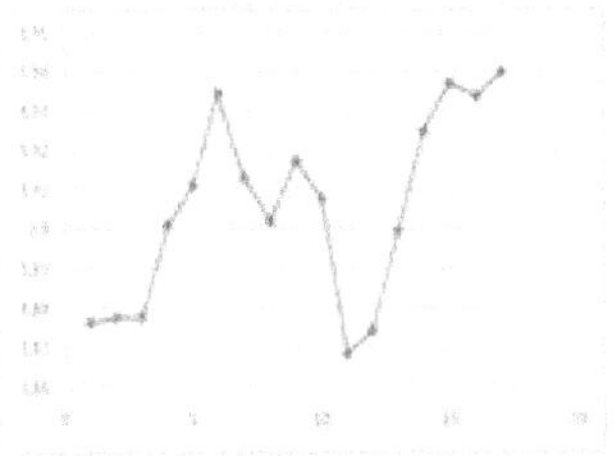

The true rate of US dollar on the day July 11th is 2.8123 (Fig.11).

Figure 11. The true rate of US dollar for day 17 (July 11th)
Source: author, 2014

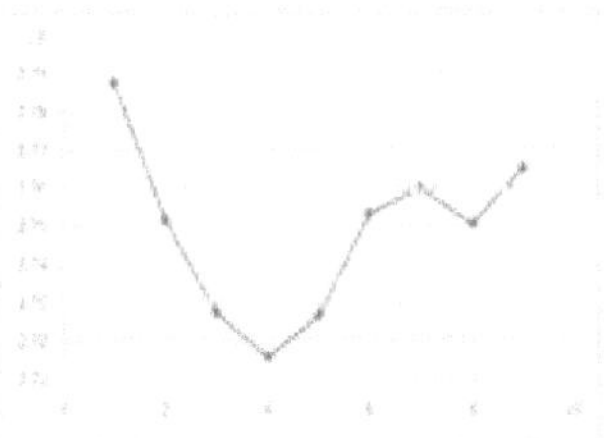

In author's opinion, extrapolated values 2.7650 (15) and 2.7672 (16) and anticipated rates in example 1 preserve the increasing trend and they are good enough to be one of the factors for making a decision of buying or selling the currency. Anticipated values, calculated by MHR method, are applied in the process of risk analysis and decision making: to follow the action or not, to do one thing or another. Extrapolated values can be used to make a decision in many branches of science and economics.

DISCUSSION OF EXAMPLES:

This chapter goes to first step of application for MHR method in point extrapolation and decision making. Figures and calculations show the trends of value: increasing or decreasing. This information could be used in analyzing risk and decision making, for example buying or selling when risk analysis is very difficult and very important.

The Length Estimation

Selection of the nodes is a key factor in the process of interpolation and extrapolation. The length of a curve is significant feature. Also the length estimation depends on the nodes. Having nodes (x_1,y_1), $(x_2,y_2),\ldots,(x_n,y_n)$ in MHR method (algorithm 1), it is possible to compute as many curve points as we want for any parameter $\alpha \in [0;1]$.

Assume that L is the number of reconstructed points plus n nodes. So curve consists of L points that could be indexed (x_1',y_1'), $(x_2',y_2'),\ldots,(x_L',y_L')$, where $(x_1',y_1') = (x_1,y_1)$ and $(x_L',y_L') = (x_n,y_n)$.

The length of curve, consists of L points, is estimated:

$$d(L) = \sum_{i=1}^{L-1} \sqrt{(x_{i+1}'-x_i')^2 + (y_{i+1}'-y_i')^2} \,. \tag{33}$$

For any accuracy of length estimation $\varepsilon > 0$, it is possible to use MHR method (algorithm 1) with suitable number and location of n nodes and reconstruct curve consists of L and L_1 points, where

$$|d(L) - d(L_1)| < \varepsilon \,.$$

So there is no need to compute

$$\int_{x_1}^{x_n} \sqrt{1+[f'(x)]^2}\, dx \tag{34}$$

as the length of curve f. Formula (33) has lower computational cost then (34).

SOLUTIONS AND RECOMMENDATIONS

The chapter deals with the problem of data interpolation and anticipation. None of the methods, that are used nowadays, applies the orthogonal basis for unknown value extrapolation and foreseeing, but proposed MHR method does as the solution. Orthogonality means a very important feature for stability in calculations. Also MHR point interpolation and extrapolation via simple computations with matrices is something new in the problem of anticipation, risk analysis and decision making. Extrapolated values show increasing or decreasing trend during the process of analyzing risk and decision making. MHR method as the solution is recommended in artificial intelligence and operational research, for example in scheduling, time-tabling or planning.

FUTURE RESEARCH DIRECTIONS

Future works with MHR method are connected with the formula of extrapolated values. Also estimation of object area in the plane, using nodes of object contour, will be possible by MHR interpolation. Length of the curve and object area are significant features in many economic problems. Future works are dealing with smoothing the curve, parameterization of whole curve and possibility to apply MHR method to three-dimensional curves. Also case of equidistance nodes must be studied with all details. Another future research direction is to apply MHR method in artificial intelligence and operational research, for example scheduling, time-tabling, planning, identification of a specific person's face or fingerprint, character recognition (Lamdan, Schwartz & Wolfson, 1990) or image restoration (Pope & Lowe, 2004). There are several specialized tasks based on recognition to consider and it is important to use the shape of whole contour (Kriegman & Ponce, 1990) for identification and detection of persons, vehicles or other objects. Other applications of MHR method will be directed to computer graphics, modeling and image processing.

CONCLUSION

The method of Hurwitz-Radon Matrices leads to curve interpolation and value extrapolation depending on the number and location of data points. No characteristic features of curve are important in MHR method: failing to be differentiable at any point, the Runge's phenomenon or differences from the shape of polynomials. These features are very significant for classical polynomial interpolations and extrapolations. MHR method gives the possibility of reconstruction a curve and anticipation the data points. The only condition is to have a set of nodes according to assumptions in MHR method. Data representation and curve extrapolation by MHR method is connected with possibility of changing the nodes coordinates and reconstruction of new data or curve for new set of nodes. The same MHR interpolation and extrapolation is valid for discrete and continues data. Main features of MHR method are: accuracy of data reconstruction depending on number of nodes; interpolation or extrapolation of a curve consists of L points is connected with the computational cost of rank $O(L)$; MHR method is dealing with local operators: average OHR operators are built by successive 4, 8 or 16 data points, what is connected with smaller computational costs then using all nodes; MHR is not an affine interpolation.

REFERENCES

Brachman, R. J., & Levesque, H. J. (2004). *Knowledge representation and reasoning*. Morgan Kaufman.

Cierniak, R. (2005). *Computed tomography*. Exit.

Citko, W., Jakóbczak, D., & Sieńko, W. (2005, September). *On Hurwitz - Radon matrices based signal processing*. Paper presented at the workshop Signal Processing at Poznan University of Technology, Poznań, Poland.

Dahlquist, G., & Bjoerck, A. (1974). *Numerical methods*. Prentice Hall.

Eckmann, B. (1999). Topology, algebra, analysis- relations and missing links. *Notices of the American Mathematical Society*, *5*(46), 520–527.

Fagin, R., Halpern, J. Y., Moses, Y., & Vardi, M. Y. (1995). *Reasoning about knowledge*. MIT Press.

Jakóbczak, D. (2006). *Application of discrete, orthogonal operator of Hurwitz - Radon in compression and reconstruction of monochromatic images' contours* (Unpublished doctoral dissertation). Polish - Japanese Institute of Information Technology, Warsaw, Poland.

Jakóbczak, D. (2007). 2D and 3D image modeling using Hurwitz - Radon matrices. *Polish Journal of Environmental Studies*, *4A*(16), 104–107.

Jakóbczak, D., & Kosiński, W. (2007). Hurwitz - Radon operator in monochromatic medical image reconstruction. *Journal of Medical Informatics & Technologies*, *11*, 69–78.

Jakóbczak, D., & Kosiński, W. (2007). Application of Hurwitz - Radon matrices in monochromatic medical images decompression. In Z. Kowalczuk & B. Wiszniewski (Eds.), *Intelligent data mining in diagnostic purposes: Automatics and informatics* (pp. 389–398). PWNT.

Kozera, R. (2004). *Curve modeling via interpolation based on multidimensional reduced data*. Silesian University of Technology Press.

Kriegman, D. J., & Ponce, J. (1990). On recognizing and positioning curved 3-D objects from image contours. *IEEE Transactions on Pattern Analysis and Machine Intelligence*, *12*(12), 1127–1137. doi:10.1109/34.62602

Lamdan, Y., Schwartz, J. T., & Wolfson, H. J. (1990). Affine invariant model-based object recognition. *IEEE Transactions on Robotics and Automation*, *5*(6), 578–589. doi:10.1109/70.62047

Latecki, L. J., & Lakaemper, R. (1999). Convexity rule for shape decomposition based on Discrete Contour Evolution. *Computer Vision and Image Understanding*, *3*(73), 441–454. doi:10.1006/cviu.1998.0738

Lowe, D. G. (1991). Fitting parameterized three-dimensional models to images. *IEEE Transactions on Pattern Analysis and Machine Intelligence*, *5*(13), 441–450. doi:10.1109/34.134043

Lowe, D. G. (1999, September). *Object recognition from local scale-invariant features*. Paper presented at the International Conference on Computer Vision, Corfu, Greece.

Lowe, D. G. (2001). *Local feature view clustering for 3D object recognition*. Paper presented at the IEEE Conference on Computer Vision and Pattern Recognition, Kauai, HI. 10.1109/CVPR.2001.990541

Markman, A. B. (1998). *Knowledge representation*. Lawrence Erlbaum Associates.

Pope, A. R., & Lowe, D. G. (2004). Probabilistic models of appearance for 3-D object recognition. *International Journal of Computer Vision*, *2*(40), 149–167.

Ralston, A. (1965). *A first course in numerical analysis*. McGraw-Hill Book Company.

Sieńko, W., & Citko, W. (2002). *Hamiltonian Neural Net based signal processing*. Paper presented at the International Conference on Signal and Electronic System ICSES, Wrocław – Świeradów Zdrój, Poland.

Sieńko, W., Citko, W., & Jakóbczak, D. (2004). Learning and system modeling via Hamiltonian Neural Networks. In L. Rutkowski, J. Siekmann, R. Tadeusiewicz, & A. Zadeh (Eds.), *Lecture notes on artificial intelligence: Artificial intelligence and soft computing - ICAISC 2004* (pp. 266–271). Springer - Verlag. doi:10.1007/978-3-540-24844-6_36

Sieńko, W., Citko, W., & Wilamowski, B. (2002). *Hamiltonian Neural Nets as a universal signal processor*. Paper presented at the 28th Annual Conference of the IEEE Industrial Electronics Society IECON, Sevilla, Spain. 10.1109/IECON.2002.1182910

Soussen, C., & Mohammad-Djafari, A. (2004). Polygonal and polyhedral contour reconstruction in computed tomography. *IEEE Transactions on Image Processing*, *11*(13), 1507–1523. doi:10.1109/TIP.2004.836159 PMID:15540458

Sowa, J. F. (2000). *Knowledge representation: logical, philosophical and computational foundations*. Brooks/Cole.

Straffin, P. D. (1993). *Game theory and strategy*. Mathematical Association of America.

Tang, K. (2005). Geometric optimization algorithms in manufacturing. *Computer-Aided Design and Applications*, *2*(6), 747–757. doi:10.1080/16864360.2005.10738338

Tarokh, V., Jafarkhani, H., & Calderbank, R. (1999). Space-Time Block Codes from orthogonal designs. *IEEE Transactions on Information Theory*, *5*(45), 1456–1467. doi:10.1109/18.771146

Ullman, S., & Basri, R. (1991). Recognition by linear combinations of models. *IEEE Transactions on Pattern Analysis and Machine Intelligence*, *10*(13), 992–1006. doi:10.1109/34.99234

Watson, J. (2002). *Strategy – an introduction to game theory*. University of California Press.

ADDITIONAL READING

Basu, S., & Bresler, Y. (2000). O($N^2 \log_2 N$) filtered backprojection reconstruction algorithm for tomography. *IEEE Transactions on Image Processing*, *9*(10), 1760–1773. doi:10.1109/83.869187 PMID:18262914

Brankov, J. G., Yang, Y., & Wernick, M. N. (2004). Tomographic image reconstruction based on a Content – Adaptive Mesh Model. *IEEE Transactions on Medical Imaging*, *2*(23), 202–212. doi:10.1109/TMI.2003.822822 PMID:14964565

Brasse, D., & Defrise, M. (2004). Fast fully 3-D image reconstruction in PET using planograms. *IEEE Transactions on Medical Imaging*, *4*(23), 413–425. doi:10.1109/TMI.2004.824231 PMID:15084067

Cetin, M., Karl, W. C., & Willsky, A. S. (2002, September). *Edge – preserving image reconstruction for coherent imaging application.* Paper presented at the IEEE International Conference on Image Processsing, Rochester, NY, USA. 10.1109/ICIP.2002.1039992

Chlebus, E., & Cholewa, M. (1999). Rapid prototyping – rapid tooling. *CADCAM Forum, 11*, 23-28.

Cormen, T. H., Leiserson, C. E., & Rivest, R. L. (1996). *Introduction to algorithms.* Massachusetts, USA: the Massachusetts Institute of Technology Press and McGraw-Hill.

Defrise, M. (2001). A short reader's guide to 3D tomographic reconstruction. *Computerized Medical Imaging and Graphics, 25*(2), 113–116. doi:10.1016/S0895-6111(00)00061-6 PMID:11137787

Dryja, M., Jankowska, J., & Jankowski, M. (1982). *Survey of numerical methods and algorithms. Part II.* WNT.

Eldar, Y. C. (2001). *Quantum Signal Processing.* (Unpublished doctoral dissertation). Massachusetts Institute of Technology, USA.

Eldar, Y. C., & Oppenheim, A. V. (2002). Quantum Signal Processing. *IEEE Signal Processing Magazine, 6*(19), 12–32. doi:10.1109/MSP.2002.1043298

Fortuna, Z., Macukow, B., & Wąsowski, J. (1982). *Numerical methods.* WNT.

Jakóbczak, D. (2005). Hurwitz - Radon matrices and their children. *Computer Science, 5*(8), 29–38.

Jankowska, J., & Jankowski, M. (1981). *Survey of numerical methods and algorithms. Part I.* WNT.

Kontaxakis, G., & Strauss, L. G. (1998). Maximum likelihood algorithms for image reconstruction in Positron Emission Tomography. *Radionuclides for Oncology – Current Status and Future Aspects, 1998*, 73-106.

Kowalczuk, Z., & Wiszniewski, B. (Eds.). (2007). *Intelligent data mining in diagnostic purposes: Automatics and informatics.* PWNT.

Kundur, D., & Hatzinakos, D. (1998). A novel blind deconvolution scheme for image restoration using recursive filtering. *IEEE Transactions on Signal Processing, 2*(46), 375–390. doi:10.1109/78.655423

Laine, A., & Zong, X. (1996). *Border identification of echocardiograms via multiscale edge detection and shape modeling.* Paper presented at the IEEE International Conference on Image Processsing, Lausanne, Switzerland. 10.1109/ICIP.1996.560486

Lang, S. (1970). *Algebra.* Addison-Wesley Publishing Company.

Le Buhan Jordan, C., Bossen, F., & Ebrahimi, T. (1997). *Scalable shape representation for content based visual data compression.* Paper presented at the International Conference on Image Processing, Santa Barbara, CA, USA. 10.1109/ICIP.1997.647962

Marker, J., Braude, I., Museth, K., & Breen, D. (2006). Contour-based surface reconstruction using implicit curve fitting, and distance field filtering and interpolation. *Volume Graphics. International Symposium on Volume Graphics, 2006*, 1–9.

Meyer, Y. (1993). *Wavelets: algorithms & applications.* Society for Industrial and Applied Mathematics.

Poggio, T., & Smale, S. (2003). The mathematics of learning: Dealing with data. *Notices of the American Mathematical Society*, *5*(50), 537–544.

Przelaskowski, A. (2005). *Data compression*. BTC.

Rutkowski, L., Siekmann, J., Tadeusiewicz, R., & Zadeh, A. (Eds.). (2004). *Lecture notes on artificial intelligence: Artificial intelligence and soft computing*. Springer - Verlag.

Vakhania, N. (1993). Orthogonal random vectors and the Hurwitz – Radon - Eckmann theorem. *Proc. of the Georgian Academy of Sciences - Mathematics, 1(1)*, 109-125.

Willis, M. (2000). *Algebraic reconstruction algorithms for remote sensing image enhancement*. Unpublished doctoral dissertation, Department of Electrical and Computer Engineering, Brigham Young University.

Xu, Fang, & Mueller, K. (2005). Accelerating popular tomographic reconstruction algorithms on commodity PC graphics hardware. *IEEE Transactions on Nuclear Science*, *3*(52), 654–661.

Zaletelj, J., & Tasic, J. F. (2003). *Optimization and tracking of polygon vertices for shape coding*. Springer - Verlag. doi:10.1007/978-3-540-45179-2_52

Zhang, J. K., Davidson, T., & Wong, K. M. (2004). Efficient design of orthonormal wavelet bases for signal representation. *IEEE Transactions on Signal Processing*, *7*(52), 1983–1996. doi:10.1109/TSP.2004.828923

KEY TERMS AND DEFINITIONS

Artificial Intelligence: Intelligence of machines and computers, as a connection of algorithms and hardware, which makes that a man – human being can be simulated by the machines in analyzing risk, decision making, reasoning, knowledge, planning, learning, communication, perception and the ability to move and manipulate objects.

Contour Modeling: Calculation of unknown points of the object contour having information about some points of the object contour.

Curve Interpolation: Computing new and unknown points of a curve and creating a graph of a curve using existing data points – interpolation nodes.

Data Extrapolation: Calculation of unknown values for the points situated outside the ranges of nodes.

Hurwitz-Radon Matrices: A family of skew – symmetric and orthogonal matrices with columns and rows that create, together with identical matrix, the base in vector spaces of dimensions $N = 2$, 4 or 8.

MHR Method: The method of curve interpolation and extrapolation using linear (convex) combinations of OHR operators.

OHR Operator: Matrix operator of Hurwitz-Radon built from coordinates of interpolation nodes.

Value Anticipation: Foreseeing next value when last value is known.

Chapter 3
Decision Making and Data Analysis:
Curve Modeling via Probabilistic Method

Dariusz Jacek Jakóbczak
https://orcid.org/0000-0002-0297-6598
Koszalin University of Technology, Poland

ABSTRACT

The proposed method, called probabilistic nodes combination (PNC), is the method of 2D curve modeling and handwriting identification by using the set of key points. Nodes are treated as characteristic points of signature or handwriting for modeling and writer recognition. Identification of handwritten letters or symbols need modeling, and the model of each individual symbol or character is built by a choice of probability distribution function and nodes combination. PNC modeling via nodes combination and parameter γ as probability distribution function enables curve parameterization and interpolation for each specific letter or symbol. Two-dimensional curve is modeled and interpolated via nodes combination and different functions as continuous probability distribution functions: polynomial, sine, cosine, tangent, cotangent, logarithm, exponent, arc sin, arc cos, arc tan, arc cot, or power function.

INTRODUCTION

Probabilistic modeling is still a developing branch of economic and computer sciences: operational research (for example probabilistic model-based prognosis) (Lorton, Fouladirad & Grall, 2013), decision making techniques and probabilistic modeling (Pergler & Freeman, 2008), artificial intelligence and machine learning. Different aspects of probabilistic methods are used: stochastic processes and stochastic model-based techniques, Markov processes (Cocozza-Thivent, Eymard, Mercier & Roussignol, 2006), Poisson processes, Gamma processes, Monte Carlo methods, Bayes rule, conditional probability and many probability distributions. In this paper the goal of a probability distribution function is to describe the position of unknown points between given interpolation nodes. Two-dimensional curve is used to

DOI: 10.4018/978-1-7998-4706-9.ch003

represent the data points and extrapolation of the unknown values enables analyzing risk and then decision making.

The chapter clarifies the significance and novelty of the proposed method compared to existing methods (for example polynomial interpolations and Bézier curves). Previous published papers of the author were dealing with the method of Hurwitz-Radon Matrices (MHR method). Novelty of this paper and proposed method consists in the fact that calculations are free from the family of Hurwitz-Radon Matrices. Problem statement of this paper is: **how to reconstruct (interpolate) missing points of 2D curve and how to anticipate (extrapolate) unknown values or data having the set of interpolation nodes (key points) and using the information about probabilistic distribution of unknown points**. For example the simplest basic (uniform) distribution leads to the easiest interpolation – linear interpolation. Apart from probability distribution, additionally there is the second factor of proposed interpolation method: nodes combination. The simplest nodes combination is zero. Thus proposed curve modeling and extrapolation is based on two agents: probability distribution and nodes combination. First trial of probabilistic modeling in MHR version was described in (Jakóbczak, 2013). Significance of this chapter consists in generalization for MHR method: the computations are done without matrices in curve fitting and data anticipation, with clear point interpolation formula based on probability distribution function (continuous or discrete) and nodes combination. The paper also consists of generalization for linear interpolation with different (non-uniform) probability distribution functions and nodes combinations. So this chapter answers the question: "Why and when should we use Probabilistic Nodes Combination (PNC) method in extrapolation and interpolation?".

Curve interpolation (Collins, 2003) represents one of the most important problems in mathematics and computer science: how to model the curve (Chapra, 2012) via discrete set of two-dimensional points (Ralston & Rabinowitz, 2001)? Also the matter of shape representation (as closed curve - contour) and curve parameterization is still opened (Zhang & Lu, 2004). Operational research in planning and scheduling, also decision making systems in risk analysis, solve the problems which are based on data modeling and extrapolation via the choice of key points. So interpolation and extrapolation is not only a pure mathematical problem but important task in economic and artificial intelligence. The paper wants to approach a problem of curve modeling by characteristic points. Proposed method relies on nodes combination and functional modeling of curve points situated between the basic set of key points and outside of this set. The functions that are used in calculations represent whole family of elementary functions with inverse functions: polynomials, trigonometric, cyclometric, logarithmic, exponential and power function. These functions are treated as probability distribution functions in the range [0;1].

An important problem in operational research and computer sciences (Ballard, 1982) is that of appropriate shape representation and reconstruction. Classical discussion about shape representation is based on the problem: contour versus skeleton. This paper is voting for contour which forms boundary of the object. Contour of the object, represented by contour points, consists of information which allows us to describe many important features of the object as shape coefficients (Tadeusiewicz & Flasiński, 1991). In the paper contour is dealing with a set of curves. Curve modeling and generation is a basic subject in many branches of industry and computer science, for example in the CAD/CAM software.

The representation of shape has a great impact on the accuracy and effectiveness of object recognition (Saber, Yaowu & Murat, 2005). In the literature, shape has been represented by many options including curves (Sebastian & Klein, 2003), graph-based algorithms and medial axis (Liu & Geiger, 1999) to enable shape-based object recognition. Digital curve (open or closed) can be represented by chain code (Freeman's code). Chain code depends on selection of the started point and transformations of the object. So Freeman's code is one of the methods how to describe and to find contour of the object. An analog (continuous) version of Freeman's code is the curve α - s. Another contour representation and reconstruction is based on Fourier coefficients calculated in Discrete Fourier Transformation (DFT). These coefficients are used to fix similarity of the contours with different sizes or directions. If we assume that contour is built from segments of a line and fragments of circles or ellipses, Hough transformation is applied to detect contour lines. Also geometrical moments of the object are used during the process of object shape representation (Choraś, 2005).

BACKGROUND

All interpolation theory is based on polynomials. But why? Many kinds of polynomials are used for interpolation: classical polynomials, trigonometric polynomials, orthogonal polynomials (Tschebyscheff, Legendre, Laguerre), rational polynomials. But what about the exceptional situations with unexpected features of curve, data or nodes. Then polynomials are not the solution, for example when:

1. The curve is not a graph of function (no matter – opened or closed curve);
2. The curve does not have to be smooth at interpolation nodes: for example curve representing symbols, signature, handwriting or other specific data;
3. Nodes are fixed and there is no possibility to choose "better" nodes as for orthogonal polynomials;
4. The curve differs considerably from any interpolation polynomial;
5. The curve fails to be differentiable at some points;
6. Between each pair of nodes we are not interested in linear interpolation (uniform probability distribution and zero nodes combination) but there ought to be some generalization (even for two nodes only) with other probability distributions and nodes combinations;
7. Interpolated points depend on some chosen nodes (two nearest nodes or more) via nodes combination $h(p_1,p_2,\ldots,p_m)$ in (1);
8. We are not interested in the formula of interpolation function (for lower computational costs) but only calculated points of modeled curve are ready to be used in numerical computations;
9. The formula of curve or function is known but from some reason (for example high computational costs or hard polynomial interpolation) the curve has to be modeled or fitted in some way for numerical calculations – the examples for PNC interpolation (in MHR version) of functions $f(x) = 2/x$ and $f(x) = 1/(1+5x^2)$ with quantified measures and experimental comparison with classical polynomial interpolation in (Jakóbczak, 2010);
10. Extrapolation problem is also a big numerical challenge and PNC interpolation enables the extension into extrapolation (Jakóbczak, 2011) with α outside of [0;1] and $\gamma = F(\alpha)$ still strictly monotonic, $F(0) = 0$, $F(1) = 1$. So for example $\gamma = \alpha^2$ is impossible for extrapolation if $\alpha < 0$ (Jakóbczak, 2011). Polynomial or other interpolations are sometimes useless for extrapolation;
11. Having only nodes the user may have "negative" information (from specific character of data): no polynomial interpolation;
12. All calculations are numerical (discrete) – even $\gamma = F(\alpha)$ is to be given in tabular (discrete) form. There is no need to build continuous function: polynomial or others;
13. Parametric version of modeled curve is to be found.

Above thirteen important and heavy individual and characteristic features of some curves and their interpolations show that there may exist the situations with unexpected assumptions for interpolation.

Why not classical interpolation? Classical methods are useless to interpolate the function that fails to be differentiable at one point, for example the absolute value function $f(x)$ =úxúat $x = 0$. If point (0;0) is one of the interpolation nodes, then precise polynomial interpolation of the absolute value function is impossible. Also when the graph of interpolated function differs from the shape of polynomial considerably, for example $f(x) = 1/x$, interpolation is very hard because of existing local extrema and the roots of polynomial. We cannot forget about the Runge's phenomenon: when nodes are equidistance then high-order polynomial oscillates toward the end of the interval, for example close to -1 and 1 with function $f(x) = 1/(1+25x^2)$ (Ralston & Rabinowitz, 2001). These classical negative cases do not appear in proposed PNC method. Experimental comparison for PNC with polynomial interpolation is to be found in (Jakóbczak, 2009).

Nowadays methods apply mainly polynomial functions in different versions (trigonometric, orthogonal, rational) and for example Bernstein polynomials in Bezier curves, splines (Schumaker, 2007) and NURBS (Rogers, 2001). But Bezier curves don't represent the interpolation method (rather interpolation-approximation method) and cannot be used for example in handwriting modeling with key points (interpolation nodes). In comparison PNC method with Bézier curves, Hermite curves and B-curves (*B-splines*) or NURBS one unpleasant feature of these curves has to be mentioned: small change of one characteristic point can result in unwanted change of whole reconstructed curve. Such a feature does not appear in proposed PNC method which is more stable than Bézier curves. Only first and last characteristic point is situated on the Bézier curve (interpolation), the rest characteristic points lay outside the Bézier curve (approximation). Numerical methods for data interpolation are based on polynomial or trigonometric functions, for example Lagrange, Newton, Aitken and Hermite methods. These methods have many weak sides (Dahlquist & Bjoerck, 1974) and are not sufficient for curve interpolation in the situations when the curve cannot be build by polynomials or trigonometric functions. Also there exists several well established methods of curve modeling, for example shape-preserving techniques (Dejdumrong, 2007), subdivision algorithms (Dyn, Levin & Gregory, 1987) and others (Kozera, 2004) to overcome difficulties of polynomial interpolation, but probabilistic interpolation with nodes combination seems to be quite novel in the area of shape modeling. Proposed 2D curve interpolation is the functional modeling via any elementary functions and it helps us to fit the curve during the computations.

This paper presents novel Probabilistic Nodes Combination (PNC) method of curve interpolation. This paper takes up new PNC method of two-dimensional curve modeling via the examples using the family of Hurwitz-Radon matrices (MHR method) (Jakóbczak, 2007), but not only this method (other nodes combinations). The method of PNC requires minimal assumptions: the only information about a curve is the set of at least two nodes. Proposed PNC method is applied in curve modeling via different coefficients: polynomial, sinusoidal, cosinusoidal, tangent, cotangent, logarithmic, exponential, arc sin, arc cos, arc tan, arc cot or power. Function for PNC calculations is chosen individually at each interpolation and it represents probability distribution function of parameter α Î [0;1] for every point situated between two interpolation knots. PNC method uses two-dimensional vectors (x,y) for curve modeling - knots $p_i = (x_i,y_i)$ Î $\boldsymbol{R}^2$ in PNC method, $i = 1,2,\ldots n$:

1. PNC needs 2 knots or more (n ³ 2);
2. If first node and last node are the same ($p_1 = p_n$), then curve is closed (contour);

3. For more precise modeling knots ought to be settled at key points of the curve, for example local minimum or maximum and at least one node between two successive local extrema.

Condition 3 means for example the highest point of the curve in a particular orientation, convexity changing or curvature extrema. So this paper wants to answer the question: how to interpolate the curve by a set of knots (Jakóbczak, 2010)?

Figure 1. Five knots of the curve before modeling
Source: author, 2014

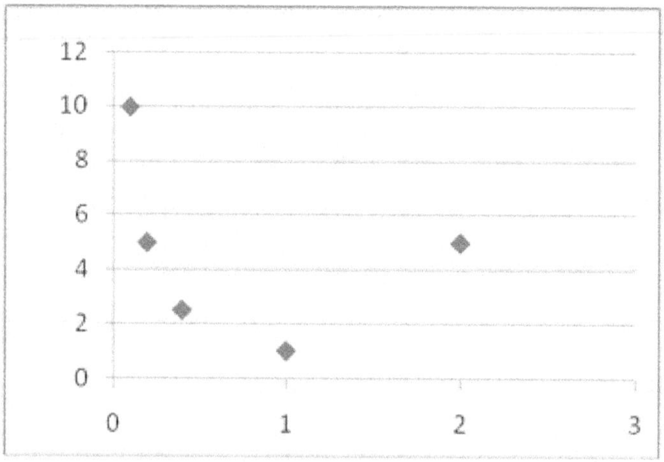

Nodes on Fig.1 represent characteristic points of the handwritten letter or symbol: if $n= 5$ then curve is opened and if $n = 6$ then curve is closed (contour). The examples of PNC curve modeling for these nodes are described later in this chapter. Coefficients for PNC curve modeling are computed using nodes combinations and probability distribution functions: polynomials, power functions, sine, cosine, tangent, cotangent, logarithm, exponent or arc sin, arc cos, arc tan or arc cot.

NOVELTY OF PROBABILISTIC INTERPOLATION AND EXTRAPOLATION

Issues

The method of PNC enables to compute points between two successive nodes of the curve: calculated points are interpolated and parameterized for real number α Î [0;1] in the range of two successive nodes. PNC method uses the combinations of nodes $p_1=(x_1,y_1)$, $p_2=(x_2,y_2)$,..., $p_n=(x_n,y_n)$ as $h(p_1,p_2,\ldots,p_m)$ and $m = 1,2,\ldots n$. Nodes combination h is defined individually for each curve to interpolate points (x,y) with second coordinate $y = y(c)$ for any first coordinate $x = c$ situated between nodes (x_i,y_i) and (x_{i+1},y_{i+1}):

$c = \alpha\times x_i + (1-\alpha)\times x_{i+1}$, $i = 1,2,\ldots n-1$,

$$y(c) = \gamma \cdot y_i + (1-\gamma)y_{i+1} + \gamma(1-\gamma) \cdot h(p_1, p_2, \ldots, p_m), \tag{1}$$

α Î [0;1], γ = *F*(α) Î[0;1], *F*:[0;1]®[0;1], F(0)=0, F(1)=1 and *F* is strictly monotonic.

PNC extrapolation requires α outside of [0;1]: α < 0 (anticipating points right of last node for $c > x_n$) or α > 1 (extrapolating values left of first node for $c < x_1$), γ = *F*(α), *F:P*®**R**, $P \supset [0;1]$, F(0)=0, F(1)=1 and *F* is still strictly monotonic for the arguments from *P*. So *c* and α represent the same – coordinate *x* of any point (*x*,*y*) between two successive nodes (x_i, y_i) and (x_{i+1}, y_{i+1}): having *c* we can calculate α and vice versa. PNC curve modeling relies on two factors: function γ = *F*(α) and nodes combination $h(p_1, p_2, \ldots, p_m)$. Function *F* is a probabilistic distribution function for random variable α Î [0;1] and parameter γ leads PNC interpolation into probabilistic modeling. Second factor, the combination of nodes *h*, is responsible for making dependent a reconstructed point on the coordinates of several nodes. The simplest case is for *h* = 0. Here are the examples of *h* computed for MHR method (Jakóbczak, 2009):

$$h(p_1, p_2) = \frac{y_1}{x_1} x_2 + \frac{y_2}{x_2} x_1$$

(only two neighboring nodes are taken for PNC calculations)or

$$h(p_1, p_2, p_3, p_4) = \frac{1}{x_1^{\,2} + x_3^{\,2}}(x_1 x_2 y_1 + x_2 x_3 y_3 + x_3 x_4 y_1 - x_1 x_4 y_3) +$$

$$+ \frac{1}{x_2^{\,2} + x_4^{\,2}}(x_1 x_2 y_2 + x_1 x_4 y_4 + x_3 x_4 y_2 - x_2 x_3 y_4)$$

(more than two neighboring nodes are used in PNC interpolation).

The examples of other nodes combinations are presented below. Formula (1) represents curve parameterization (*x*(α),*y*(α)) between two successive nodes (x_i, y_i) and (x_{i+1}, y_{i+1}) as α Î [0;1]:

$x(\alpha) = \alpha \times x_i + (1-\alpha) \times x_{i+1}$

and

$$y(\alpha) = F(\alpha) \cdot y_i + (1 - F(\alpha))y_{i+1} + F(\alpha)(1 - F(\alpha)) \cdot h(p_1, p_2, \ldots, p_m),$$

$$y(\alpha) = F(\alpha) \cdot (y_i - y_{i+1} + (1 - F(\alpha)) \cdot h(p_1, p_2, \ldots, p_m)) + y_{i+1}.$$

Proposed parameterization gives us the infinite number of possibilities for curve calculations (determined by choice of *F* and *h*) as there is the infinite number of human handwritten letters and symbols.

Nodes combination is the individual feature of each modeled curve (for example a handwritten character). Coefficient $\gamma = F(\alpha)$ and nodes combination h are key factors in PNC curve interpolation and data extrapolation.

Distribution Functions in PNC Interpolation and Extrapolation

Points settled between the nodes are computed using PNC method. Each real number c Î $[a;b]$ is calculated by a convex combination $c = \alpha \times a + (1 - \alpha) \times b$ for

$$\alpha = \frac{b-c}{b-a} \text{ Î } [0;1].$$

Key question is dealing with coefficient γ in (1). The simplest way of PNC calculation means $h = 0$ and $\gamma = \alpha$ (basic probability distribution – uniform distribution). Then PNC represents a linear interpolation. MHR method (Jakóbczak, 2010) is not a linear interpolation. MHR is the example of PNC modeling.

Each interpolation requires specific distribution of parameter α and γ (1) depends on parameter α Î [0;1]:

$\gamma = F(\alpha)$, F:[0;1]®[0;1], $F(0) = 0$, $F(1) = 1$ and F is strictly monotonic.

Coefficient γ is calculated using different functions (polynomials, power functions, sine, cosine, tangent, cotangent, logarithm, exponent, arc sin, arc cos, arc tan or arc cot, also inverse functions) and choice of function is connected with initial requirements and curve specifications. Different values of coefficient γ are connected with applied functions $F(\alpha)$. The functions (2)-(34) represent the examples of probability distribution functions for random variable α Î [0;1] and real number $s > 0$:

1. power function

$\gamma = \alpha^s$ with $s > 0$. (2)

For $s = 1$: basic version of PNC and MHR (Jakóbczak, 2010) methods when $\gamma = \alpha$.

2. sine

$\gamma = sin(\alpha^s \cdot \pi/2)$, $s > 0$ (3)

or

$\gamma = sin^s(\alpha \cdot \pi/2)$, $s > 0$. (4)

For $s = 1$: $\gamma = sin(\alpha \cdot \pi/2)$. (5)

3. cosine

$$\gamma = 1\text{-}\cos(\alpha^s \cdot \pi/2),\ s > 0 \tag{6}$$

or

$$\gamma = 1\text{-}\cos^s(\alpha \cdot \pi/2),\ s > 0. \tag{7}$$

For $s = 1$: $\gamma = 1\text{-}\cos(\alpha \cdot \pi/2)$. (8)

4. tangent

$$\gamma = \tan(\alpha^s \cdot \pi/4),\ s > 0 \tag{9}$$

or

$$\gamma = \tan^s(\alpha \cdot \pi/4),\ s > 0. \tag{10}$$

For $s = 1$: $\gamma = \tan(\alpha \cdot \pi/4)$. (11)

5. logarithm

$$\gamma = \log_2(\alpha^s + 1),\ s > 0 \tag{12}$$

or

$$\gamma = \log_2^s(\alpha + 1),\ s > 0. \tag{13}$$

For $s = 1$: $\gamma = \log_2(\alpha + 1)$. (14)

6. exponent

$$\gamma = \left(\frac{a^\alpha - 1}{a - 1}\right)^s,\ s > 0 \text{ and } a > 0 \text{ and } a^1\ 1. \tag{15}$$

For $s = 1$ and $a = 2$: $\gamma = 2^\alpha - 1$. (16)

7. arc sine

$\gamma = 2/\pi \cdot arcsin(\alpha^s),\ s > 0$ (17)

or

$\gamma = (2/\pi \cdot arcsin\ \alpha)^s,\ s > 0.$ (18)

For $s = 1$: $\gamma = 2/\pi \cdot arcsin(\alpha)$. (19)

8. arc cosine

$\gamma = 1\text{-}2/\pi \cdot arccos(\alpha^s),\ s > 0$ (20)

or

$\gamma = 1\text{-}(2/\pi \cdot arccos\ \alpha)^s,\ s > 0.$ (21)

For $s = 1$: $\gamma = 1\text{-}2/\pi \cdot arccos(\alpha)$. (22)

9. arc tangent

$\gamma = 4/\pi \cdot arctan(\alpha^s),\ s > 0$ (23)

or

$\gamma = (4/\pi \cdot arctan\ \alpha)^s,\ s > 0.$ (24)

For $s = 1$: $\gamma = 4/\pi \cdot arctan(\alpha)$. (25)

10. cotangent

$\gamma = ctg(\pi/2 - \alpha^s \cdot \pi/4),\ s > 0$ (26)

or

$\gamma = ctg^s\ (\pi/2 - \alpha \cdot \pi/4),\ s > 0.$ (27)

For $s = 1$: $\gamma = ctg(\pi/2 - \alpha \cdot \pi/4)$. (28)

11. arc cotangent

$$\gamma = 2 - 4/\pi \cdot arcctg(\alpha^s),\ s > 0 \tag{29}$$

or

$$\gamma = (2 - 4/\pi \cdot arcctg\ \alpha)^s,\ s > 0. \tag{30}$$

For $s = 1$: $\gamma = 2 - 4/\pi \cdot arcctg(\alpha)$. (31)

Functions used in γ calculations (2)-(31) are strictly monotonic for random variable $\alpha\hat{I}[0;1]$ as $\gamma = F(\alpha)$ is probability distribution function. Also inverse function $F^{-1}(\alpha)$ is appropriate for γ calculations. Choice of function and value s depends on curve specifications and individual requirements.

Proposed (2)-(31) probability distributions are continuous, but of course parameter γ can represent discrete probability distributions, for example: $F(0.1)=0.23$, $F(0.2)=0.3$, $F(0.3)=0.42$, $F(0.4)=0.52$, $F(0.5)=0.63$, $F(0.6)=0.69$, $F(0.7)=0.83$, $F(0.8)=0.942$, $F(0.9)=0.991$. What is very important in PNC method: two curves (for example a handwritten letter) may have the same set of nodes but different h or γ results in different interpolations (Fig.2-10).

Algorithm of PNC interpolation and modeling (1) looks as follows:

Step 1: Choice of knots p_i at key points.

Step 2: Choice of nodes combination $h(p_1,p_2,\ldots,p_m)$.

Step 3: Choice of distribution $\gamma = F(\alpha)$: (2)-(31) or others (continuous or discrete).

Step 4: Determining values of α: $\alpha = 0.1, 0.2\ldots0.9$ (nine points) or $0.01, 0.02\ldots0.99$ (99 points) or others.

Step 5: The computations (1).

These five steps can be treated as the algorithm of PNC method of curve modeling and interpolation (1). Without knowledge about the formula of curve or function, PNC interpolation has to implement the coefficients γ (2)-(31), but PNC is not limited only to these coefficients. Each strictly monotonic function F between points (0;0) and (1;1) can be used in PNC modeling.

Data Modeling and Curve Fitting

Curve knots $p_1 = (0.1;10)$, $p_2 = (0.2;5)$, $p_3 = (0.4;2.5)$, $p_4 = (1;1)$ and $p_5 = (2;5)$ from Fig.1 are used in some examples of PNC method in handwritten character modeling. Fig.2-9 represent PNC as MHR interpolation (Jakóbczak, 2011) with different γ. Points of the curve are calculated with no matrices ($N = 1$) and $\gamma = \alpha$ in example 1 and with matrices of dimension $N = 2$ in examples 2-8 for $\alpha = 0.1, 0.2,\ldots,0.9$.

Example 1

PNC curve interpolation (1) for $\gamma = \alpha$ and

$$h(p_1, p_2) = \frac{y_1}{x_1}x_2 + \frac{y_2}{x_2}x_1:$$

Figure 2. PNC character modeling for nine reconstructed points between nodes
Source: author, 2014

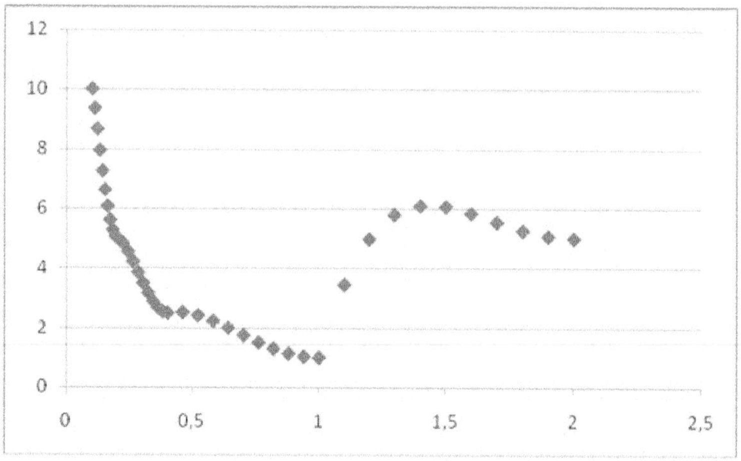

For $N = 2$ (examples 2 – 8) MHR version (Jakóbczak, 2011) as PNC method gives us

$$h(p_1, p_2, p_3, p_4) = \frac{1}{x_1^{\,2} + x_3^{\,2}}(x_1x_2y_1 + x_2x_3y_3 + x_3x_4y_1 - x_1x_4y_3) +$$

$$+\frac{1}{x_2^{\,2} + x_4^{\,2}}(x_1x_2y_2 + x_1x_4y_4 + x_3x_4y_2 - x_2x_3y_4).$$

Example 2

PNC sinusoidal interpolation with $\gamma = sin(\alpha \cdot \pi/2)$.

Figure 3. Sinusoidal modeling with nine reconstructed curve points between nodes
Source: author, 2014

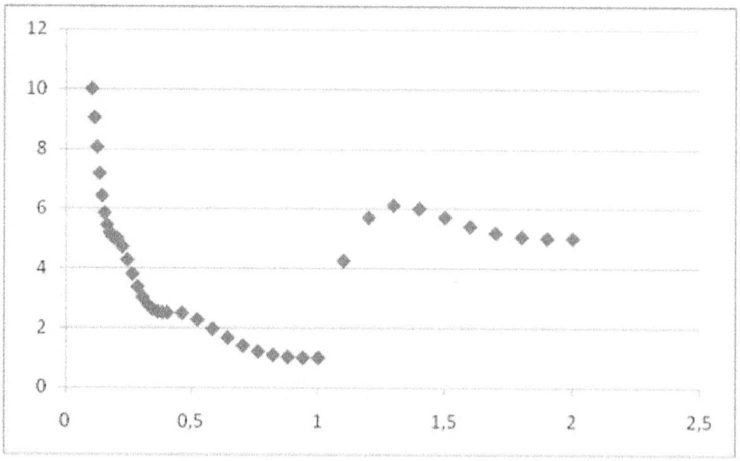

Example 3

PNC tangent interpolation for $\gamma = tan(\alpha \cdot \pi/4)$.

Figure 4. Tangent character modeling with nine interpolated points between nodes
Source: author, 2014

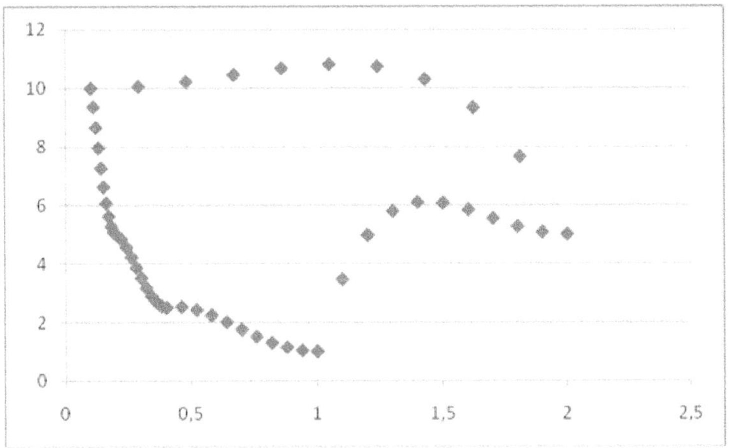

Example 4

PNC tangent interpolation with $\gamma = tan(\alpha^s \cdot \pi/4)$ and $s = 1.5$.

Figure 5. Tangent curve modeling with nine recovered points between nodes
Source: author, 2014

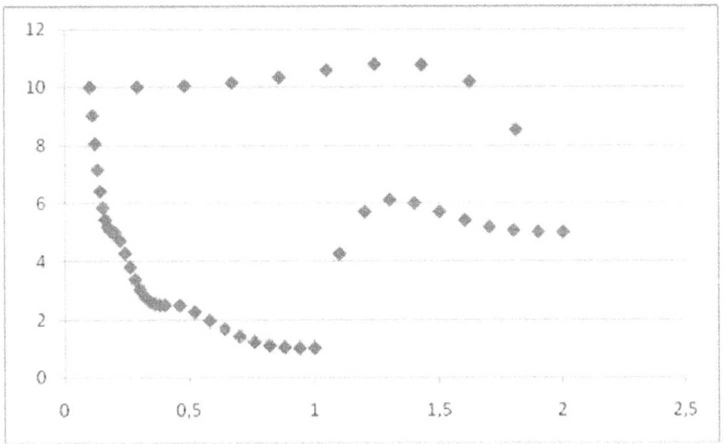

Example 5

PNC tangent curve interpolation for $\gamma = tan(\alpha^s \cdot \pi/4)$ and $s = 1.797$.

Figure 6. Tangent symbol modeling with nine reconstructed points between nodes
Source: author, 2014

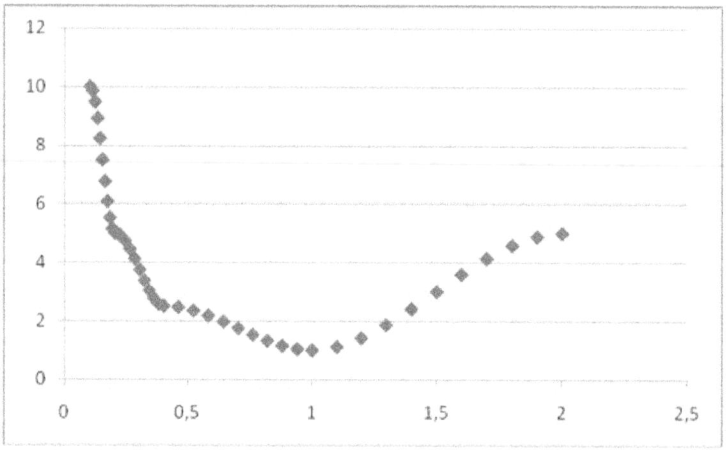

Example 6

PNC sinusoidal interpolation with $\gamma = sin(\alpha^s \cdot \pi/2)$ and $s = 2.759$.

Figure 7. Sinusoidal modeling with nine interpolated curve points between nodes
Source: author, 2014

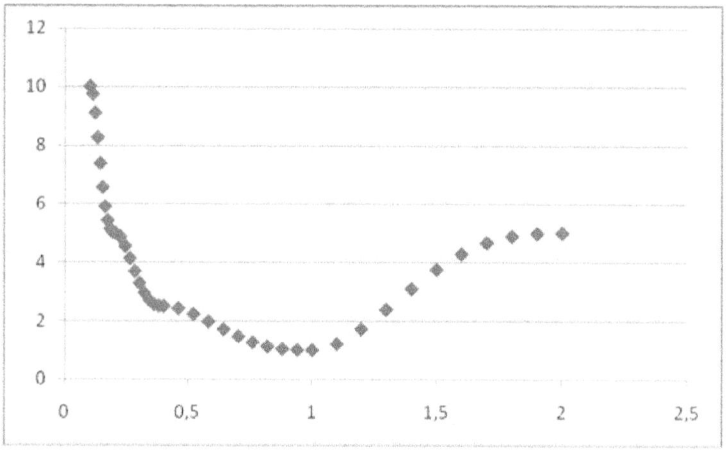

Example 7

PNC power function modeling for $\gamma = \alpha^s$ and $s = 2.1205$.

Figure 8. Power function curve modeling with nine recovered points between nodes
Source: author, 2014

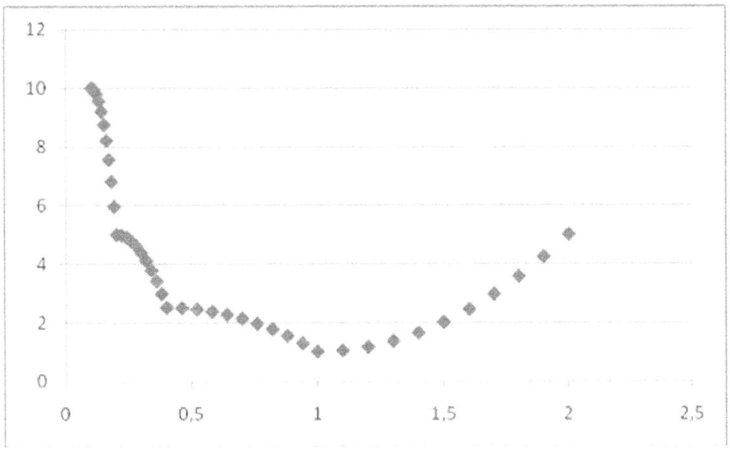

Example 8

PNC logarithmic curve modeling with $\gamma = log_2(\alpha^s + 1)$ and $s = 2.533$.

Figure 9. Logarithmic character modeling with nine reconstructed points between nodes
Source: author, 2014

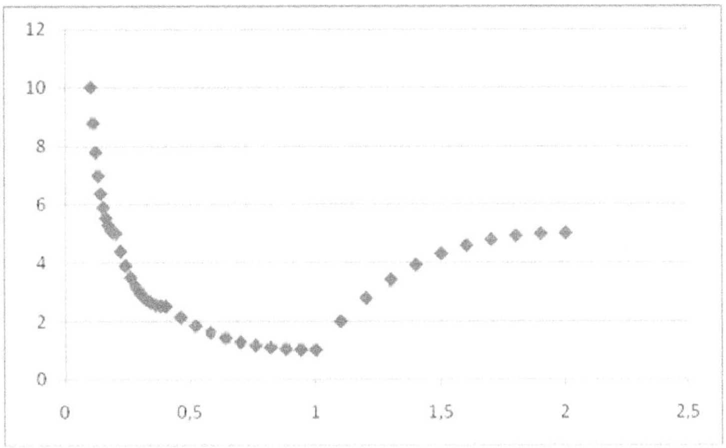

These eight examples demonstrate possibilities of PNC curve interpolation and handwritten character modeling for key nodes in MHR version. And here are other examples of PNC modeling (but not MHR):

Example 9

PNC for $\gamma = \alpha^2$ and $h(p_1, p_2) = x_1 y_1 + x_2 y_2$:

Figure 10. Quadratic symbol modeling with nine reconstructed points between nodes
Source: author, 2014

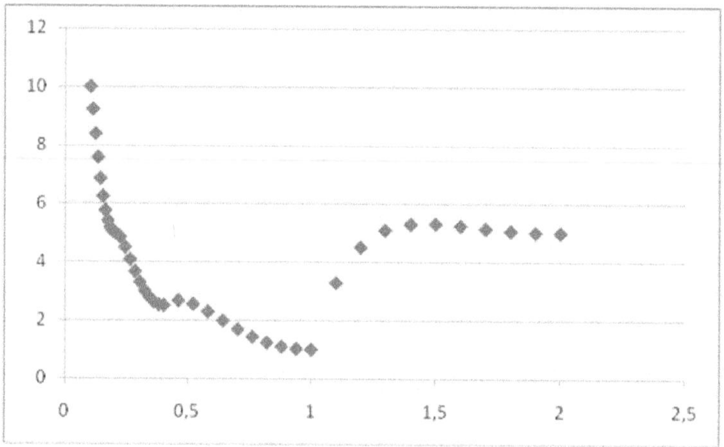

Example 10

PNC for $\gamma = \alpha^3$ and $h(p_1, p_2) = x_1 y_1 + x_2 y_2$:

Figure 11. Cubic character modeling with nine reconstructed points between nodes
Source: author, 2014

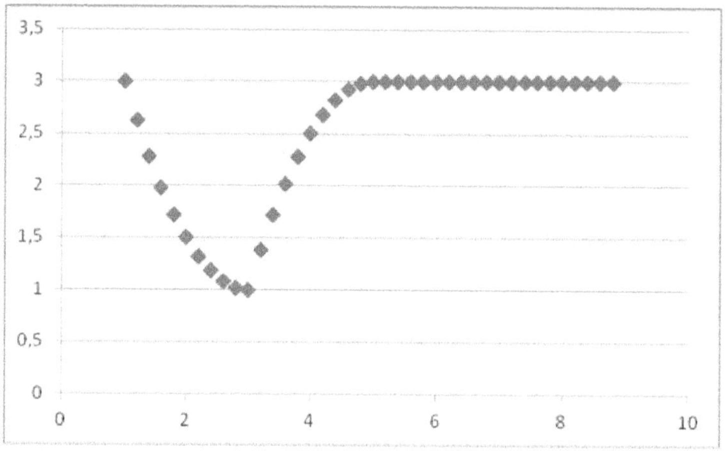

If there is considered Fig.1 as closed curve (contour) with the node $p_6 = p_1 = (0.1;10)$ then examples 9 and10 give the shapes:

Example 11

PNC for $\gamma = \alpha^2$ and $h(p_1, p_2) = x_1 y_1 + x_2 y_2$:

Figure 12. Quadratic contour modeling with nine reconstructed points between nodes
Source: author, 2014

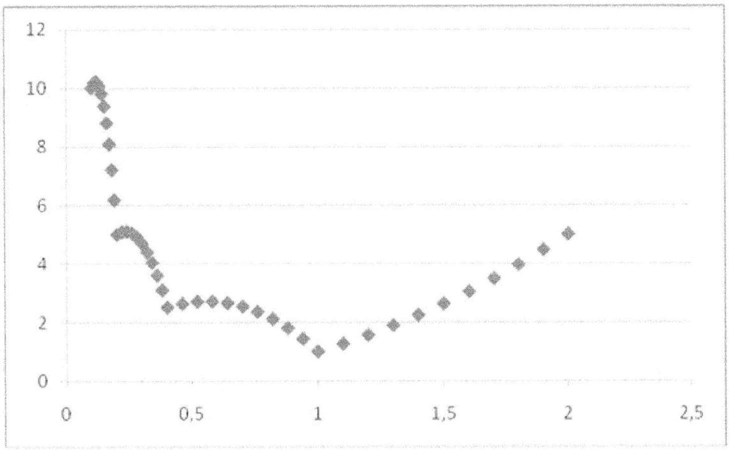

Example 12

PNC for $\gamma = \alpha^3$ and $h(p_1, p_2) = x_1 y_1 + x_2 y_2$:

Figure 13. Cubic shape modeling with nine reconstructed points between nodes
Source: author, 2014

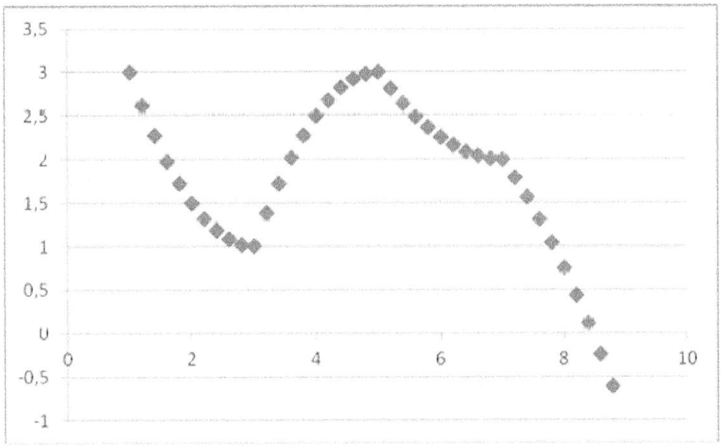

Every man has individual style of handwriting. Recognition of handwritten letter or symbol need modeling and the model of each individual symbol or character can be built by choice of γ and h in (1).

PNC modeling via nodes combinations h and parameter γ as probability distribution function enables curve interpolation for each specific letter or symbol.

Number of reconstructed points depends on a user by value α. If for example $\alpha = 0.01, 0.02, \ldots, 0.99$ then 99 points are interpolated for each pair of nodes. Reconstructed values and interpolated points, calculated by PNC method, are applied in the process of curve modeling. Every curve can be interpolated by some distribution function as parameter γ and nodes combination h. Parameter γ is treated as probability distribution function for each curve.

Beta Distribution

Considering nowadays used probability distribution functions for random variable αÎ[0;1] - one distribution is dealing with the range [0;1]: beta distribution. Probability density function f for random variable α Î [0;1] is:

$$f(\alpha) = c \cdot \alpha^{s} \cdot (1-\alpha)^{r},\ s^{3}\ 0,\ r^{3}\ 0. \tag{32}$$

When $r = 0$ probability density function (32) represents $f(\alpha) = c \cdot \alpha^{s}$ and then probability distribution function F is like (2), for example $f(\alpha) = 3\alpha^{2}$ and $\gamma = \alpha^{3}$. If s and r are positive integer numbers then γ is the polynomial, for example $f(\alpha) = 6\alpha(1-\alpha)$ and $\gamma = 3\alpha^{2} - 2\alpha^{3}$. So beta distribution gives us coefficient γ in (1) as polynomial because of interdependence between probability density f and distribution F functions:

$$f(\alpha) = F'(\alpha),\ F(\alpha) = \int_{0}^{\alpha} f(t)dt. \tag{33}$$

For example (33):

$$f(\alpha) = \alpha \cdot e^{\alpha} \text{ and } \gamma = F(\alpha) = 1 - (1-\alpha)e^{\alpha}. \tag{34}$$

Basic (uniform) distribution ($\gamma = \alpha$) with nodes combination $h = 0$ turns PNC interpolation (1) to linear interpolation. What about PNC in the case of yet another distribution on the range [0;1]: beta distribution (32)? Power functions as γ used in examples 1, 7 and 9-12 are also connected with beta distribution. Here are the examples of PNC modeling for beta distribution with nodes combination $h = 0$.

Example 13

PNC for $\gamma = 3\alpha^2\text{-}2\alpha^3$ and $h(p_1, p_2) = 0$:

Figure 14. Beta distribution in handwritten character modeling
Source: author, 2014

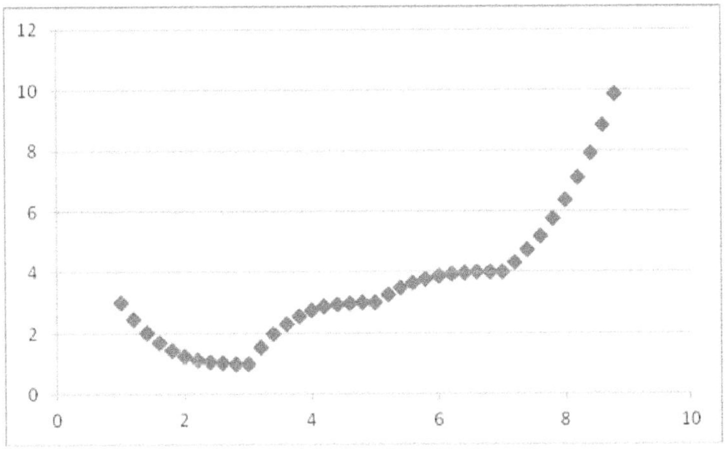

Example 14

PNC for $\gamma = 4\alpha^3\text{-}3\alpha^4$ and $h(p_1, p_2) = 0$:

Figure 15. Beta distribution in handwritten symbol modeling
Source: author, 2014

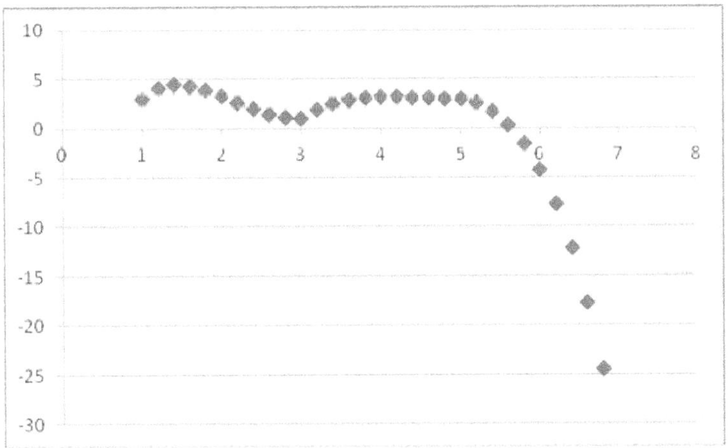

Example 15

PNC for $\gamma = 2\alpha\text{-}\alpha^2$ and $h(p_1, p_2) = 0$:

Figure 16. Beta distribution in handwritten letter modeling
Source: author, 2014

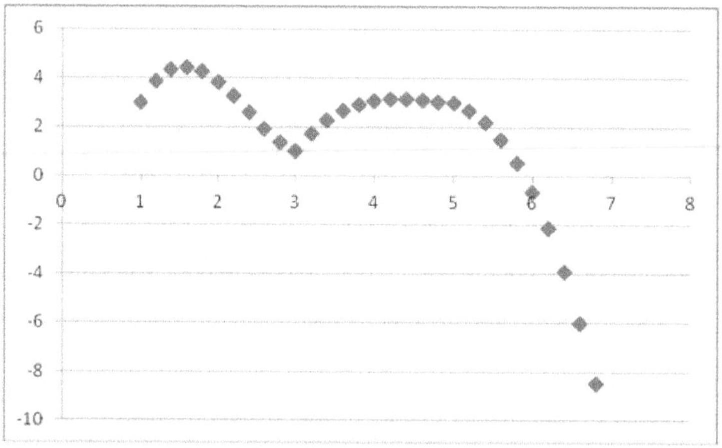

Examples 9-12 represent beta distribution with $h(p_1, p_2) = x_1y_1 + x_2y_2$.

Exponential Distribution

Exponential distribution is dealing with random variable ³ 0, but in PNC interpolation random variable αÎ[0;1]. Then exponential distribution is represented by distribution function (34):

$$\gamma = F(\alpha) = 1 - (1 - \alpha)e^{\alpha}.$$

Example 16

PNC for $\gamma = 1-(1-\alpha)e^{\alpha}$ and $h(p_1, p_2) = 0$:

Figure 17. Exponential distribution in handwritten character modeling
Source: author, 2014

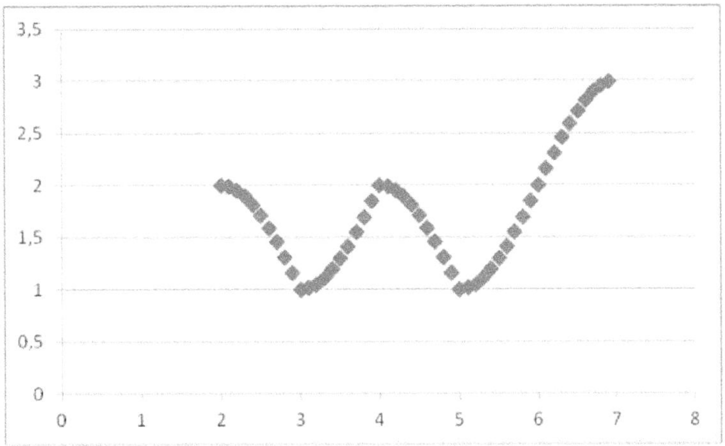

Example 17

PNC for $\gamma = 1-(1-\alpha)e^{\alpha}$ and $h(p_1, p_2) = \dfrac{y_2}{y_1} + \dfrac{x_2}{x_1}$:

Figure 18. Exponential distribution in handwritten symbol modeling
Source: author, 2014

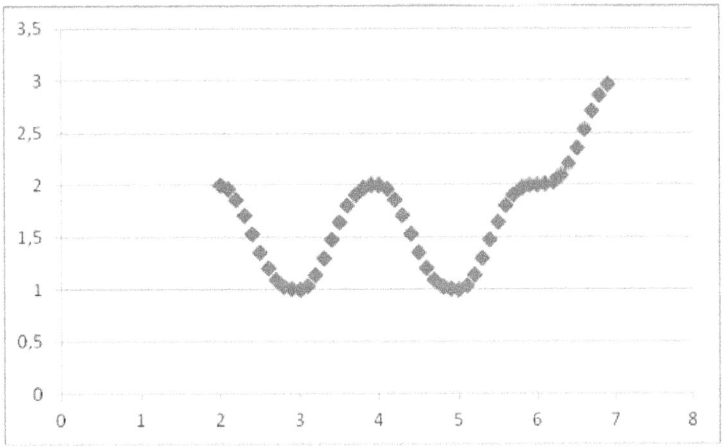

These examples show the variety of possibilities in curve modeling via the choice of nodes combination and probability distribution function for interpolated points.

PNC Extrapolation as the Support in Planning

Unknown data, important for planning or decision making, are modeled (interpolated or extrapolated) by the choice of nodes, determining specific nodes combination and characteristic probabilistic distribution function. Less complicated models take $h(p_1,p_2,\ldots,p_m) = 0$ and then the formula of interpolation (2) looks as follows:

$$y(c) = \gamma \cdot y_i + (1-\gamma) y_{i+1} \cdot$$

It is linear interpolation for basic (uniform) probability distribution ($\gamma = \alpha$).

Example 1

Nodes (1;3), (3;1), (5;3), (7;3) and $h = 0$, $\gamma = F(\alpha) = \alpha^2$. This function F requires $\alpha > 1$ and extrapolation is computed with (4)-(5):

Figure 19. PNC modeling for nine interpolated points between successive nodes and nine extrapolated points right of the last node
Source: author, 2014

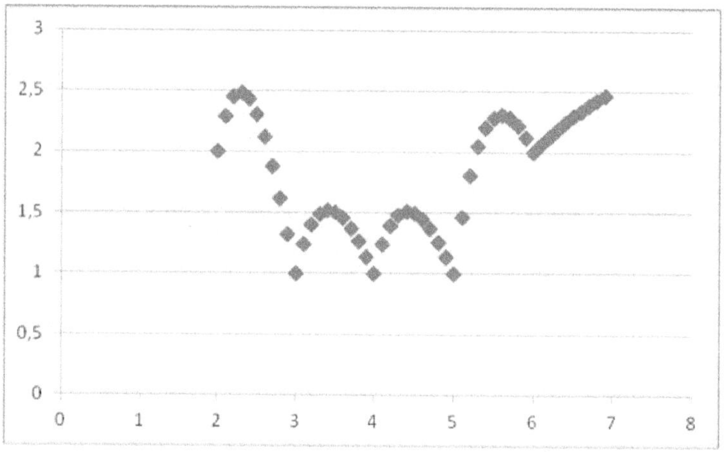

Extrapolated points: (7.2;3), (7.4;3), (7.6;3), (7.8;3), (8;3), (8.2;3), (8.4;3), (8.6;3), (8.8;3).

Example 2

Nodes (1;3), (3;1), (5;3), (7;2) and $h = 0$, $\gamma = F(\alpha) = \alpha^2$. This function F requires $\alpha > 1$ and extrapolation is computed with (4)-(5) too:

Figure 20. PNC modeling with nine interpolated points between successive nodes and nine extrapolated points right of the last node
Source: author, 2014

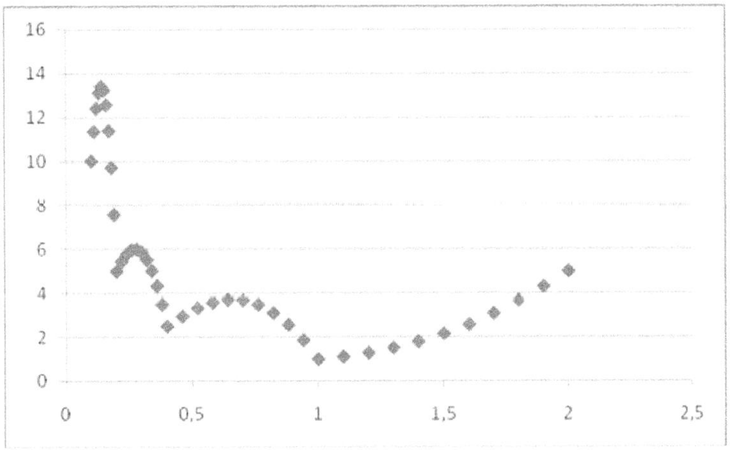

Extrapolated points: (7.2;1.79), (7.4;1.56), (7.6;1.31), (7.8;1.04), (8;0.75), (8.2;0.44), (8.4;0.11), (8.6;-0.24), (8.8;-0.61).

Example 3

Nodes (1;3), (3;1), (5;3), (7;4) and $h = 0$, $\gamma = F(\alpha) = \alpha^3$:

Figure 21. PNC modeling for nine interpolated points between successive nodes and nine extrapolated points right of the last node
Source: author, 2014

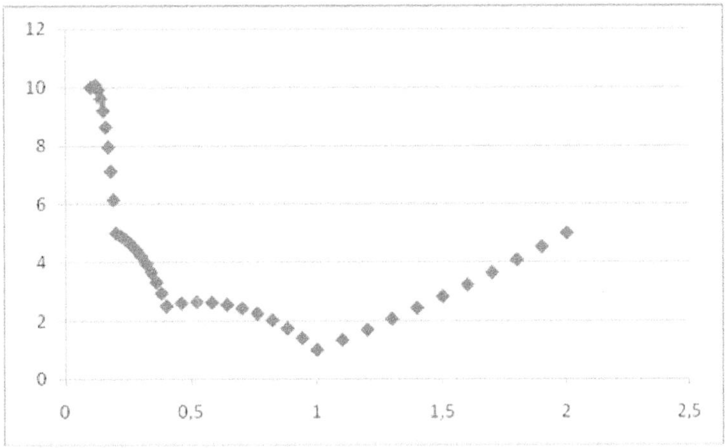

Extrapolated points: (7.2;4.331), (7.4;4.728), (7.6;5.197), (7.8;5.744), (8;6.375), (8.2;7.096), (8.4;7.913), (8.6;8.832), (8.8;9.859).

These three examples 1-3 (Fig.1-3) with nodes combination $h = 0$ differ at fourth node and probability distribution functions $\gamma = F(\alpha)$. Much more possibilities of modeling are connected with a choice of nodes combination $h(p_1,p_2,\ldots,p_m)$. MHR method uses the combination (3) with good features because of orthogonal rows and columns at Hurwitz-Radon family of matrices:

$$h(p_i, p_{i+1}) = \frac{y_i}{x_i} x_{i+1} + \frac{y_{i+1}}{x_{i+1}} x_i$$

and then (2):

$$y(c) = \gamma \cdot y_i + (1-\gamma) y_{i+1} + \gamma(1-\gamma) \cdot h(p_i, p_{i+1}).$$

Here are two examples 4 and 5 of PNC method with MHR combination (3).

Example 4

Nodes (1;3), (3;1), (5;3) and $\gamma = F(\alpha) = \alpha^2$. This function F requires $\alpha > 1$ and extrapolation is computed with (4)-(5):

Figure 22. PNC modeling with nine interpolated points between successive nodes and nine extrapolated points right of the last node
Source: author, 2014

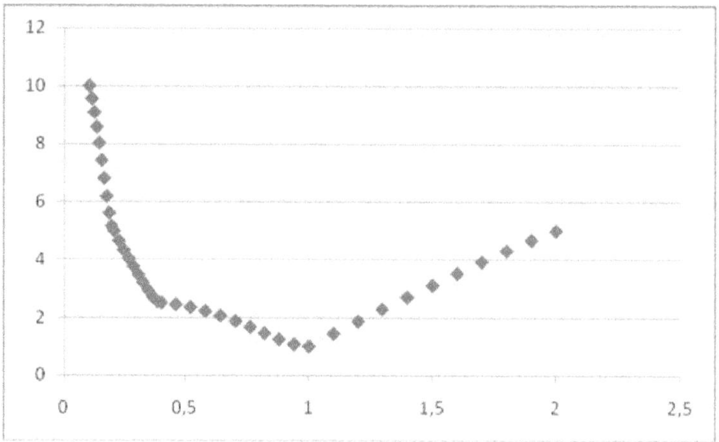

Extrapolated points: (5.2;2.539), (5.4;1.684), (5.6;0.338), (5.8;-1.603), (6;-4.25), (6.2;-7.724), (6.4;-12.155), (6.6;-17.68), (6.8;-24.443).

Example 5

Nodes (1;3), (3;1), (5;3) and $\gamma = F(\alpha) = \alpha^{1.5}$. This function F requires $\alpha > 1$ and extrapolation is computed with (4)-(5):

Figure 23. PNC modeling with nine interpolated points between successive nodes and nine extrapolated points right of the last node
Source: author, 2014

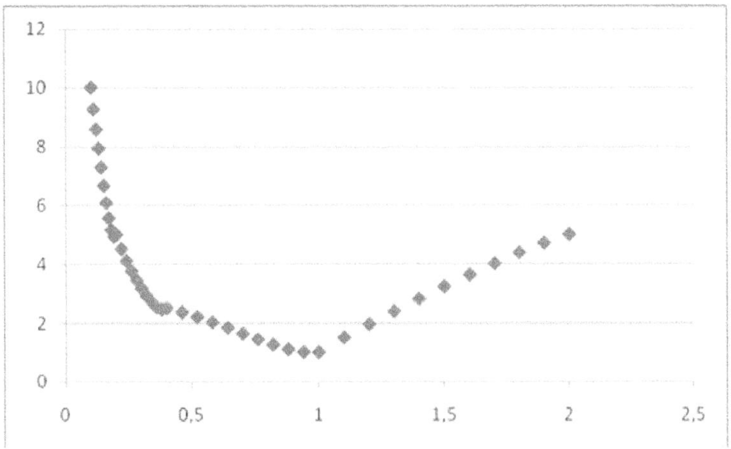

Extrapolated points: (5.2;2.693), (5.4;2.196), (5.6;1.487), (5.8;0.543), (6;-0.657), (6.2;-2.136), (6.4;-3.915), (6.6;-6.016), (6.8;-8.461).

Now let us consider PNC method with other functions F than power functions, $\alpha < 0$ for extrapolation (1)-(2) and nodes combination $h = 0$.

Example 6

Nodes (2;2), (3;1), (4;2), (5;1), (6;2) and $\gamma = F(\alpha) = sin(\alpha \cdot \pi/2)$, $h = 0$:

Figure 24. PNC modeling with nine interpolated points between successive nodes and nine extrapolated points right of the last node
Source: author, 2014

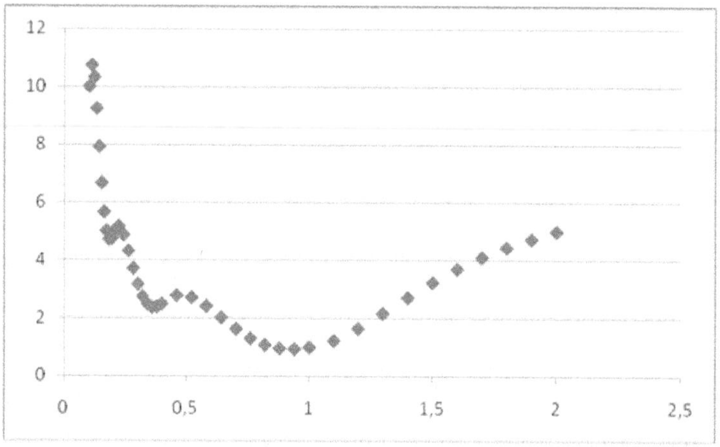

Extrapolated points: (6.1;2.156), (6.2;2.309), (6.3;2.454), (6.4;2.588), (6.5;2.707), (6.6;2.809), (6.7;2.891), (6.8;2.951), (6.9;2.988).

Example 7

Nodes (2;2), (3;1), (4;2), (5;1), (6;2) and $\gamma = F(\alpha) = sin^3(\alpha \cdot \pi/2)$, $h = 0$:

Figure 25. PNC modeling with nine interpolated points between successive nodes and nine extrapolated points right of the last node
Source: author, 2014

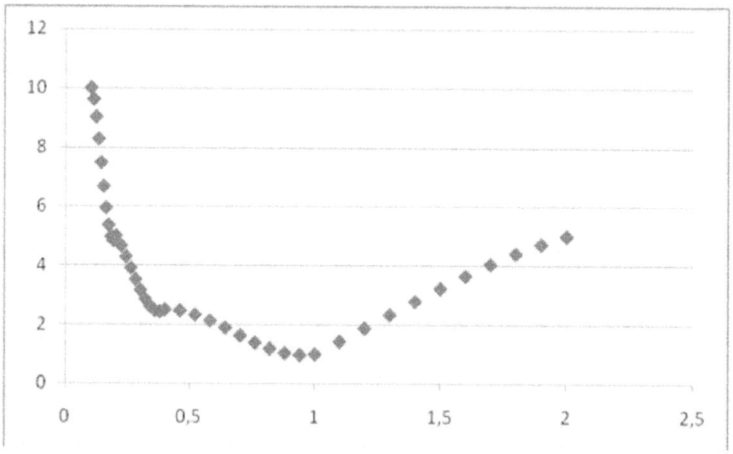

Extrapolated points: (6.1;2.004), (6.2;2.03), (6.3;2.094), (6.4;2.203), (6.5;2.354), (6.6;2.53), (6.7;2.707), (6.8;2.86), (6.9;2.964).

These two examples 6 and 7 (Fig.6-7) with nodes combination $h = 0$ and the same set of nodes differ only at probability distribution functions $\gamma = F(\alpha)$. Fig.8 is the example of nodes combination h as (3) in MHR method.

Example 8

Nodes (2;2), (3;1), (4;1), (5;1), (6;2) and $\gamma = F(\alpha) = 2^{\alpha} - 1$:

Figure 26. PNC modeling with nine interpolated points between successive nodes and nine extrapolated points right of the last node
Source: author, 2014

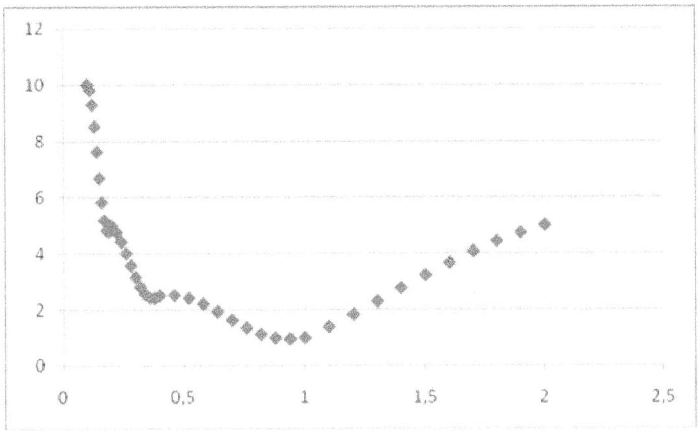

Extrapolated points: (6.1;2.067), (6.2;2.129), (6.3;2.188), (6.4;2.242), (6.5;2.293), (6.6;2.34), (6.7;2.384), (6.8;2.426), (6.9;2.464).

Examples that are calculated above have one function $\gamma = F(\alpha)$ and one combination h for all ranges between nodes. But it is possible to create a model with functions $\gamma_i = F_i(\alpha)$ and combinations h_i individually for a range of nodes $(p_i;p_{i+1})$. Then it enables very precise modeling of data between each successive pair of nodes. Each data point is interpolated or extrapolated by PNC via three factors: the set of nodes, probability distribution function $\gamma = F(\alpha)$ and nodes combination h. These three factors are chosen individually for each data, therefore this information about modeled points seems to be enough for specific PNC curve interpolation and extrapolation. Function γ is selected via the analysis of known points before extrapolation, we may assume $h = 0$ at the beginning and after some time exchange h by more adequate.

These eight examples illustrate the extrapolation of some important values in planning process, for example anticipation of some costs or expenses and foreseeing the prices or other significant data in the process of planning.

SOLUTIONS AND RECOMMENDATIONS

Proposed method, called Probabilistic Nodes Combination (PNC), is the method of 2D curve interpolation and extrapolation using the set of key points (knots or nodes). Nodes can be treated as characteristic points of data for modeling and analyzing. The model of data can be built by choice of probability distribution function and nodes combination. PNC modeling via nodes combination and parameter γ as probability distribution function enables value anticipation in risk analysis and decision making. Two-dimensional curve is extrapolated and interpolated via nodes combination and different functions as discrete or continuous probability distribution functions: polynomial, sine, cosine, tangent, cotangent, logarithm, exponent, arc sin, arc cos, arc tan, arc cot or power function. Novelty of the paper consists of two generalizations: generalization of previous MHR method with various nodes combinations and generalization of linear interpolation with different (no uniform) probability distribution functions and nodes combinations.

FUTURE RESEARCH DIRECTIONS

Future trends will go to various directions: how to fix the best probability distribution function for the nodes, how to calculate the most appropriate nodes combination and what extrapolation is the most valuable in decision making and risk analysis.

CONCLUSION

Planning process requires the anticipation of unknown values or data and foreseeing some important factors. The method of Probabilistic Nodes Combination (PNC) enables interpolation and extrapolation of two-dimensional curves using nodes combinations and different coefficients γ: polynomial, sinusoidal, cosinusoidal, tangent, cotangent, logarithmic, exponential, arc sin, arc cos, arc tan, arc cot or power function, also inverse functions. Function for γ calculations is chosen individually at each planning and it is treated as probability distribution function: γ depends on initial requirements and data specifications. PNC method leads to point extrapolation and interpolation via discrete set of fixed knots. Main features of PNC method are:

1. The smaller distance between knots the better;
2. PNC method develops a linear interpolation and extrapolation into other functions as probability distribution functions;
3. PNC is a generalization of MHR method via different nodes combinations;
4. Interpolation and extrapolation of L points is connected with the computational cost of rank $O(L)$ as in MHR method;
5. Nodes combination and coefficient γ are crucial in the process of data probabilistic anticipation and foreseeing.

Future works are going to: application of PNC method in decision making, choice and features of nodes combinations and coefficient γ, implementation of PNC in computer vision and artificial intelligence: shape geometry, contour modelling, object recognition and curve parameterization.

Why and when should we use PNC method? Interpolation methods and curve fitting represent so huge problem that each individual interpolation is exceptional and requires specific solutions. PNC method is such a novel tool with its all pros and cons. The user has to decide which interpolation method is the best in a single situation. The choice is yours if you have any choice. Presented method is such a new possibility for curve fitting and interpolation when specific data (for example handwritten symbol or character) starts up with no rules for polynomial interpolation. This paper consists of two generalizations: generalization of previous MHR method with various nodes combinations and generalization of linear interpolation with different (non-uniform) probability distribution functions and nodes combinations.

The method of Probabilistic Nodes Combination (PNC) enables interpolation and modeling of two-dimensional curves using nodes combinations and different coefficients γ: polynomial, sinusoidal, cosinusoidal, tangent, cotangent, logarithmic, exponential, arc sin, arc cos, arc tan, arc cot or power function, also inverse functions. This probabilistic view is novel approach a problem of modeling and interpolation. Computer vision and pattern recognition are interested in appropriate methods of shape representation and curve modeling. PNC method represents the possibilities of shape reconstruction and curve interpolation via the choice of nodes combination and probability distribution function for interpolated points. It seems to be quite new look at the problem of contour representation and curve modeling in artificial intelligence and computer vision.

Function for γ calculations is chosen individually at each curve modeling and it is treated as probability distribution function: γ depends on initial requirements and curve specifications. PNC method leads to curve interpolation as handwriting modeling via discrete set of fixed knots. So PNC makes possible the combination of two important problems: interpolation and modeling. Main features of PNC method are:

6. The smaller distance between knots the better;
7. Calculations for coordinates close to zero and near by extremum require more attention because of importance of these points;
8. PNC interpolation develops a linear interpolation into other functions as probability distribution functions;
9. PNC is a generalization of MHR method via different nodes combinations;
10. Interpolation of L points is connected with the computational cost of rank $O(L)$ as in MHR method;
11. Nodes combination and coefficient γ are crucial in the process of curve probabilistic parameterization and interpolation: they are computed individually for a single curve.

Future works are going to: application of PNC method in signature and handwriting recognition, choice and features of nodes combinations and coefficient γ, implementation of PNC in computer vision and artificial intelligence: shape geometry, contour modelling, object recognition and curve parameterization.

REFERENCES

Ballard, D. H. (1982). *Computer Vision*. Prentice Hall.

Chapra, S. C. (2012). *Applied Numerical Methods*. McGraw-Hill.

Choraś, R. S. (2005). *Computer Vision*. Exit.

Cocozza-Thivent, C., Eymard, R., Mercier, S., & Roussignol, M. (2006). Characterization of the Marginal Distributions of Markov Processes Used in Dynamic Reliability. *Journal of Applied Mathematics and Stochastic Analysis*, 1–18.

Collins, G. W. II. (2003). *Fundamental Numerical Methods and Data Analysis*. Case Western Reserve University.

Dahlquist, G., & Bjoerck, A. (1974). *Numerical Methods*. Prentice Hall.

Dejdumrong, N. (2007). A Shape Preserving Verification Techniques for Parametric Curves. *Computer Graphics, Imaging and Visualization. CGIV, 2007*, 163–168.

Dyn, N., Levin, D., & Gregory, J. A. (1987). A 4-Point Interpolatory Subdivision Scheme for Curve Design. *Computer Aided Geometric Design, 4*, 257–268.

Jakóbczak, D. (2007). 2D and 3D Image Modeling Using Hurwitz-Radon Matrices. *Polish Journal of Environmental Studies, 4A*(16), 104–107.

Jakóbczak, D. (2009). Curve Interpolation Using Hurwitz-Radon Matrices. *Polish Journal of Environmental Studies, 3B*(18), 126–130.

Jakóbczak, D. (2010). Shape Representation and Shape Coefficients via Method of Hurwitz-Radon Matrices. *Lecture Notes in Computer Science, 6374*, 411–419.

Jakóbczak, D. (2010). Object Modeling Using Method of Hurwitz-Radon Matrices of Rank k. In W. Wolski & M. Borawski (Eds.), *Computer Graphics: Selected Issues* (pp. 79–90). University of Szczecin Press.

Jakóbczak, D. (2011). Curve Parameterization and Curvature via Method of Hurwitz-Radon Matrices. *Image Processing & Communications-. International Journal (Toronto, Ont.), 1-2*(16), 49–56.

Jakóbczak, D. (2011). Data Extrapolation and Decision Making via Method of Hurwitz-Radon Matrices. *Lecture Notes in Computer Science, 6922*, 173–182.

Jakóbczak, D. (2011). Curve Extrapolation and Data Analysis using the Method of Hurwitz-Radon Matrices. *Folia Oeconomica Stetinensia, 9*(17), 121-138.

Jakóbczak, D. (2013). Probabilistic Modeling of Signature using the Method of Hurwitz-Radon Matrices. *Global Perspectives on Artificial Intelligence, 1*(1), 1–7.

Kozera, R. (2004). *Curve Modeling via Interpolation Based on Multidimensional Reduced Data*. Silesian University of Technology Press.

Liu, T., & Geiger, D. (1999). Approximate tree matching and shape similarity. *Int. Conf. Computer Vision*.

Lorton, A., Fouladirad, M., & Grall, A. (2013). A Methodology for Probabilistic Model-based Prognosis. *European Journal of Operational Research*, *225*, 443–454.

Pergler, M., & Freeman, A. (2008). Probabilistic Modeling as an Exploratory Decision-Making Tool. *McKinsey Working Papers on Risk, 6*, 1-18.

Ralston, A., & Rabinowitz, P. (2001). *A First Course in Numerical Analysis* (2nd ed.). Dover Publications.

Rogers, D. F. (2001). *An Introduction to NURBS with Historical Perspective*. Morgan Kaufmann Publishers.

Saber, E., Yaowu, X., & Murat Tekalp, A. (2005). Partial shape recognition by sub-matrix matching for partial matching guided image labeling. *Pattern Recognition*, *38*, 1560–1573.

Schumaker, L. L. (2007). *Spline Functions: Basic Theory*. Cambridge Mathematical Library.

Sebastian, T. B., & Klein, P. N. (2003). On aligning curves. *IEEE Transactions on Pattern Analysis and Machine Intelligence*, *25*(1), 116–124.

Tadeusiewicz, R., & Flasiński, M. (1991). *Image Recognition*. PWN.

Zhang, D., & Lu, G. (2004). Review of Shape Representation and Description Techniques. *Pattern Recognition*, *1*(37), 1–19.

KEY TERMS AND DEFINITIONS

Artificial Intelligence: Intelligence of machines and computers, as a connection of algorithms and hardware, which makes that a man – human being can be simulated by the machines in analyzing risk, decision making, reasoning, knowledge, planning, learning, communication, perception and the ability to move and manipulate objects.

Curve Interpolation: Computing new and unknown points of a curve and creating a graph of a curve using existing data points – interpolation nodes.

Data Extrapolation: Calculation of unknown values for the points situated outside the ranges of nodes.

Value Anticipation: Foreseeing next value when last value is known.

Section 2

Dual Approach of Data Analytics and Machine Learning Modelling in Real Case Scenarios

Chapter 4

Patient Arrival to Public OPDs: Analysis and Use of Statistical Distribution for Improving Performance Indicators in Rural Hospitals

Ahan Chatterjee
https://orcid.org/0000-0001-5217-4457
The Neotia University, India

Swagatam Roy
https://orcid.org/0000-0002-8012-5529
The Neotia University, India

Trisha Sinha
The Neotia University, India

ABSTRACT

The main objective of this chapter is to take a deeper look into the infrastructural condition of the hospitals across the districts of West Bengal, India. There is a liaison between various variables and the infrastructural growth of the public healthcare centres. In this chapter, the authors have formed a panel data from the year 2004 2017, consisting of 17 districts across West Bengal. They have assessed the random effect model on the data to choose their respective hypothesis. A Bayesian risk analysis had also been carried out on the mortality rate of the patients on which factors it depends. Next, a Poisson distribution model is being fit to get some insights into the data. Afterward, they predicted the number of patients who will arrive in 2020 and the shortfall of hospitals is also being projected. The remedies to these have also been suggested in that section. At last, they carried out an econometric analysis in the healthcare domain and took a closer look at how healthcare expenditure affects our focus variables performance.

DOI: 10.4018/978-1-7998-4706-9.ch004

INTRODUCTION

It is health that is real wealth and not pieces of gold and silver

- Mahatma Gandhi

As said by Mahatma Gandhi, health is our real wealth that should be taken care of. In this paper, we will use the statistical approach to analyze the current infrastructure of the hospitals in the rural part of West Bengal, India. We have created a panel data taking from the year 2004 -2017. Our aim in this paper to analyze and improve the health indicators in the hospitals. We have applied predictive forecasting to predict the rough number of patients arriving at OPDs in every district for years to come. (Raghupathi & Ragupathi, 2015) Using those figures we can analyze the indicator results and can suggest the changes needed to be done for better performances. Further, we have applied the Bayesian Model on the data and analyzed those on that basis. Poisson Model also has been applied for analysis. The model which yields better performance indications is taken into account. We know Indian healthcare is a three-tiered system and there is an acute shortfall of healthcare centres and we can predict how much is being required for the near. In this limelight, we also see the growth from the econometric perspective where the lion's share of GDP is being invested in the healthcare sector of India's total GDP, and how it is hitting an all-time low in recent past. In this backdrop, we will analyze how we can make a growth in the GDP through proper infrastructure and attracting people to public OPD centres for their healthcare access and make a contribution in the growth of country's GDP. Here we will suggest the changes required in those parts of the infrastructure which is most significant and those are measured using hypothesis testing and fitting random effect regression model on the panel data which we have created to analyze the data. Making required changes in processes of care and service delivery is essential for the improvement of healthcare. This analysis is complicated by the existence of natural variation though process performance. This is measured to determine if these changes are having the desired beneficial effects. Lion's share of the total budget of a country is being spending on the healthcare unit, it is thus essential for the hospitals to provide essential services. In this paper, we have taken 14 years of data and the indicators are Postnatal Care, Antenatal Care, Child Delivery, Deaths Reported, Child Care, Laboratory Testing, and GDP. Here district GDP is being taken to analyze the growth in economics measured through patients arriving at the public OPD's. (Gajewski et al., 2008)

THE CONTRIBUTION OF HELATH IN THE PROCESS OF HUMAN DEVELOPMENT

Health is being considered as the one of the most important pillar in the process of development of human race. A healthy human is capable to perform better in each and every aspect of life over a non-healthy person. In spite of these facts the health care infrastructure as well as healthcare service in the developing countries like India is not up to the mark. In spite investing a lion's share of total GDP of India into healthcare service still the basic needs are not fulfilled, in support of this statement the fact is still now in India every second young child across the nation is still malnourished. With such facts we can safely claim our statements as true across the length and breadth of country. A report was published in late 2000's as Millennium Development Goal Report, it stated that target in healthcare section is to reduce Infant Mortality Ratio (IMR) and Maternal Mortality Ratio (MMR) will be reduced to 1 per 1000, live birth. And this target was scheduled to be achieved by the year 2012. But in reality still now we are far behind that target. From these statistics we can observe that the healthcare service is not performing up to the mark. Through various reports we have formed an observation based on the health performance indicators across the international level and it's evident that India lags much behind to achieve those targets. Table 1, shows some of the healthcare indicators across the globe.

Table 1. A Demographic Picture across the globe on the parameter health

Indicators	West Bengal	India	Less Developed Countries	Developing Countries	Industrialized Countries	World
Population (in millions)	85.2	1112.2	785.4	5358.2	969.9	6577.2
Crude Birth Rate	18.4	23.5	36.0	23.0	11.0	21.0
Crude Death Rate	6.4	7.5	13.0	8.0	9.0	9.0
Infant Mortality Rate	38.0	57.0	96.0	54.0	5.0	49.0
Under Five Mortality Rate	59.6	74.3	140.0	79.0	6.0	72.0
Total Fertility Rate	2.0	2.9	4.7	2.8	1.7	2.6
Life Expectancy	69.6	67.0	55.0	66.0	79.0	68.0

Source: The state of the World's children 2008 (UNICEF)

From the statistics presented in Table 1, we can say that the overall performing health infrastructure is in mess. India definitely has high quality medical options but those are highly costly and it's only limited to the higher class of peoples. In India where a staggering 23% of people are under *Below Poverty Line* can't certainly access those high cost medical treatments. In such condition we can the health care situation in the rural or underdeveloped parts of West Bengal rather India is extremely poor. The health infrastructure immediately needs change in order to provide better life to the masses. (Marchal et al., 2010)

BASIC INTRODUCTION OF THREE-TIERED HEALTH INFRASTRUCTURE IN INDIA

Now-a-days it's very common among people to visit the OPD for checkup and access healthcare services. The healthcare structure of India is basically a three-tier structure. The topmost tier is the Community Healthcare Centres, the middle tier is known as Primary Health Centres and the lowest tier is the Sub-Centres. As per recent report from The District Level Household & Facility Survey (*DLHS*) of year 2014-15 the stats shows that for per 5,000 population in general area there exist 1 Sub-Centre, and this figure goes to 3,000 population per Sub-Centre in hilly and tribal areas of India.

The figure for Primary Health Centres (PHC) goes as 1 PHC per 30,000 populations in general areas and in hilly and tribal areas it's 1 PHC per 20,000 populations. And for the top tier of healthcare infrastructure Community Health-Care Infrastructure (CHC), the count goes as 1,20,000 populations for unit CHC in general areas and for 1 CHC in hilly and tribal areas this stat goes for 80,000 population.

Figure 1. Graph showing status of Sub-Centres in India, India from year 1981-2015
Source: DLHS Report

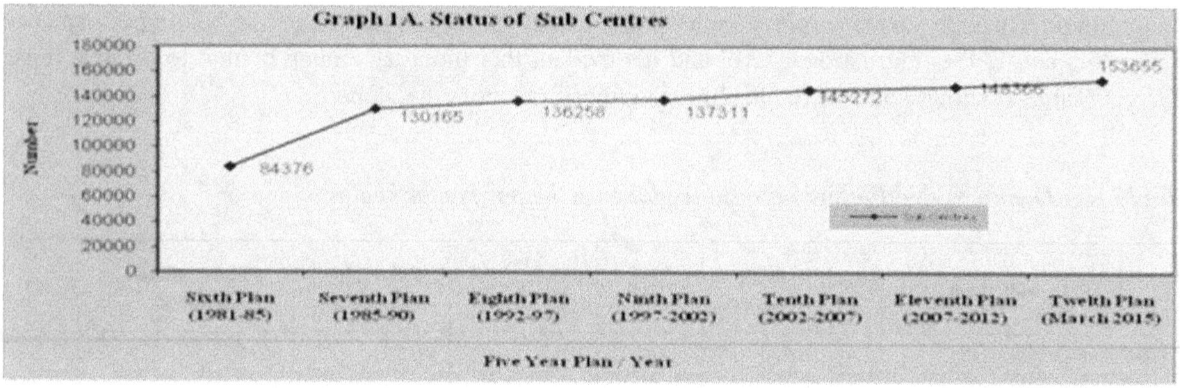

After statistics of the three-tiered healthcare infrastructure we will look into the overall statistics of the healthcare in India. As per the Rural Healthcare Statistics report we see there are 1,53,655 Sub-Centres (SCs), 25,308 Primary Health Centres (PHCs), 5,396 Community Health Centres (CHCs) available in our country.

After calculation we see there is a shortfall of 33,145 SCs, 6566 PHCs, and 1022 CHSs. This is equivalent to 20% shortfall in Sub-Centres, 22%, and 32% shortfall for Primary Health Centres and Community Health Centres respectively across our nation. After seeing this kind of statistics for the country we can easily say there is a need of lots of improvements needed in this healthcare arena for making India a better place to live. Extending this we can highlight the econometric view of the country from this side. The allotted healthcare expenditure is 4 percent of India's GDP and it took a low of 1 percent as per latest reports shown. This strongly reflects the need of improvement of the health indicators. In this backdrop, this paper shows a statistical approach to improve the health indicators across the rural part of West Bengal, India. (Mukherjee, 2017)

Figure 2. Graph showing status of Primary Health Centres in India, India from year 1981-2015
Source: DLHS Report

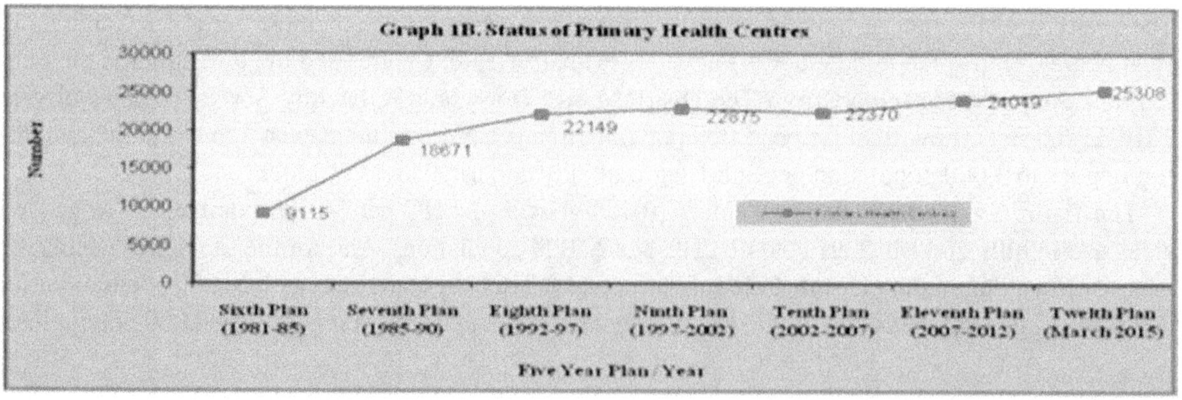

Figure 3. Graph showing status of Community Health Centres in India, India from year 1981-2015
Source: DLHS Report

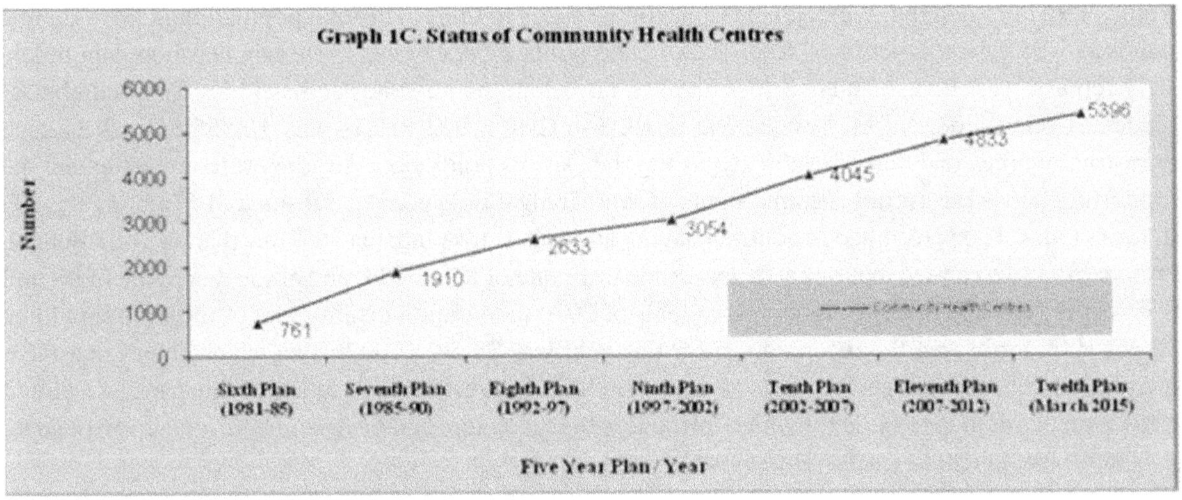

In a typical manner every OPD's take the registration of the patients in advance and there is a buffer of waiting time for each patient to visit the respective doctor. If we can minimize the model of waiting time there is betterment in the performance indications.

Deaths reported is another important indicator which is being used, as when this figure will go less we can conclude that the OPD are working fine. Along with that availability of the hospital facility also highly contribute in the performance indicators. (Singh et al., 2013)

Figure 4. Bar Plot showing shortage of Three-Tired Health Centres in India, India from year 1981-2015
Source: DLHS Report

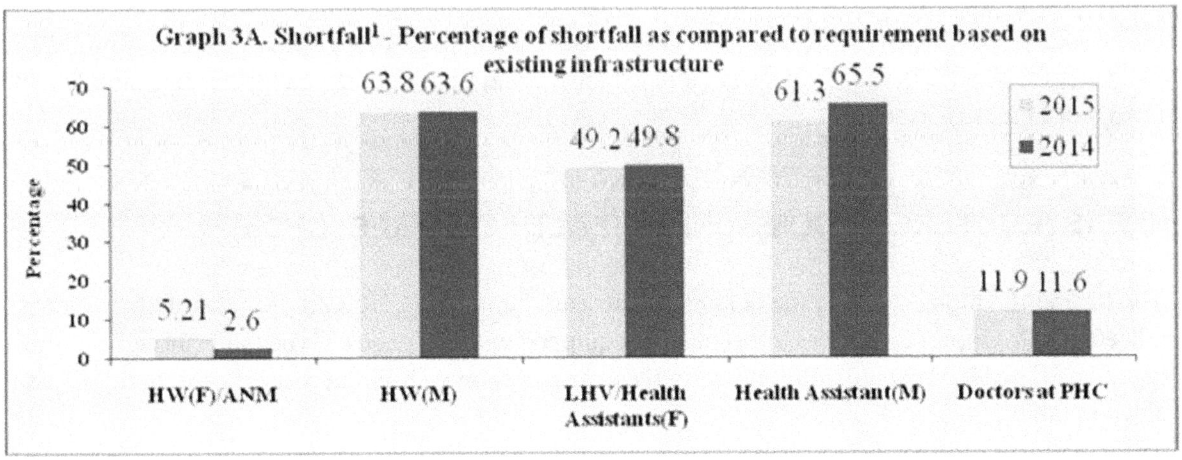

CURRENT HEALTH INFRASTRUCTURE OF WEST BENGAL IN A NUTSHELL

Among 29 States in India, West Bengal is one of the states in India. The total population in West Bengal stands as 9.03 crores as per latest Census, 2011. As per latest statistics, on Human Development Index, 2017 reports West Bengal stands 28th in the list, far behind major states across India. Total number of Medical Colleges present in West Bengal is 19, 346 CHCs, 922 PHC's, and 10,356 SCs. With such infrastructure the condition of health across the state seems pretty poor. In support to our statement the supporting fact is the overall aneamia status of West Bengal is so poor that it stands 19th among the 29th states in India. The percentage of children having aneamia across India is 78% low than of West Bengal. From table 1, we can see that this state has crude birth rate of about 19.8 and crude death rate of around 6.3. As per National Socio-Demographic Goals of 2010 the infant mortality rate of the state should be 30 per 1000 births, but the recorded stat for that year was 38 per 1000 births, which shows that there is a dip in the healthcare performances across the OPDs in West Bengal, India. In this paper we aim to bring a limelight to the key factor which influences the indicators and bring some insights from the data to better these statistical goals. (Hossein & Gerard, 2013)

HEALTH INFRASTRUCTURE OF WEST BENGAL

As defined earlier in the section 1.b. we know that the health infrastructure of India is three-tiered. There consists the Sub Health Centres, Primary Health Centres, and Community Health Centres. In a similar way West Bengal also follows this health infrastructure. In 1948, as recommended by Bhore Committee that for every 1000 population present in the country there should a bed available in public hospitals, still after 71 years we have failed to provide this facility. (Hakkinen et al., 2013)

As per national norm of India there should be at least 1 Community Health Centres per 1 lakh population present but the scenario in West Bengal seems not good as there is 1 CHCs per 3 lakh population present thus indicating a very poor healthcare system. This causes overcrowdings in the hospitals and this leads to poor healthcare services and reduces their efficiencies in performances. Low performance leads many patients to private hospital to get health support this certainly leads to less income of government from the public healthcare centres. Huge numbers of patients are being referred to the Medical Colleges and District Hospitals from lower tier as they don't have sufficient infrastructure to provide good treatment to them. This is again another major cause of overcrowding in those places.

The statistic for private medical support at West Bengal stands as 10,461 beds in 3558 hospitals. But the healthcare service in those places is quite expensive and poor people across the state simply can't afford those kinds of expenses.

INFRASTRUCTURE AT COMMUNITY HEALTH CENTRES (CHCs)

CHC lies on the topmost tier in the healthcare infrastructure. The CHCs are designed to provide specialized health support to the mass. It should be equipped with Surgeons, Gynecologists etc. and those should be 30 bedded hospitals. But in reality the statistics show a different story regarding this. This is shown in table 2.

Table 2. Comparison between available CHC and capable CHC

Total Number of CHCs in West Bengal	Number of CHCs Capable for Critical Treatment
269	72

Source: Created by Author, based on dataset

From this we can clearly see that the present scenario is far below the average healthcare infrastructure, this can be considered as poor. (Butler et al., 2013)

The specialized CHC have many specialized outdoors, among the districts in West Bengal the worst conditions is seen in North 24 Parganas, followed by South 24 Parganas, they have the worst facilities available. Among the districts with high population density or population pressure Hooghly, Howrah, Burdwan also have equally poor facilities open. In the table 3, we have put some stats regarding how bad the current scenario is:

INFRASTRUCTURE AT PRIMARY HEALTH CENTRES (PHCs)

The CHCs are being operated at a district level while the PHCs are being operated at the block level across the districts. According to national norms of India there should be a Primary Health Centres available for every 30,000 population present in plain areas and for hilly or tribal areas this stat goes down to 20,000 population. In West Bengal there is 1 PHCs available for 90,000 populations in plain areas and for hilly areas it's 60,000. This shows how poor the availability of PHCs across the state and districts. PHCs act as referral unit for six SCs. Total numbers of PHCs across West Bengal is 909. And to claim our statement of poor service in PHCs we present the statistics in table 4. (Geweke et al., 2003)

Table 3. Comparison between available CHC with different parameters

CHCs with Parameter	Total Number of CHCs
Total Number of CHCs in West Bengal	269
CHCs with Blood Storage Facility	5
CHCs with Gynecologist Facility	3
CHCs with Anesthetist Facility	35
CHCs with New Born Baby Facility	40

Source: Created by Author, based on dataset

INFRASTRUCTURE AT SUB HEALTH CENTRES (SCs)

The SCs belong to the lowest tier in the three-tiered health infrastructure, it acts as the first preliminary contact point in between PHCs and Community. This is build to support medical issues at village level, at grassroots level. Patients from SCs will only be transferred to PHCs and CHCs when they can't provide medical treatments. In every 100 villages there is 94 SCs present, and the availability of SCs is

Table 4. Comparison between available CHC with different parameters

PHCs with Parameter	Total Number of PHCs
Total number of PHCs in West Bengal	909
Proper available PHCs in Villages	45
Regular Electricity available in PHCs in Villages	318
Presence of at least 4 beds in PHCs	227
PHCs having experience of 10 deliveries	91
PHCs operating for 24 hours	229

Source: Created by Author, based on dataset

better observed in the districts of Howrah, Jalpaiguri. The health performance indicator in West Bengal is pretty good, and we can state that West Bengal has done a significant improvement in this section at rural areas, in grassroots level the development have been done. To support our claim the next table 5 shows some stats regarding the same.

Table 5. Comparison between available CHC with different parameters

SCs with Parameter	Total Number of SCs
Total Number of SCs in West Bengal	10,369
SCs with equipped Health Service	10,161
SCs equipped with Auxiliary Nurse	9228
SCs equipped with Drug Reserve Facility	9332

Source: Created by Author, based on dataset

ANALYTICS FRAMEWORK

Now, coming to the notion of the health infrastructure of West Bengal, India, the support of the health service is the most basic need of the citizens of a country belongs to the third world, and a booming economy, India. Our research in this paper involves the use of analytics in the healthcare domain of India, and we will showcase a framework of health analytics which can provide us better health indicators result, our model is based on the general statistical modeling and business intelligence. The steps included in our work are namely, data collection, data cleaning, data transformation, and data analytics. Figure 5 shows our analytics framework.

DATA COLLECTION

For the analysis, the data have been collected from the open data source of Indian Government. All the data have been collected from data.gov.in website. The data is being validated genuine by none other

Figure 5. Flowchart showing the Analytics Framework
Source: Created by the Author

than Indian Government. We have collected all the data and merged them into panel data form to fit our models and carry out our analysis. (Benneyan et al., 2003)

HYPOTHESIS GENERATION AND PERFORMANCE INDICATOR

Table 6. Defining the Health Indicators

Indicators	Meaning
Antenatal Care	Care received during pregnancy
Postnatal Care	Care given to baby and mother after birth.
Child Deliveries	Child born.
Child Death Case	Child death cases registered.
RTI/ STI Case	Sexually transmitted diseases.
Child Immunization	Immunization provided to the children.
Laboratory Testing	Blood or other forms of lab test.
District GDP	Gross Domestic Product rate for districts.
Death Case Reported	The toll of deaths reported.

Source: Created by the Author

In table 6, we have defined the health indicators we have taken to analyze the performance of the OPD's across the state, West Bengal. All the indicators may don't have significance for such we will be using hypothesis testing.

There are in total of 9 health indicators we have taken in our analysis. Not all of them contribute significantly in determining whether the factor does measure the performance level of a hospital. Thus we take help of the *hypothesis testing*, from the descriptive statistics to find the factors with significance. (Aven & Eidesen, 2007)

In total we take 9 (nine) null hypothesis for each factor and same number of alternate hypothesis.

H_0: Null Hypothesis is represented
H_1: Alternate Hypothesis is represented.
$H_{0(Antenatal\ Care)}$: Change in figures of Antenatal Care doesn't affect the performance indication.
$H_{1(Antenatal\ Care)}$: Change in figure of Antenatal Care does affect the performance indication.

Similarly, for all the other factors we took the same hypothesis, to draw our conclusion on the most significant performance indicators. Our data set spans from the year 2004-2017. As this is a time series data we took the help of panel data to analyze and frame our data.

PANEL DATA ANALYSIS AND RANDOM EFFECT MODELING

In our panel dataset the panel variable was, District Name and the Year variable is being declared as the time variable which was yearly, and our dataset was a balanced dataset. We have applied *Random Effect Model* and regressed our model keeping the variable "Patient coming to OPD's" as the dependent variable and the other nine performance indicators as independent variable. The *Random effect* model is also known as the variance components model. Here, the model parameters are taken as random variables and those don't have any fixed effect on the model. It's a kind of hierarchal linear model, which analyzes the data drawn from different populations who have a difference. (Alkhatib et al., 2016)

Implementing the *Random Effect Model,* on the panel data we get the value of *Ptest,* this value decides whether we can accept or reject the null hypothesis.

If the value of $P_{Test} < 0.05$, we can accept our alternate hypothesis true, and can safely reject the null hypothesis. While for $P_{Test} \geq 0.05$ the null hypothesis stands true and we would accept that, and can conclude that those parameters don't have any effect on the target variable.

Table 7. Results of Random Effect Model on Panel Data

Performance Overall Average	Coef.	Std. Err.	z	P>|z|	[95% Conf. Interval]	
Antenatal Care Services	.0113013	.0230261	0.49	0.00	-.033829	.0564316
Deliveries	-.155958	.0701845	-2.22	0.02	-.2935172	-.0183989
Child Death Reported	.3516577	.2432727	1.45	0.03	-.125148	.8284634
Postnatal Care	.012186	.0092801	1.31	0.51	-.0060027	.0303748
Details of deaths reported	-.0140501	.0622569	-0.23	0.03	-.1360713	.1079711
RTI/ STI Cases	-.0525188	.0632639	-0.83	0.44	-.1765138	.0714763
Child Immunization	-.0112014	.0108885	-1.03	0.00	-.0325425	.0101396
Laboratory Testing	-.0034249	.0026659	-1.28	0.04	-.00865	.0018001
GDP	-2258.756	1209.395	-1.87	0.03	-4629.128	111.6148

Source: Created by the Author

In Table 7, we show the result received from implementing the random effect model in the panel data and from that we also get the significance value by which we can decide the fate of hypothesis.

After fitting random effect model on our panel data we can conclude that the most significant factors which contribute the performance indicator, accepting the alternate hypothesis. The following table 8 lists all the indicators which have significance. We will accept only those indicators which have $P_{Test} < 0.05$.

RISK ANALYSIS THROUGH BAYESIAN HIEARARCHICAL MODEL

Table 8. Most Significant Indicator names being listed

Significant Indicator Names
Antenatal Care Services
Child Deliveries
Child Death Reported
Details of deaths reported
Child Immunization
Laboratory Testing
GDP

Source: Created by the Author

In recent time, the use of Bayesian Hierarchical Model (BHM) gained a lot of grounds in statistical modeling and analysis. Using this we can predict the uncertainty in the terms of both input and output. Now-a-days Bayesian Model is widely used to create a framework for risk analysis in healthcare section and management. In this paper we have used this model to analyze and to get insights regarding the risks related to the key health indicators. Through BHM analysis of risk associated with a parameter can be better understood, over traditional subjective predictions. (Paoli et al., 2019)

BHM represents and can solve hierarchical, complex and multilevel data structures. Through this model we can easily predict the death chances and can analyze and reduce the numbers of deaths occurred due to poor facility or poor infrastructure so that we can upgrade those in order to reduce the rates and make the state a better place to live. (Nallala et al., 2015)

Through our panel data model and the taking important key factors, we can analyze the results and produce better result for our analysis reports. We consider a Probabilistic Risk Analysis (PRA) to analyze a specific parameter.

Let, *p* be the probability of deaths occurred, i.e. an unsuccessful event occurs.

In order to determine the value of *p* we will implement models like event trees and fault trees.*p,* is being computed using some parameterized function *f,* with parameters *q*. This is stated in equation 1.

$$p = f\left(q\right) \tag{1}$$

Figure 6. Bayesian Network
Source: Created by Author

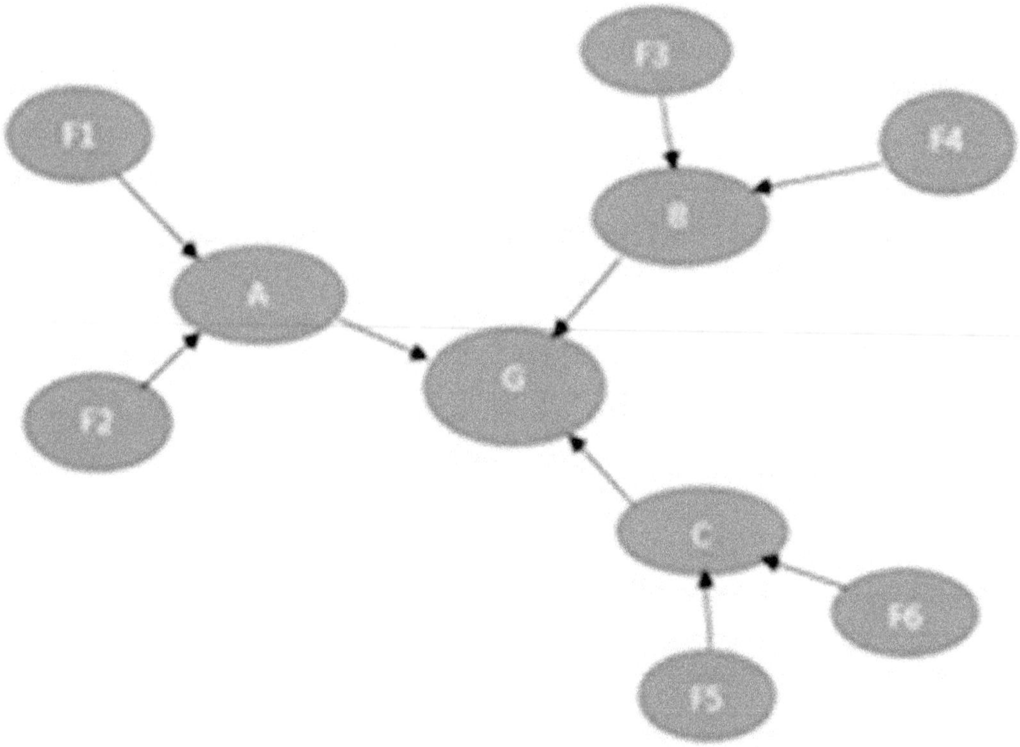

Here, q is a vector which includes the probabilities for human failure or human mistake and infrastructural error or failure. This model is represented by the true parameters between p and q. Now, we will be analyzing death stats using Bayesian Model. Figure 6, visualize the general Bayesian network.

Let, a patient *L*, suffering from a specific disease

Treatment A: High success rate, but partial recovery

Treatment B: Moderate success rate, but full recovery expected

We can say with the results and hospital infrastructure L should chose treatment A.

Table 9. Risk Factors Associated with the Model

Code	Factor
F_1	Blood Facility in CHCs/PHCs/SCs
F_2	Specialized Doctor Availability
F_3	Functioning CHCs/PHCs/SCs
F_4	Doctors ability
F_5	Patients Condition

Source: Created by the Author

Table 10. Probability of deaths based on risk factors

Year	Coefficient	F_1	F_2	F_3	F_4	F_5
2017	0.6	0.1				
2016	0.8	0.2			0.1	
2015	0.8		0.32	0.5	0.1	0.15
2014	1	0.15			0.1	0.1
2013	0.4		0.12	0.1		0
2012	0.8					
2011	0.8	0.1	0.14	0.1		0.4
2010	0.17				0.1	
2009	0.78		0.05	0.1		0.1
2008	0.40					0.1
2007	0.76	0.2		0.05	0.2	
2006	0.88		0.6			
2005	0.65				0.34	
2004	0.96	0.1				
Probability		0.132	0.215	0.476	0.073	0.104

Source: Created by the Author, based on the dataset

In this section we will analyze the risk associated in various districts across West Bengal, India through panel data model where year ranging from 2004 – 2017. (Staggs & Gajewski, 2017)

In table 9 we have enlisted the risk factors associated with the model.

We have tabulated average death model in yearly form, and only overall West Bengal data has been taken into account and not in district spread.

The Bayesian Model has been applied to infer the risk probabilities associated with the factors through which we can set our statement. (Mohapatra, 2019)

The risk factor was done by applying the univariate analysis by Bayesian MNMC approach. The model gives us the output as '0' for death and '1' for survived in a Bayesian model which in case was used to find the risk factors associated and how much associated. In Table 11, a specification of the model which was used is being stated. (Rohleder et al., 2011)

Table 11. Parameters used in Bayesian MNMC Model Deployment

Parameters	Value
Mean	0
Variance	10,000
Number of Parallel Chains	3
Number of Iterations	500000
p-value for likehood test	0.25

Source: Created by the Author, based on the dataset

Figure 7. Bayesian Analysis Result
Source: Created by Author, based on the dataset

Model	DIC	Deviance	AUC (95% CI)	Brier score	H-L statistic (p)	SMR (95% CI)
Bayesian						
M1	695.4	635.7	0.809 (0.747, 0.862)	0.113	9.06 (p = 0.3373)	0.909 (0.626, 1.277)
M2	695.0	634.8	0.802 (0.739, 0.855)	0.118	8.53 (p = 0.3835)	0.862 (0.593, 1.211)
M3	696.0	644.4	0.792 (0.728, 0.846)	0.113	18.06 (p = 0.0208)	0.952 (0.655, 1.337)
M4	719.4	643.4	0.807 (0.744, 0.860)	0.117	12.37 (p = 0.1354)	1.013 (0.697, 1.423)
Frequentist						
M1	-	670.56	0.810 (0.748, 0.862)	0.113	3.97 (p = 0.8598)	0.858 (0.590, 1.205)
M2	-	670.43	0.804 (0.741, 0.857)	0.117	8.47 (p = 0.389)	0.817 (0.562, 1.147)
M3	-	680.63	0.792 (0.729, 0.847)	0.114	14.84 (p = 0.0623)	0.882 (0.607, 1.238)
M4	-	684.77	0.806 (0.744, 0.859)	0.116	8.92 (p = 0.3491)	0.931 (0.641, 1.307)

Figure 8. Mortality Rate
Source: Created by Author, based on the dataset

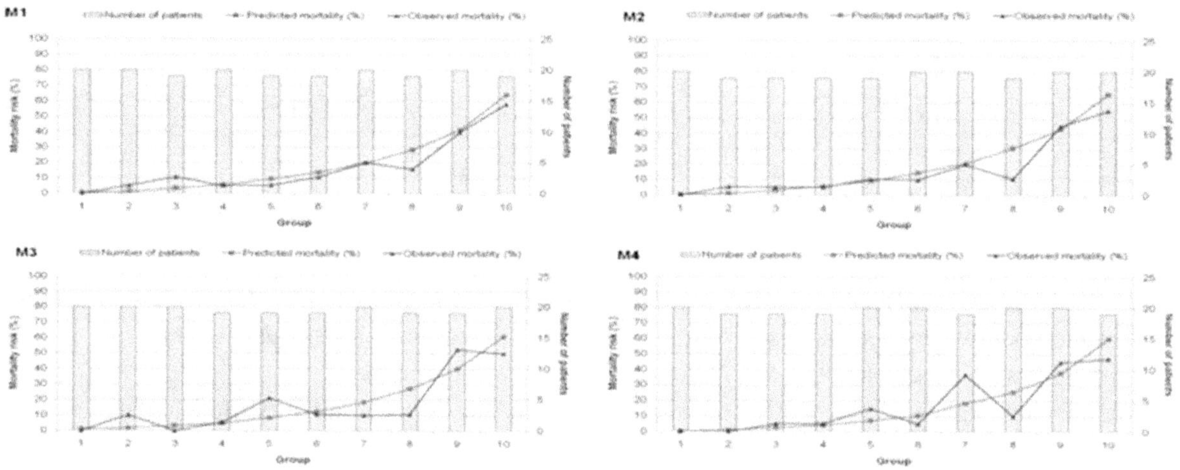

In figure 7, the values obtained from applying Bayesian model is being shown. Values of 95% Confidence Interval (CI) and other parameters are shown.

The mortality rate graph has been represented in figure 8. The predicted or posterior mean graph along with the original trend has been plotted and the significance of the risk factors has been noted in this limelight.

From the results of risk factors and CI interval value for each district we see that hospital infrastructure contribute most in the survival rate of a patient, thus in table 12 we have stated district with best and worst hospital infrastructure based on the results of Bayesian model.

Table 12. District with best and worst risk analyzed hospital infrastructure

District Name	Condition
Howrah	Best in terms of Hospital Infrastructure
North 24 Parganas	Worst in terms of Hospital Infrastructure

Source: Created by the Author, based on the dataset

ANALYSIS BASED ON POISSON MODELING

The Poisson model follows the probability mass function; it belongs to the generalized family of regression model and regression analysis model is being used. It's mainly used to model count data and in contingency tables analysis. Sometimes, Poisson model is considered as log-linear model for its nature and spread.

When the input in Poisson model is θ and x as input vector the mean predicted associated with Poisson distribution is given in equation 2.

$$E\left(\frac{Y}{X}\right) = e^{\theta x} \tag{2}$$

In this section we will fit the Poisson model to estimate the effect of our health parameters tabulated in table 7, on the rate of health performances recorded. Then we have compared the results with the random effect model. In this section we have used graphical model to estimate our models.

Figure 9. Poisson Model Fit Result
Source: Created by Author, based on the dataset

PerformanceOverallAverage	Coef.	Std. Err.	z	P>|z|	[95% Conf.	Interval]
M1AnteNatalCareServicesANC	1.33e-07	8.60e-10	154.41	0.000	1.31e-07	1.34e-07
M2Deliveries	-3.10e-08	3.60e-09	-8.60	0.000	-3.81e-08	-2.39e-08
M3NumberofChildDeathCasesR	1.08e-06	9.57e-09	113.10	0.000	1.06e-06	1.10e-06
M4PostNatalCare	7.00e-08	3.83e-10	182.81	0.000	6.92e-08	7.07e-08
M5Detailsofdeathsreportedu	-4.86e-07	2.20e-09	-220.60	0.000	-4.90e-07	-4.81e-07
M5RTISTICases	-5.39e-07	2.77e-09	-194.66	0.000	-5.45e-07	-5.34e-07
M7CHILDIMMUNIZATION	-7.61e-08	4.17e-10	-182.66	0.000	-7.69e-08	-7.53e-08
M8LaboratoryTesting	-3.13e-08	1.05e-10	-297.18	0.000	-3.15e-08	-3.11e-08
M9GDP	-.0155158	.0000414	-374.44	0.000	-.0155971	-.0154346
_cons	11.93214	.0638207	186.96	0.000	11.80705	12.05723
/lnalpha	-2.6704	.3391111			-3.335045	-2.005754
alpha	.0692246	.0234748			.035613	.134558

From figure 7, we can see the results obtained from Poisson Model, and the factors which highly influence the performance indicators come out same as the tabulated in figure 7. Separate analysis has been done on each of the factors in the following sections.

The normal distribution graph has been plotted against the time and performance in OPDs it can be visualized in figure 8.

Figure 8 shows the Poisson distribution for our fitted panel data model and in table 13 the results obtained from the model is being show:

Figure 10. Poisson Model Graph
Source: Created by Author, based on the dataset

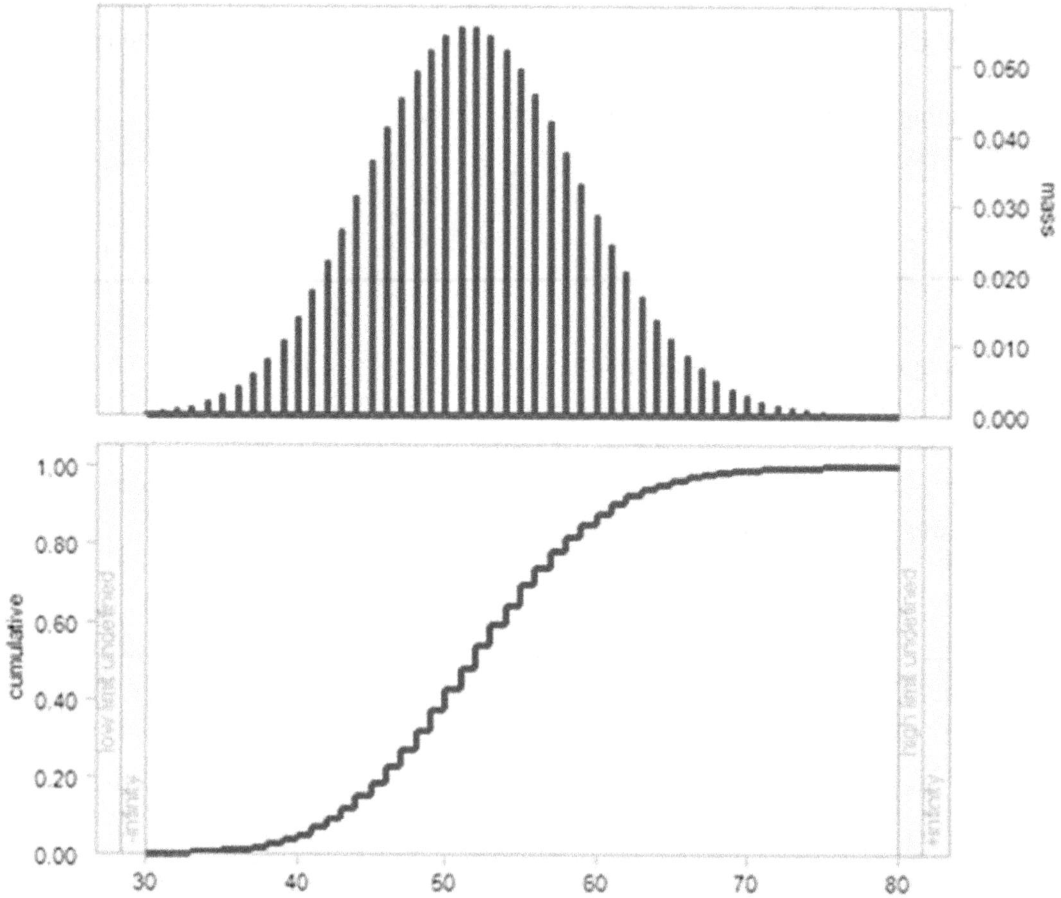

ANALYSIS OF THE INDICATORS

Table 13. Poisson model graph parameter data

Parameter Name	Value
Standard Deviation	7.211
Variance	52
CofV	0.1387
Skewness	0.1373
Kurtosis	3.0192
P Spread	1
Mode	51

Source: Created by the Author, based on the dataset

From the analysis we can raise questions to better our performance indicators such as:

- Has there been a reduction in the count of death cases registered during 2004-2017?
- Has the service of Antenatal Care increased during 2004-2017?
- Has the child death rate decreased during 2004-2017?
- Has the effective result of laboratory testing increased during the tenure of 2004-2017?
- Has the district GDP figure gone high from the health cost side during 2004-2017?

PREDICTIVE FORECASTING OF PEOPLE VISITING OPDs IN 2020

We have forecasted the number of people which will visit OPDs for healthcare facility using a linear regression model. In this case we have taken longitudinal data instead of panel data.

Regression Model

Regression model or analysis is one of the statistical methods to study the relationship between two or more variable in a dataset. The regression model comes under the supervised learning part as the dependent and independent values are known and numeric in nature. It's generally used to find effect of one variable on other. For example, the effect of price inflation of vegetables upon low crop yields.

Figure 11. Linear regression model
Source: (Chatterjee & Sinha, 2018)

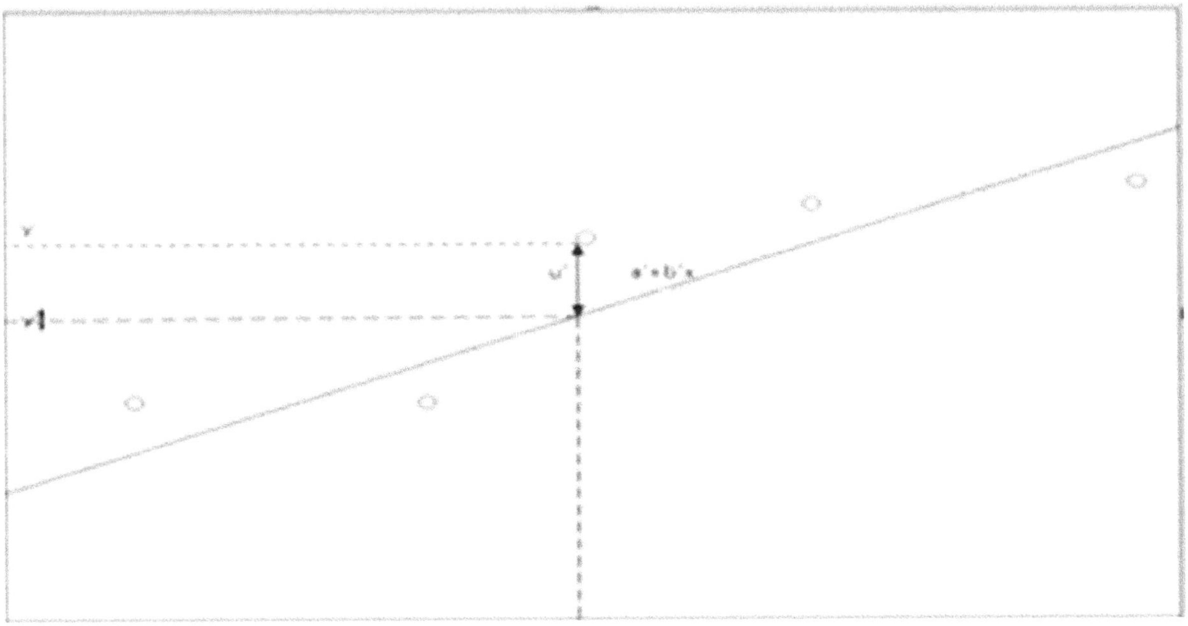

Table 14. Predicted Values of Patients coming to OPDs in 2018 in West Bengal, India

District Name		Patients Coming to OPD (*Predicted Value*)
1. Alipurduar		356901
2. Bankura		231737
3. Birbhum		31111
4. Dakshin Dinajpur		131715
5. Darjeeling		98986
6. Howrah		112788
7. Hooghly		126301
8. Jalpaiguri		128708
9. Koch Bihar		160507
10. Malda		116323
11. Murshidabad		77203
12. North 24 Parganas		53251
13. Paschim Medinipur		148633
14. Purba Medinipur		102848
15. Purulia		45703
16. South 24 Parganas		40367

Source: Created by the Author

In table 14, the average patients arriving to OPDs are mentioned district wise for the state West Bengal, India. From this table we have the insights how much the indicators have to perform to accommodate the crowd. (Chatterjee & Sinha, 2018)

MORTALITY RATE INDICATOR ANALYSIS

Mortality is considered as one of the most important key factors for a country's health indicator. Low mortality rate clearly indicates possession of high quality infrastructure through which illness can be cured. But on the contrast with high birth rate and low death rate can result in huge population burst, thus we have to take care of those factors also. (Paddock, 2014)

In India as well as in West Bengal we observe there is a high birth rate with a low death rate which is causing huge population growth across the state as well as in the country.

Table 15. Decrement of IMR across West Bengal, India

Year	Percentage Decreased
1961 - 1991	50
1991 – 2010	30
2010 - 2020	18 (*Predicted through regression model)*

Source: Created by the Author, based on the dataset

Table 16. Districts with highest IMR across West Bengal, India

Districts with Highest Infant Mortality Rates
Maldah
Darjeeling
Murshidabad

Source: Created by the Author, based on the dataset

During our taken tenure of year 2004-2017, we observe that the average death rate across West Bengal, India during 2004-2016 is being just 2.8% more than the deaths recorded in the year 2017. Our aim is to increase that percentage so that death toll decreases with the coming year. To maintain the flow in the year 2018, and the indicator to be increased we suggest the percentage should be close to 3.4% - 3.8% as per our predicted numbers. In figure 11 we visualize the death toll reported graphically, with each district visualized separately the index number can be tallied from table 4. The graph has been taken from year 2004 – '17.

Another significant indicator in this limelight can be taken as the child death rates or we can call it as Infant Mortality Rate (IMR). Table 15 shows how much IMR rate varied over the years and future targets of the government in this issue. (Cacace et al., 2013)

Due to poor infrastructure in the hilly areas and presence of people from socially backward classes led to poor education background and those factors contributed in comparatively high IMR around those

Figure 12. Time Series Plot for Death Cause Reported from 2004-2017
Source: Created by Author, based on the dataset

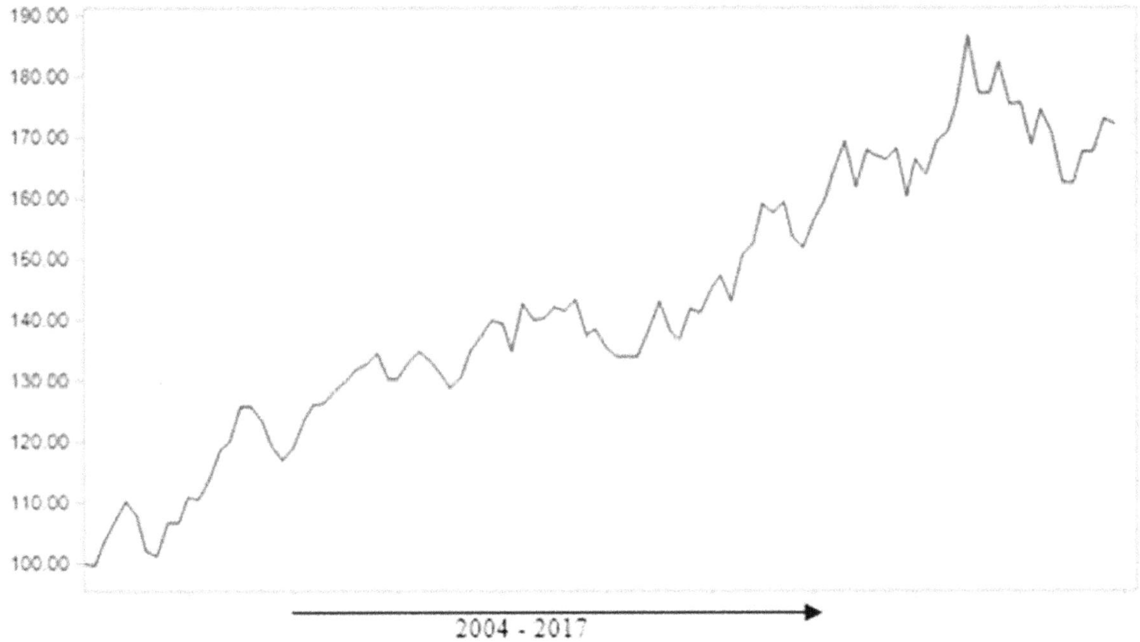

Figure 13. Graphical view of death registered during the tenure 2004 - 2017
Source: Created by the Author

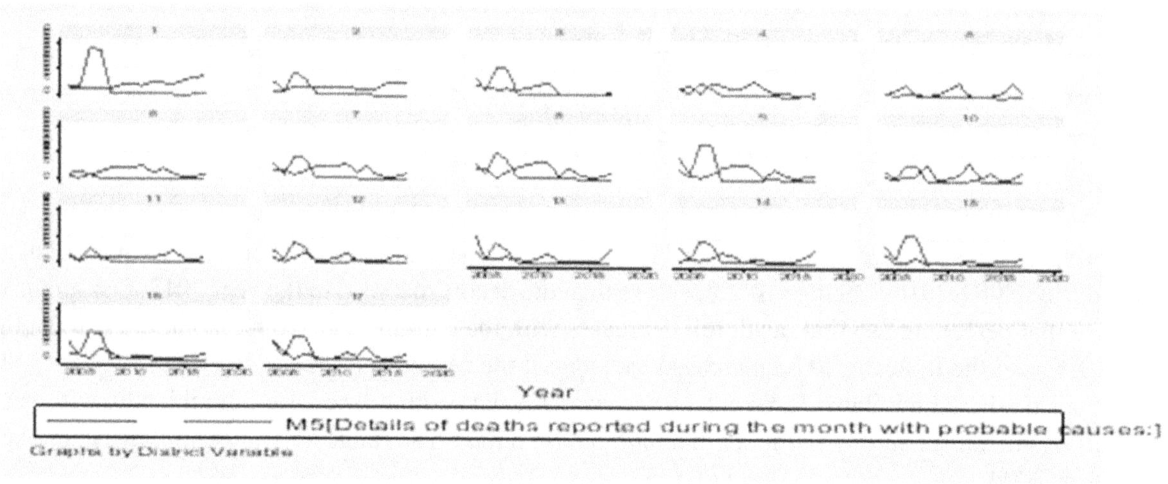

areas. In table 16 we see districts in West Bengal, India with high IMR. Factors like literacy and related unconsciousness for recent things contribute as indirect indicators affecting healthcare unit. (Balov, 2016)

Figure 10, shows how the death rates across West Bengal, India varied from the year 2004 to 2017. Graphical approach is taken in this case.

ANTENATAL CARE SERVICE INDICATOR ANALYSIS

The growth of a country highly depends on the youth of the country and to produce youths which can contribute to the growth of country at first those pregnant women should be taken care off. Antenatal Care Service is such a kind of care unit which is given to the pregnant women's. Rural or the local health centres are given the duties to perform such task, but unfortunately in West Bengal, India the performance in this unit is very poor indeed as most of the PHCs don't have proper infrastructure to support the medical issues as shown statistically in table 3.

Studies show that woman takes more antenatal care services with higher rate of education.

Table 17. Percentage of Woman taking ANC across various stages in West Bengal, India

Percentage of Women	Stage of Antenatal Care Service (ANC)
4	No ANC taken
30	Only 1st Stage ANC taken
47	Completes only 1st,2nd,3rd Stage ANC
20	Completes full ANC Checkup

Source: Created by the Author, based on the dataset

Table 18. Districts with best ANC Service across West Bengal, India

Districts with Best ANC Treatment Service
Hooghly
Howrah
Dakshin Dinajpur

Source: Created by the Author, based on the dataset

Table 19. Neonatal Mortality Rate Comparison

Zone	Neonatal Mortality Rate (in percentage)
India	4.4
West Bengal	3.3

Source: Created by the Author, based on the dataset

- Study shows that for one standard higher education among females lead to 0.55 percentage increased chance for antenatal care service, thus here also literacy rate acts as one of the indirect health indicators. This result can be mathematically expressed as it's done in equation 3. (Breiman, 2001)

$$\textit{Chance to take Anntenatal Care Service} \propto \textit{Literacy Rate among female} \quad (3)$$

Table 17 shows percentage of woman visits the antenatal care service for their regular checkups.

Table 18, shows district names which have recorded best ANC treatment across the year 2004 – 2017. The SCs are designated to provide this kind of treatment to the pregnant woman. So those districts which have better result we can conclude that SC infrastructure at those districts are better than the rest and we can suggest the local authority of the other districts to take up facilities like those districts having.

CHILD DELIVERIES AND INFANT MORTALITY RATE INDICATOR ANALYSIS

One of the most important and key health indicators is how much safely child delivery is taking place across the districts of West Bengal, India. In a medical institutional organization child delivery should

Table 20. Best and worst child deliveries recorded

Districts with Best Rate	Districts with Worst Rate
Darjeeling	Purulia
Nadia	Purba Midnapur
Howrah	Dinajpur

Source: Created by the Author, based on the dataset

Figure 14. Graphical view of child delivered and child death cased registered during the tenure 2004 - 2017
Source: Created by the Author

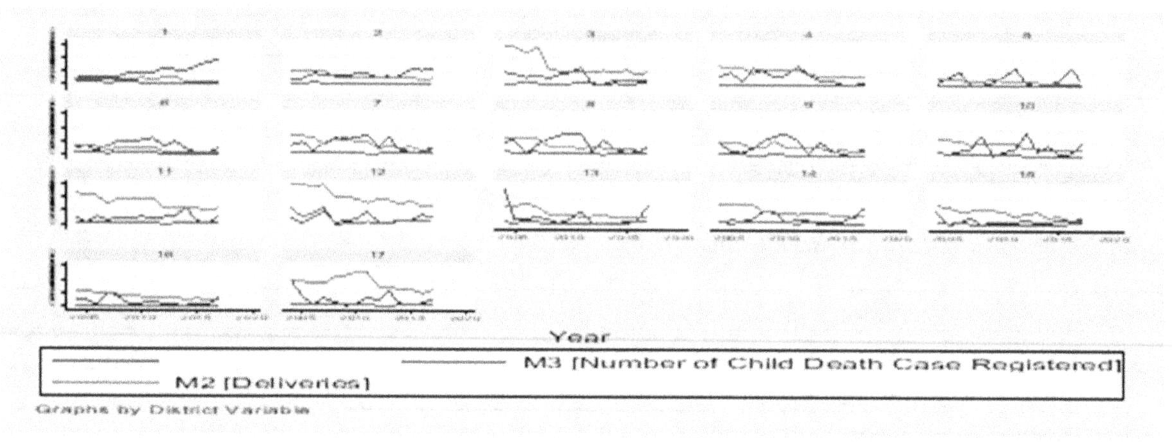

be done under supervision of trained professional. As per stats over one million newborn babies die before completing their one month lifecycle. This factor is called as Neonatal Mortality Rate. In table 19 a comparative NNR is given of India and West Bengal.

Some significant facts and stats regarding this is around 50% child delivery happens in home across the rural part of India. Illiteracy shows significant effect in this case as huge numbers of people believe to give birth to their child in home to take care of their age old traditional believes and holding onto them. Here also we can see that illiteracy comes as one of the most significant indirect health indicators. Table 20, tabulates districts with best and worst successful child deliveries across the state. (Boulkedid et al., 2011)

During our taken tenure from year 2004 – 2017, we at first observe the average child deliveries from 2004-2016 is around 1.856% greater than the stats registered in the year 2017. Thus this performance indicator is let us down and the performance is registered as a negative remark. With our prediction of 2018, people coming to OPDs we suggest that this percentage figure should go high as 3.432% to accommodate this as counting parameter and the performance is increasing in nature. In figure 8 we visualize the child deliveries reported graphically, with each district visualized separately the index number can be tallied from table 4. The graph has been taken from year 2004 – '17. (Forde et al., 2013)

On other hand we have child death reported and the percentage is 1.123% less in the year 2017 than the average spanning of 2004 – 2017. Thus for coming year this percentage should go as 0.872% to accommodate the capacity. In figure 8 we visualize the child death reported and child deliveries reported simultaneously graphically, with each district visualized separately the index number can be tallied from table 4. The graph has been taken from year 2004 – '17.

In Figure 12, the factors contributing in health care performance indicators are shown.

In figure 13, we see a composite scoring rank has been created based on the data from Indian Institute of Population Science. All the data have been normalized and Principal Component Analysis (PCA) has been applied and a certain total value is received and based on the final scores each districts have been ranked in order of best health indicators in terms of performance parameters. This study and ranking has been done by researchers namely (Hati & Majumdar, 2011.)

Figure 15. Factors Contributing in HealthCare Performance Indicator
Source: (Hati & Majumdar, 2011)

ANALYSIS AND PREDICTIVE FORECASTING ON NEED OF HEALTH CENTRES

As we have said earlier the health infrastructure of India is a three-tiered structure and lot of health indicators directly depends on the availability and current conditions of the CHCs, PHCs, and the SCs. Thus in table 21 we will go through how much shortfall can be there in the year 2020. (Klazinga et al., 2011)

The growth of population is taking a huge scale in West Bengal, and number of patient arriving at OPDs for healthcare access is also pretty high thus there will be need of more numbers of healthcare centres to accommodate and support such increasing populations. According to national norms of India we predict how many numbers of CHCs, PHCs, SCs should be there in West Bengal to perfectly cater healthcare service.

HEALTH CARE INDICATOR ANALYSIS ON THE BACKDROP OF HEALTHCARE EXPENDITURE AND GDP

In this section we will drop a light and have a closer look in the close relation between the two focus variable *viz.* public health expenditure and its respective implementations in the field of public healthcare infrastructure development. Initially a model has been setup on the backdrop of data received from the panel data to link the two focus variables and further analysis also has been done in this section. (Singh et al., 2018)

As study by Mukherjee (2018) in the field of economics it suggests some inter relationship between the focus variables. The statements and observations are stated below.

Figure 16. Ranking based on health performance indicators
Source: (Hati & Majumdar, 2011)

District	IMR	Percentage of Women Who/Had		Percentage of		Composite Health Outcome Score	Rank
		Received Full ANC	Safe Delivery	Child Received Full PNC	Morbidity		
Darjiling	40.2	30.3	73.9	85.5	5.3	88.3	2
Jalpaiguri	40.2	17.9	49.2	78.6	4.2	70.2	9
Koch Bihar	40.2	13.6	47.2	77.2	5.1	63.8	13
Uttar Dinajpur	41.5	8.2	30.3	54.1	2.9	40.4	18
Dakshin Dinajpur	41.5	27.3	42.4	87.9	4.7	79.1	6
Maldah	41.5	14.5	29.8	68.8	3.6	54.2	17
Murshidabad	41.5	18.2	43.7	63.0	4.7	54.6	16
Birbhum	41.5	15.6	50.5	63.6	5.3	56.2	15
Barddhaman	30.8	21.1	58.3	66.7	3.9	65.3	12
Nadia	41.5	25.9	70.6	89.2	5.8	86.2	3
North 24 Parganas	30.8	24.6	65.8	81.6	3.1	82.5	4
Hugli	30.8	32.6	84.9	98.0	3.6	100.0	1
Bankura	40.2	17.5	64.4	92.7	6.3	82.0	5
Puruliya	40.2	14.5	46.9	78.4	5.0	68.0	11
Pachim Medinipur	41.5	21.3	49.6	83.0	3.5	69.7	10
Haora	30.8	27.8	69.8	72.2	2.4	77.2	7
South 24 Parganas	30.8	20.5	36.7	70.8	2.7	63.6	14
Purab Medinipur	38.8	22.0	45.0	84.6	2.4	72.4	8
West Bengal	**38.0**	**19.6**	**51.5**	**75.7**	**3.7**	**68.2**	---

Table 21. Predicted Shortfall in Health Infrastructures

Type of Health Care Centres	Current Shortfall (in percentage)	Predicted Shortfall (in percentage)
CHCs	14.45	15.68
PHCs	24.27	27.67
SCs	18.45	22.32

Source: Created by the Author, based on the dataset

Figure 17. Visualized Health Infrastructure Shortfall
Source: Created by the Author

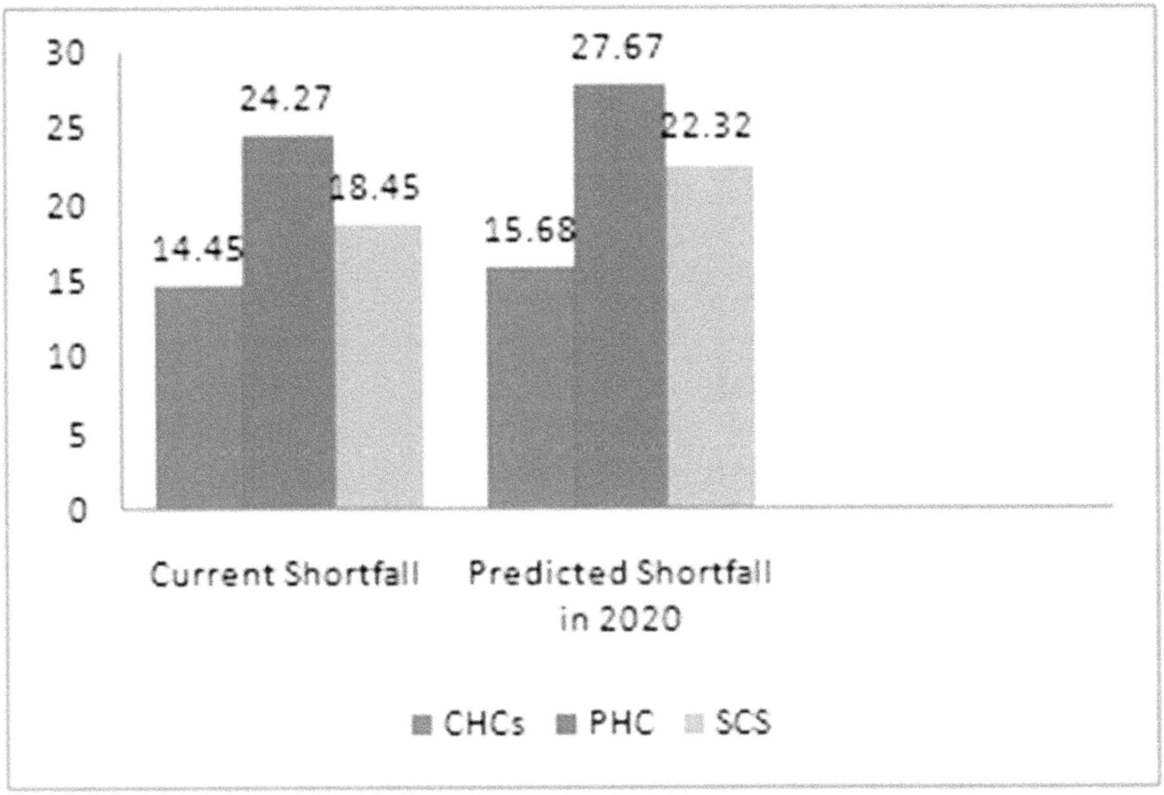

1. Healthcare spending depends on income but varies across different population.
2. Income growth highly depends on the current health status

From, the study Mukherjee also suggested an empirical model or formula to calculate the health status of a country and the earnings related to this in this particular field.

Let y_t be the earning; h_t be the health status and s_t be the spending at an age t. Then the earning and health status can be represented as:

Figure 16, suggests the empirical structure through which the two focus variables are being correlated to each other is shown.

Table 22. Number of health facility required in 2020 (Predicted)

Facility Type	Number Required on 2021 *(Predicted)*
CHCs	430
PHCs	1214
SCs	13,338

Source: Created by the Author, based on the dataset

Figure 18. Empirical Formula showing earning and health status
Source: (Mukherjee, 2018)

$$y_t = \left[y_0 - \left(\frac{a+bg}{1-bk}\right)\right]\left(\sqrt{bk}\right)^t + \left(\frac{a+bg}{1-bk}\right)$$

$$h_t = \left[h_0 - \left(\frac{ak+g}{1-bk}\right)\right]\left(\sqrt{bk}\right)^t + \left(\frac{ak+g}{1-bk}\right)$$

To compare the results and growth rate we have taken two end points of our data set in order to compute the difference across the model. The two end points are taken as 2004-05 and 2016-17. The key dimensions taken are CHCs, PHCs, SCs, and the current hospital infrastructure in mind.

A Calculation model is being proposed in this section to find out the indexing values and condition of health infrastructure:

Step 1: Each parameter is being normalized. In this data we have taken four parameters.

$$Parameterized\,Value = \frac{Observed - Minimum\,Value\,Recorded}{Maximum\,Value\,Recorded - Minimum\,Value\,Recorded}$$

Figure 19. Economic Model on Health Infrastructure
Source: Created by the Author

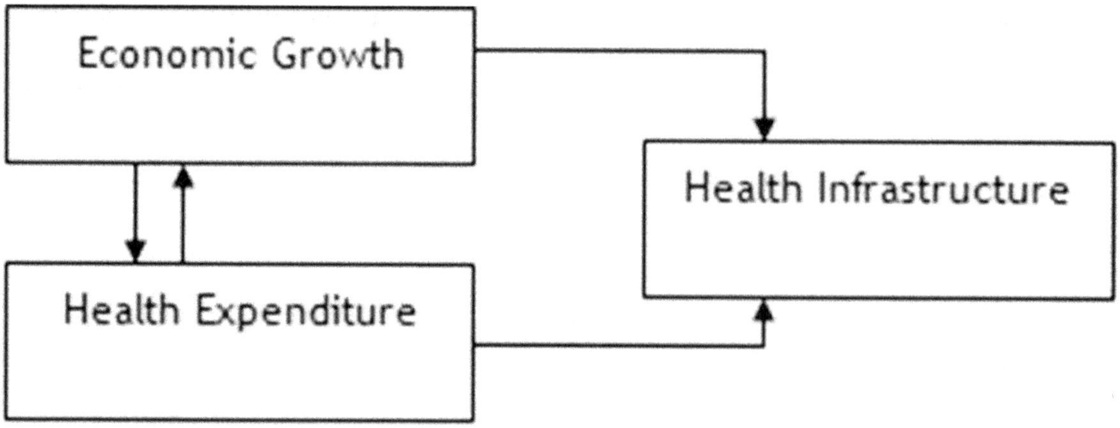

Table 23. Investments Planned in various Plan Period

Plan Period	Total Planned Investment	Family Welfare (*in percentage*)	Health Sector (*in percentage*)
Eighth Plan (1992-97)	434100	1.5	3.2
Ninth Plan (1997-02)	859200	1.76	4.09
Tenth Plan (2002-07)	1484131	1.83	3.97
Eleventh Plan (2007-12)	2156571	6.31	6.49

Source: Created by the Author, based on the dataset

Figure 20. Trend of GDP allocation
Source: (Mukherjee, 2018)

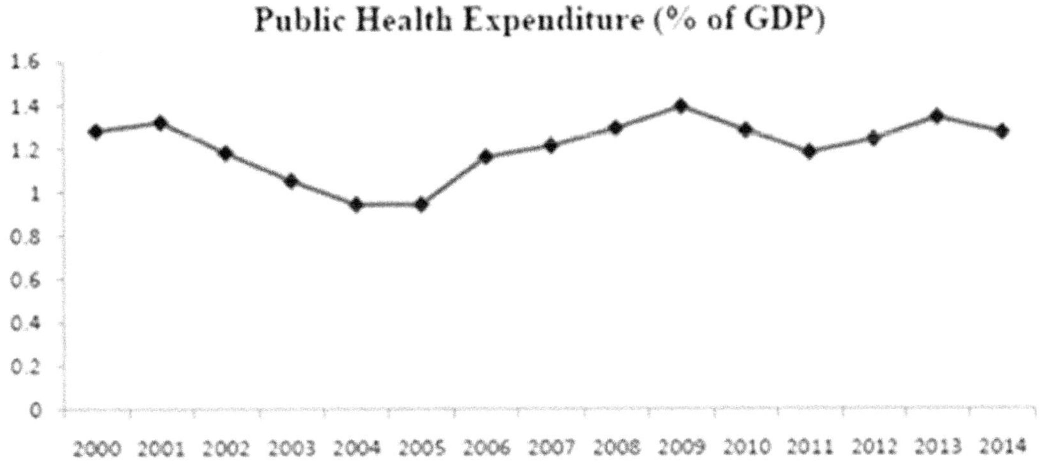

This step gives us result in binary form i.e. 0 and 1. Here 0 represents worst case and 1 represents best case scenario. (Niens & Brouwer, 2013)

Step 2: Summation of Weights with all four parameters taken into account.

In this step distance of each parameterized value is calculated from the best case scenario and least square model has been implemented to get the minimum error in this case.

Table 24. Index Value Calculation

State Name	Index Value Calculated (2004-05)	Index Value Calculated (2016-17)	Rank wrt. Other states (2004-05)	Rank wrt. Other States (2016-17)
West Bengal	0.411	0.365	7	8

Source: Created by the Author, based on the dataset

Figure 21. Graphical view of GDP distribution during the tenure 2004 - 2017
Source: Created by the Author

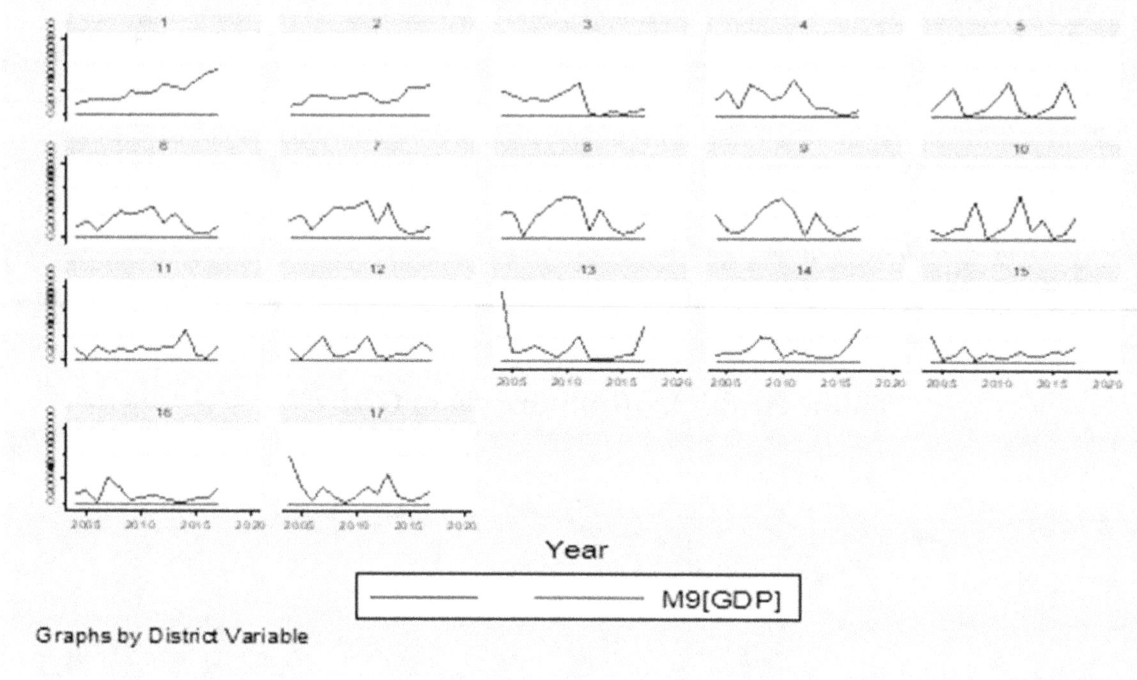

$$Index\,Value = \sqrt{\frac{\left[(1-SC)^2+(1-CHC)^2+(1-PHC)^2+(1-Inf)^2\right]}{4}}$$

The higher Index value will be there from this calculation the better infrastructure can be said on the backdrop of econometric perspective. Table 23 shows the economic investment from the planning commission over the years. (Lahariya, 2018)

India's healthcare expenditure directly contributes to the infrastructure level and infrastructure directly contributes to the performance of the health indicators. The central as well as the state government should allocate funds in the growth of number of PHCs, SCs, and CHCs across the state as well as country for better condition for living. According tour study and for most significant indicators government should invest more in the Antenatal Care Service, Postnatal Care Service, Laboratory Testing as well for a better growth. Currently with this kind of investments there is a serious shortage of CHCs, PHCs, and SCs across the country and in coming year 2020 the shortfall will get increased as predicted by us in table 18, thus the government should think to invest much more in order to tackle those problems in upcoming years. (Pillai, 2016)

The results clearly validates the current condition of the healthcare system is strictly not up to the mark. Analysis showing states spending more in infrastructure have a comparatively better result than the others. The private sector hospitals in current scenario plays a tricky role as they are attracting huge number of patients by visualizing the better infrastructural facilities and making the families believing that health is more important and charging huge money from the patient families to push the total health sector condition.

During our taken tenure of 2004 – 2016 the average GDP is much higher than the GDP recorded in the year 2017, as in the year 2017 GDP took low hit. This section goes under the macro economics part. In figure 9 we visualize GDP variation across districts graphically, with each district visualized separately the index number can be tallied from table 4. The graph has been taken from year 2004 – '17.

INDIRECT FACTORS INFLUECING HEALTH INDICATOR PERFROMANCES

The direct factors which influence the healthcare indicators are discussed above, with those there are also some indirect factors. Among which illiteracy stands as one of the most significant indicators. We have seen this in case of Mortality Analysis and ANC Service analysis where illiteracy led to lower number of pregnant women opting for ANC Service. Poor knowledge in the field often leads to death without any treatment.

Along with this we have also seen around 50% child deliveries happens in homes which can cause fatal results, here also illiteracy is one of the primary cause.

CONCLUDING REMARKS AND SCOPE FOR FUTURE STUDY

In this paper we have only statistical method to predict the increase in the number of patients that will arrive in the OPDs for healthcare access. On this context we have given a comparative study of the indicators that how much they should perform, or how much the performance should be increased to meet the desired result what can increase the performances.

Through panel data analysis we have find the key factors or the most significant factors which contribute largely in the healthcare performance. We have opted for Random Variable Effect model as the hypothesis for Fixed Effect model has been ruled out by Haussmann Test.

We have analyzed the risk factors and key indicators on which the survival rates of the patient depends through Bayesian Hierchichal Modeling, from which we observed blood facility in hospitals with active number of presence of doctors play a key role in this.

Along with that to find an analyzed result regarding the count variables like overall performance, we have applied Poisson model to get the probabilities of function ability.

In the next section we have computed the current shortfall in the count of hospitals across West Bengal, India and predicted the same for upcoming years and how much the count should be increased to reach the national norms prescribed totals.

The next is an econometric analysis on how much health expenditure is being allotted and for that how much the conditions of hospitals are now.

In a nutshell, we can say that the health care condition in West Bengal as well as in India is erratic. The hospitals and plans are being unfunded resulting in overcrowding in the hospitals. Lack of access in the rural areas also plays a key role in this pathetic condition. The poor healthcare system is one of the most celebrated failures of India where common people cannot get proper treatment. One of the missing parts in our analysis is the private health care giants. We have failed to collect data from those giant sectors thus we cannot conclude about them.

The public health sectors now needs a serious observation and planned infrastructure to make India a better place to live and they there is a need to upgrade their infrastructures, specially those which have been suggested in the paper as those verticals attract most number of patients and the flow of patients from private healthcare to public healthcare needs that change.

Although the growth of Indian economy is being highly supported by the private health sectors, are the positions of the public health centres being justified? The answer is inconclusive.

REFERENCES

Alkhatib, M., Talaei-Khoei, A., & Ghapanchi, A. (2016). *Analysis of research in healthcare data analytics.* arXiv preprint arXiv:1606.01354

Aven, T., & Eidesen, K. (2007). A predictive Bayesian approach to risk analysis in health care. *BMC Medical Research Methodology*, *7*(1), 38. doi:10.1186/1471-2288-7-38 PMID:17714597

Balov, N. (2016, August). *Bayesian hierarchical models in Stata. In 2016 Stata Conference (No. 30).* Stata Users Group.

Benneyan, J. C., Lloyd, R. C., & Plsek, P. E. (2003). Statistical process control as a tool for research and healthcare improvement. *BMJ Quality & Safety*, *12*(6), 458–464. doi:10.1136/qhc.12.6.458 PMID:14645763

Boulkedid, R., Abdoul, H., Loustau, M., Sibony, O., & Alberti, C. (2011). Using and reporting the Delphi method for selecting healthcare quality indicators: A systematic review. *PLoS One*, *6*(6), e20476. doi:10.1371/journal.pone.0020476 PMID:21694759

Breiman, L. (2001). Statistical modeling: The two cultures (with comments and a rejoinder by the author). *Statistical Science*, *16*(3), 199–231. doi:10.1214s/1009213726

Butler, J., Foot, C., Bomb, M., Hiom, S., Coleman, M., Bryant, H., Vedsted, P., Hanson, J., & Richards, M.ICBP Working Group. (2013). The international cancer benchmarking partnership: An international collaboration to inform cancer policy in Australia, Canada, Denmark, Norway, Sweden and the United Kingdom. *Health Policy (Amsterdam)*, *112*(1-2), 148–155. doi:10.1016/j.healthpol.2013.03.021 PMID:23693117

Cacace, M., Ettelt, S., Mays, N., & Nolte, E. (2013). Assessing quality in cross-country comparisons of health systems and policies: Towards a set of generic quality criteria. *Health Policy (Amsterdam)*, *112*(1-2), 156–162. doi:10.1016/j.healthpol.2013.03.020 PMID:23628482

Chatterjee, A., & Sinha, T. (2019). Correlation between Absence, Interest in the Field and Grades in an Organization using Regression Model. *International Journal of Engineering and Advanced Technology*, *8*(6), 1436–1441. doi:10.35940/ijeat.F8118.088619

Forde, I., Morgan, D., & Klazinga, N. S. (2013). Resolving the challenges in the international comparison of health systems: The must do's and the trade-offs. *Health Policy (Amsterdam)*, *112*(1-2), 4–8. doi:10.1016/j.healthpol.2013.01.018 PMID:23434265

Gajewski, B. J., Mahnken, J. D., & Dunton, N. (2008). Improving quality indicator report cards through Bayesian modeling. *BMC Medical Research Methodology, 8*(1), 77. doi:10.1186/1471-2288-8-77 PMID:19017399

Geweke, J., Gowrisankaran, G., & Town, R. J. (2003). Bayesian inference for hospital quality in a selection model. *Econometrica, 71*(4), 1215–1238. doi:10.1111/1468-0262.00444

Häkkinen, U., Iversen, T., Peltola, M., Seppälä, T. T., Malmivaara, A., Belicza, É., Fattore, G., Numerato, D., Heijink, R., Medin, E., & Rehnberg, C. (2013). Health care performance comparison using a disease-based approach: The EuroHOPE project. *Health Policy (Amsterdam), 112*(1-2), 100–109. doi:10.1016/j.healthpol.2013.04.013 PMID:23680074

Hati, K. K., & Majumder, R. (2011). *Health for development: a district level study in West Bengal.* Academic Press.

Hossein, Z., & Gerard, A. (2013). Trends in cost sharing among selected high income countries—2000–2010. *Health Policy (Amsterdam), 112*(1-2), 35–44. doi:10.1016/j.healthpol.2013.05.020 PMID:23809913

Klazinga, N., Fischer, C., & Ten Asbroek, A. (2011). Health services research related to performance indicators and benchmarking in Europe. *Journal of Health Services Research & Policy, 16*(2_suppl), 38-47.

Lahariya, C. (2018). 'Ayushman Bharat' program and universal health coverage in India. *Indian Pediatrics, 55*(6), 495–506. doi:10.1007/13312-018-1341-1 PMID:29978817

Marchal, B., Dedzo, M., & Kegels, G. (2010). Turning around an ailing district hospital: A realist evaluation of strategic changes at Ho Municipal Hospital (Ghana). *BMC Public Health, 10*(1), 1–16. doi:10.1186/1471-2458-10-787 PMID:21184678

Mohapatra, S. (n.d.). *Public Health Expenditure and its Effect on Health Outcomes: A New Methodological approach in the Indian Context.* Academic Press.

Mukherjee, S. (2017). Anatomy and Significance of Public Healthcare Expenditure and Economic Growth Nexus in India: Its Implications for Public Health Infrastructure Thereof. In Social, Health, and Environmental Infrastructures for Economic Growth (pp. 120-144). IGI Global.

Nallala, S., Swain, S., Das, S., Kasam, S. K., & Pati, S. (2015). Why medical students do not like to join rural health service? An exploratory study in India. *Journal of Family & Community Medicine, 22*(2), 111. doi:10.4103/2230-8229.155390 PMID:25983608

Niëns, L. M., & Brouwer, W. B. F. (2013). Measuring the affordability of medicines: Importance and challenges. *Health Policy (Amsterdam), 112*(1-2), 45–52. doi:10.1016/j.healthpol.2013.05.018 PMID:23827263

Paddock, S. M. (2014). Statistical benchmarks for health care provider performance assessment: A comparison of standard approaches to a hierarchical Bayesian histogram-based method. *Health Services Research, 49*(3), 1056–1073. doi:10.1111/1475-6773.12149 PMID:24461071

Paoli, F., Schmidt, I., Wigzell, O., & Ryś, A. (2019). An EU approach to health system performance assessment: Building trust and learning from each other. *Health Policy (Amsterdam), 123*(4), 403–407. doi:10.1016/j.healthpol.2019.02.004 PMID:30777300

Pillai N, V. (2016). *Panel data analysis with stata part 1 fixed effects and random effects models*. Academic Press.

Raghupathi, V., & Raghupathi, W. (2015). *Benchmarking hospital performance using health analytics*. Academic Press.

Rohleder, T. R., Lewkonia, P., Bischak, D. P., Duffy, P., & Hendijani, R. (2011). Using simulation modeling to improve patient flow at an outpatient orthopedic clinic. *Health Care Management Science*, *14*(2), 135–145. doi:10.100710729-010-9145-4 PMID:21152989

Singh, P., Hashmi, G., & Swain, P. K. (2018). High prevalence of cesarean section births in private sector health facilities-analysis of district level household survey-4 (DLHS-4) of India. *BMC Public Health*, *18*(1), 613. doi:10.118612889-018-5533-3 PMID:29747609

Singh, S., Shekhar, C., Acharya, R., Moore, A. M., Stillman, M., Pradhan, M. R., ... Sundaram, A. (2018). The incidence of abortion and unintended pregnancy in India, 2015. *The Lancet. Global Health*, *6*(1), e111–e120. doi:10.1016/S2214-109X(17)30453-9 PMID:29241602

Staggs, V. S., & Gajewski, B. J. (2017). Bayesian and frequentist approaches to assessing reliability and precision of health-care provider quality measures. *Statistical Methods in Medical Research*, *26*(3), 1341–1349. doi:10.1177/0962280215577410 PMID:25788482

Chapter 5
An Econometric Overview on Growth and Impact of Online Crime and Analytics View to Combat Them

Swagatam Roy
https://orcid.org/0000-0002-8012-5529
The Neotia University, India

Ahan Chatterjee
https://orcid.org/0000-0001-5217-4457
The Neotia University, India

Trisha Sinha
The Neotia University, India

ABSTRACT

In this chapter, the authors take a closer look into the economic relation with cybercrime and an analytics method to combat that. At first, they examine whether the increase in the unemployment rate among youths is the prime cause of the growth of cybercrime or not. They proposed a model with the help of the Phillips curve and Okun's law to get hold of the assumptions. A brief discussion of the impact of cybercrime in economic growth is also presented in this paper. Crime pattern detection and the impact of bitcoin in the current digital currency market have also been discussed. They have proposed an analytic method to combat the crime using the concept of game theory. They have tested the vulnerability of the cloud datacenter using game theory where two players will play the game in non-cooperative strategy in the Nash equilibrium state. Through the rational state decisions of the players and implementation MSWA algorithm, they have simulated the results through which they can check the dysfunctionality probabilities of the datacenters.

DOI: 10.4018/978-1-7998-4706-9.ch005

INTRODUCTION

Cybercrime is the greatest threat to every company in the world.

-Ginni Rommety

Crime and criminality have a high correlation with man from very beginning. It is now very difficult to catch and bust a gut to hide itself in featuring the development of different fields. It is the crime committed in the cyberspace which in this tech-savvy world is spreading like wild fire and as Ronald Reagan says that information is oxygen of modern era, the main target is digital information. The primary effect of cybercrime is the financial section and economy of a country.

The modern technology is inextricable from human life. The most integral part of this technology is the internet. The internet consists of millions of hyper-connected devices and is the chief source of knowledge for novices as well as for experts. In this era of developing technology, cybercrime is one of the most challenging aspects of the recent decade, nations and multinational companies are facing the threat of attack which costs around some billion and millions of dollars. Crime is one of the major impediments to the GDP growth of a country and we can reduce it if we can analyze the root cause of it.

Our aim in section 1 of this paper is to analyze the number of cyber attacks affected due to unemployment rate, inflation rate and GDP's growth or fall per capita. We have framed a cross-sectional time series panel data to analyze the situation and to validate our hypothesis regarding the same. Our data is from the year 2011- 2018. Various analysis models have been taken such as OLS Regression model and Auto Distribution lag model. (An & Kim, 2018)

The growing rate of cyber crime now-a-days is similar to breeding rate of the rabbits. According to a report the number of cybercrime reported in 2016, is near about 1 billion where most of the cases are malware based. "Data" is referred to as modern gold in recent times, and most of the company's store this data into cloud systems and those cloud servers are there in a datacenters, a place from where all the servers are controlled. Data breaching is one of the most celebrated type of cyber crime, and that can be easily done by getting hold the data from the datacenters, with such humungous amount of data the concept of big data also comes into the game.

In the section 2, we will take a closer look into the cloud storage system in the datacenters, and their security issues or challenges that those datacenters face, and we will form a game theory approach to detect and protect datacenter vulnerability.

DIFFERENT TYPES OF CYBER CRIME PREVAILING

Cyber criminals are experts who are adept with all the nitty-gritty of this network system who works in a collaborative manner to perform crimes which affect the economy of a country in an adverse manner. There are various types of cyber crimes that may occur in this advanced 21st century. Some of the most common cyber crimes are fraud, hacking, identity theft, scamming, computer viruses, ransomware, DDoS attack, botnets, spamming, phishing, malvertising, cyber-stalking, software piracy, cyber-bullying, etc.

Few of the above mentioned terms are described in detail below.

Fraud: This term is generally used to describe a cyber crime which is carried out to gain some important data or information by deceiving a person. It is done to either alter or suppress any kind of important detail to secure unlawful gain.

Hacking: This involves unlawful acquisition of the applications and functions within a system or network. This is a way of attacking corporate or government accounts that ultimately affects the economy of the country. (Abdelhamid et al., 2017)

Identity Theft: This means stealing of personal data including bank account and credit card details which again can hamper economic growth of a country.

Ransomware: One of the most disastrous malware-based attacks, which enter the computer network and encrypt files and information through public-key encryption.

Cyber-stalking: This includes anonymously following a person online.

Cyber-bullying: Any social mal-practitioner demands ransom to permanently stop bullying over the internet.

These are few of the major cyber crimes prevalent in the present societal condition. These crimes bring about a financial setback for the country. From a recent survey, it has been revealed that India incurs an average economic loss of US$10.3 million in cyber attacks. Hence, these attacks and crimes must be curbed in every way possible using the same technology in a more pragmatic and constructive manner.

DATA VISUALIZATION AND STATISTICAL MODELING

Cyber crime has increased considerably with advancing technology. This technology is readily available to the people of all age groups and hence, crimes using these technologies are growing day-by-day.

Figure 1. Showing the number of cyber crime cases

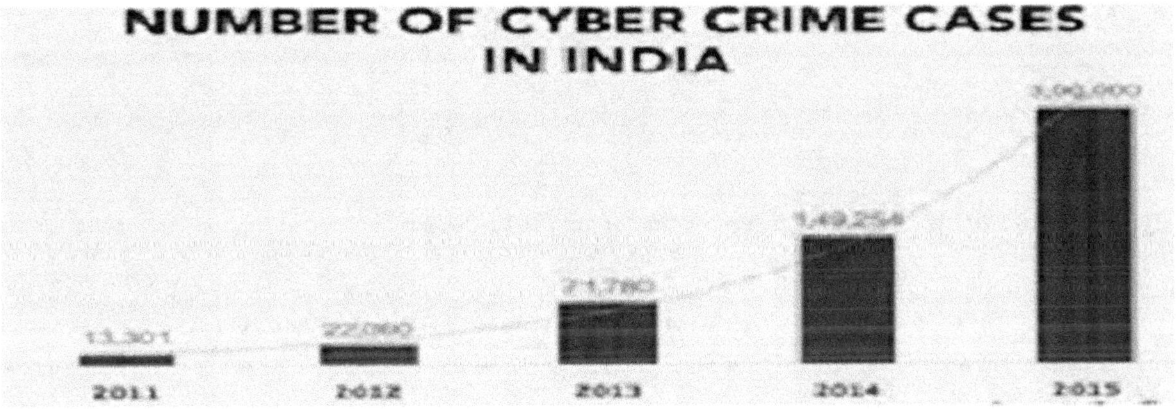

In a report published in 2012, it has been predicted that cyber attacks is one of the top five risks in the world for government and business sector. With ever increasing popularity of online banking and online shopping, where intricate details of an individual's bank account are needed, cyber crime has been quite common and inevitable.

According to the statistic released by NCRB(National Crime Report Bureau), cyber crime in India almost doubled in the year 2017. The report states that, "During 2017, 56% of cyber-crime cases reg-

Figure 2. Statistics of cyber crime in different states
Source: Jagranjosh

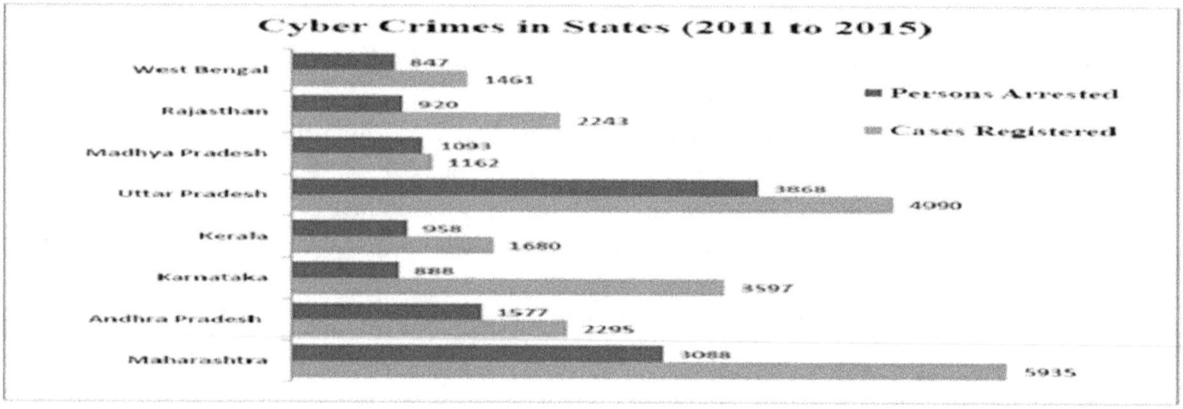

istered were for the motive of fraud (12,213 out of 21,796 cases) followed by sexual exploitation with 6.7% (1,460 cases) and causing disrepute with 4.6% (1,002 cases)". Cyber crime relating social media had upped from 155 in 2016 to 328 in 2017. The highest number of cases registered in 2017 under "Violation of privacy" was in Assam. Also, there were 13 registered cases under "Cyber Terrorism" across the entire country in the same year. 11,592 cyber crime cases were registered in the year 2015. The Information Technology (IT) Act was passed by the Parliament of India in May 2000. It aimed to reduce cyber crime and secure e-commerce transactions. Hacking is a rapidly growing cyber crime. In 2011, 157 cases of hacking were reported, whereas the Figure went up to 435 in 2012. So, the percentage of variation was nearly 177.1%

Figure 3. Cases of cyber crime registered under it act (2011-2015)
Source: Factly

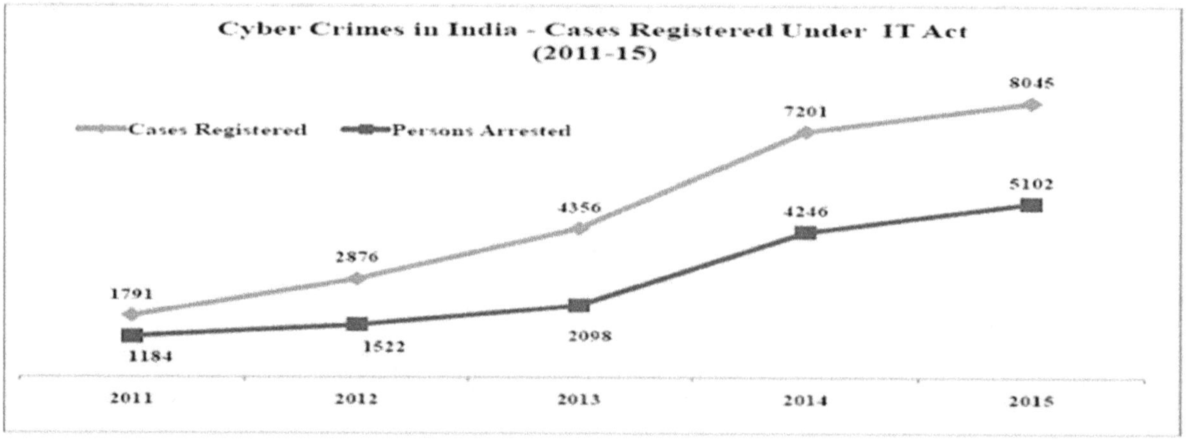

UNDERGROUND ECONOMY OF CYBER CRIME

The threats imposed by cyber attacks have increased with time and individuals as well as organizations have worked towards finding ways to defend them. Cyber attacks that take place globally are carried out by extremely organized criminal groups. Even the national-level attacks are executed by trained criminals. They buy and sell hacking equipments and the required services from the online underground market and this transaction thus supports the underground cybercrime economy. The attackers form an underground market which is type of black market and this form underground community and *CaaS*(Crimeware as a Service) remains invisible to authority. *CaaS* is basically a type of business model which is prevalent in the underground market to provide cyber attackers or more specifically, the underground buyers with illegal services to carry out cyber crime such as attacks and frauds in an electronic manner. Also, these services can be provided by a cyber criminal to other cyber criminals as needed. These products benefit both the sellers and the buyers. The sellers are benefitted in a way that they get quick and instant money by selling these equipments and the buyers benefit in a way that they get unique methods and techniques to carry out malicious activities that they can implement almost immediately. (Basu, 2014, pp. 215-242)

The equipments mainly include personal information of people collected by spammers through mass emails and through financial frauds, the internal and confidential data of an organization, username and password combinations that are illegally acquired or stolen, etc.

The services provided by the underground black markets are given at a fixed rate and for a definite period of time. The sellers acquire large amount of revenue and the buyers get access of the malicious tools for a fair amount of time which aids in fulfilling their vicious intentions. One of the major service among these is Distributed Denial of Service or frequently known as DDoS. In computer networking, DDoS is a type of cyber-attack in which the cyber-criminals tend to make a network unavailable to its definite users by temporarily or permanently creating problems in the services of the host network. Also, exploit kits are used to get access to the exploit toolkit leased with a monthly rate. Similar to these, the transfer of illegally acquired funds through bank accounts is also one of the services offered in the above mentioned underground markets.

According to the reports of a leading global organisation against cyber crime, in 2011, 403 new variants of malware were created which increased by 43% than it was in 2010. It also stated that web based attacks leaped by a percentage of 36% with new attacks ranging over 4500 each day.

ECONOMETRIC RELATION BETWEEN CYBERCRIME, UNEMPLOYMENT, AND INFLATION RATE IN INDIA

In the era of exponential growth of technological advancement we can see there is also a growth of crime, especially cybercrime. In last few years there is a growth of unemployment, and with that there is also a growth of inflation rate in the country for this reason unemployed peoples find difficult to meet both the ends. In this section we examine the major cause behind the growth of cybercrime through an econometric point of view.

The methods of cybercrime is not fixed there are various types which comes under this banner, there are online harassing, online trolling to credit card frauds, bank frauds to the extremes like terrorism, human trafficking. While committing the crimes those criminals ensure to protect their anonymity. Figure 4 shows some statistics regarding cybercrime in India.

Figure 4. Graph showing cybercrime cases in 2016, and 2017
Source: Statista

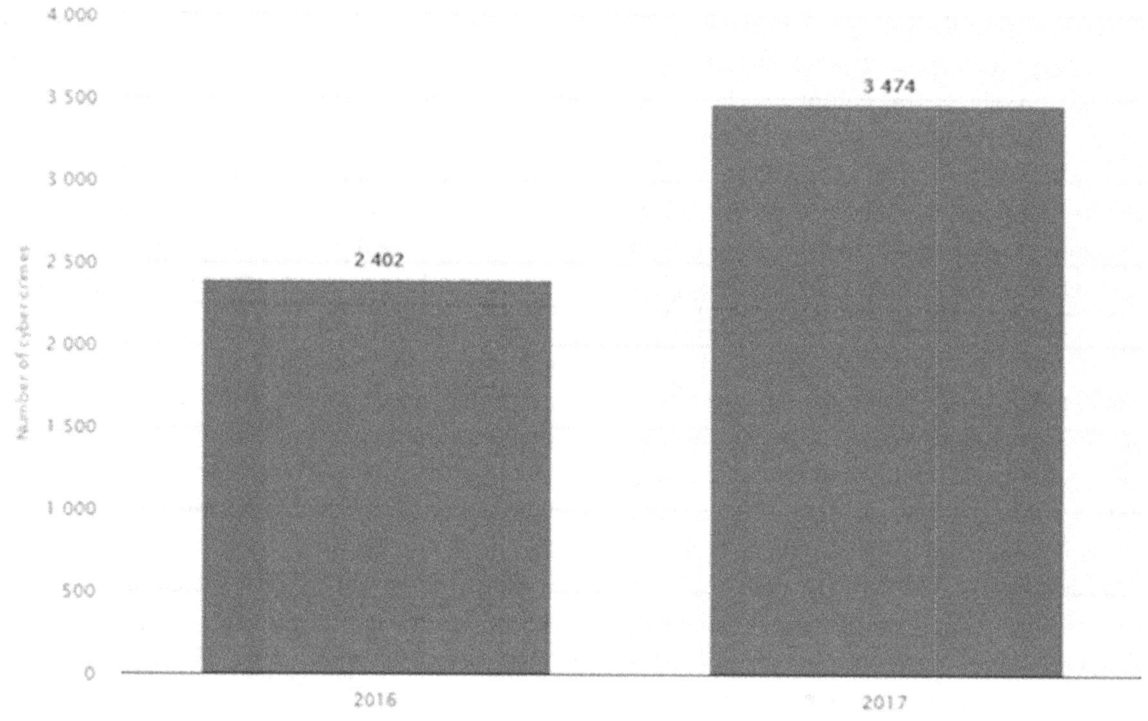

One of the most dependable pillar of economics or economy of a country goes by the name of unemployment. Unemployment is defined as a state of a person who is actively searching for employment but unable to find the work. It generally gives a measure of health of economy, and its major indicator is unemployment rate. High rate of unemployment signals a poor economic condition of a country. An economy with high unemployment has a lower output with a share decline in the need for consumption. Majorly unemployment can be divided into 4 parts namely, *frictional, cyclical, structural, and institutional.*

Frictional Unemployment

This state of unemployment rises when a person is in between a job, the person is changing in between jobs thus the unemployment tenure is the minimum and it's least harmful from the economic standpoint.

Cyclical Unemployment

It's defined as the variation in the numbers of unemployed workers during a season or up and down of an economy, that is there is an increase of employment when the economy is high and similarly a fall when low.

Structural Unemployment

This comes through the technological advancements, technological advancements often takes many jobs of workers through which there is an increment of unemployment.

Institutional Unemployment

This type of unemployment comes as effects of long term or nearly permanent institutional factors and incentives in economy.

India's current year unemployment rate is the highest in last 45 years, and we can consider this as government's celebrated failure. From such scenarios we can easily say that there can be an increment of cybercrime due to this. Figure5 shows the rate from 1998-2018; in 2019 it was the highest.

Figure 5. Graph showing Unemployment variation from 1998-2018
Source: Statista

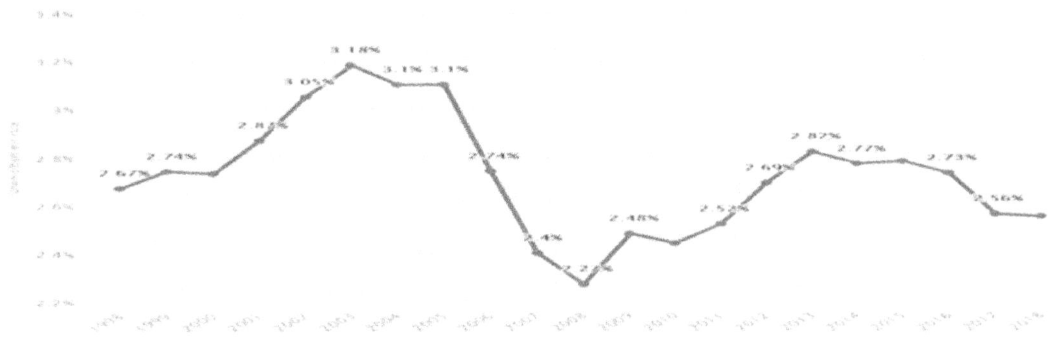

Inflation is another pillar of measure of economics. It's defined as a quantitative measure of rate at which average price level of goods increase in a time period. Increase in inflation percentage rate indicates a decreasing in purchasing power of persons in a country. Rising Price is the root of inflation there are mainly 3 causes namely, *Demand-Pull Inflation,* here demands goods exceeds the production capacity thus inflation occurs. Next, is *Cost-Push,* here the cause is increment of production cost then *Built-In,* where price and wages both increases to maintain the living. Figure6 shows inflation rate from 1984-2016.

RELATION BETWEEN UNEMPLOYMENT AND INFLATION RATE:

These two factors can be connected using Phillips Curve which says, "Inflation and Unemployment have an inverse stable relationship." With increasing inflation rate in a country there will be lower employability.

The stability of Philips Curve is challenged in 1970's where the policy makers observed that the inflation rate is not related to unemployment. Then explanation came as there is two models of Phil-

Figure 6. Graph showing Inflation Rate variation from 1984-2016
Source: Statista

lips curve, one is *Short Run Phillips Curve,* and the other is *Long Run Phillips Curve.* There will be no tradeoff factor in the Long Run Phillips Curve. (Bhatt et al., 2018)

During the period of 1993-2008, the unemployment has seen several lows but the inflation rate didn't rise much as predicted by Phillips Curve thus the trade-off factor is being justified.

Figure 7. Graphical representation of phillips curve
Source: Created by the Author

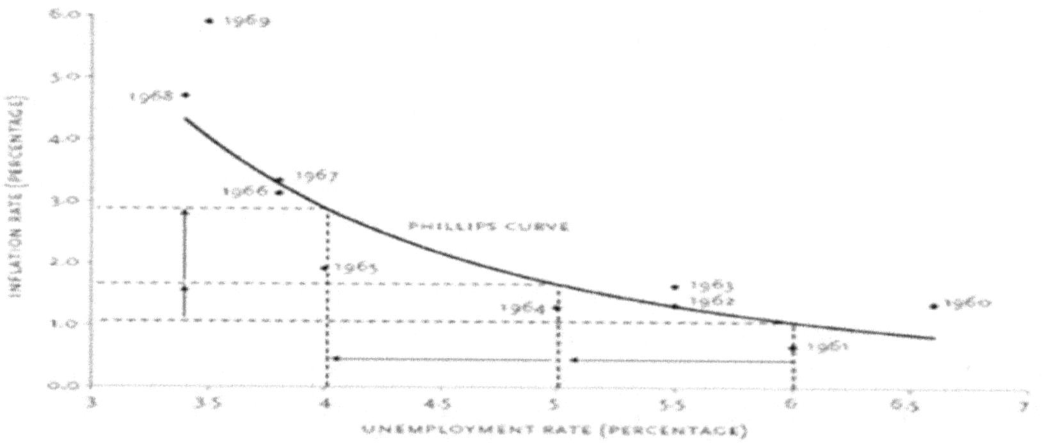

The movement of economy moves to the upward zone of SAS Curve, it indicates a higher price level with a general reduction of unemployment.

Let U' be the NRU

Let U actual employment level

Thus, the employment gap is $U - U'$

Therefore, trade-off between unemployment with –

$$W_{t+1} = W_t \left[1 - \epsilon \left(U - U'\right)\right]^2 \tag{1}$$

Let N' be the full employment level

Let N actual employment level

$$U - U' = \frac{N - N'}{N'} \tag{2}$$

Unemployment rate is the fraction of full-employment labor force, and N which is not employed.

$$g_W = \frac{(W_{t+1} - W_t)}{W_t} \tag{3}$$

Phillips Curve is represented in equation 4.

$$\pi = \pi_e - \beta\left(U - U'\right) + \delta \tag{4}$$

Where:

π = Inflation

π_e = Expected Inflation

β = Parameter which measures the response of inflation

$\left(U - U'\right)$ = Unemployment

δ = Supply Shocks

From the relation in equation 3 we can write Phillips Curve as represented in equation 5.

Figure 8. Graphical Representation Unemployment and Inflation Rate from 2000-2015
Source: ONS LF 2Q

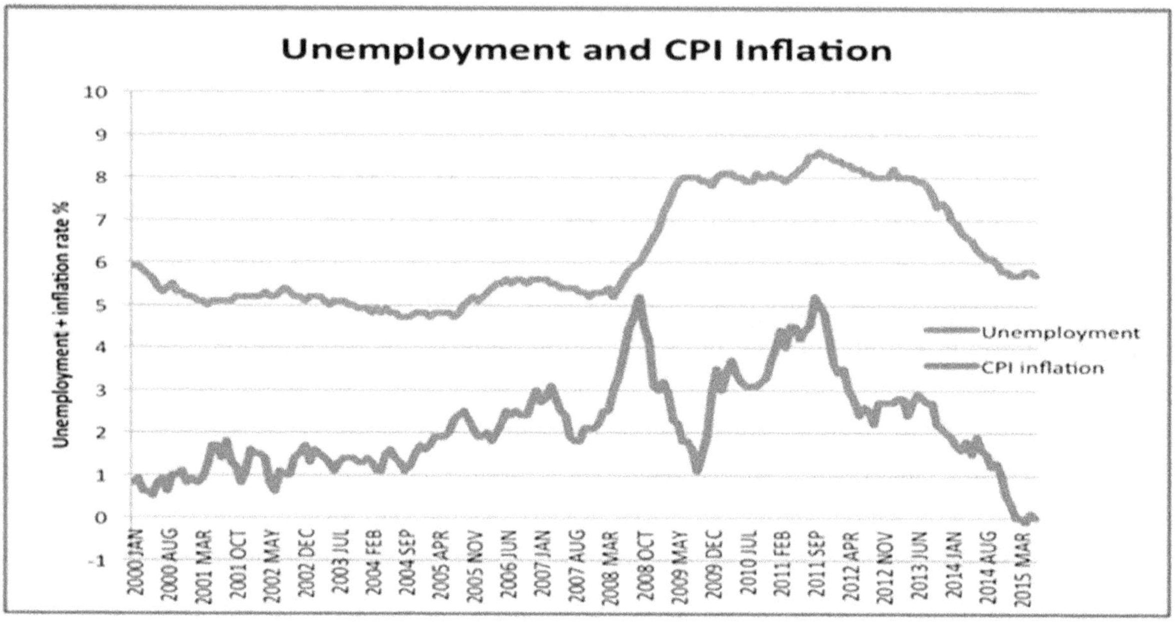

$$g_W = -\epsilon\left(\left(U - U'\right)\right) \tag{5}$$

$$g_W = -\epsilon\left(\frac{N - N'}{N'}\right) \tag{6}$$

ϵ is the responsiveness of wages to unemployment.

Putting the value of equation 6 in equation 4:

$$W_{t+1} = W_t\left[1 + \epsilon\left(\frac{N - N'}{N'}\right)\right] \tag{7}$$

Equation 7 represents the wage employment relation.

PROPOSED MODEL REGARDING CYBER CRIME AND UNEMPLOYMENT

According to reports there is a jump of around 35% in the count of cybercrimes. Campeau and Higgins proposed a social cognition theory regarding the connection of cybercrime with personal factors and behavior in 1995. Study of Kigerl, 2012 showed that magnitude of persons spamming the internet is more in those places where unemployment is high.

Hypothesis in the Model:

H_0: Count of Cybercrime doesn't affect due to unemployment rate
$H_{1:}$ Count of Cybercrime increases with the increase of unemployment Figureure

The dataset regarding cybercrime is being created from the cases registered against cyber offenders. It's collected from The IT and Electronics Department report of Government of India. Major classes which taken into account are Cyber Terrorism, Prank (Online), Credit Card Fraud, Bank Fraud, Online Trolling. (Das & Das, 2017)

METHODOLOGY

We have created a cross-sectional time series panel data to analyze the data and to get some insights from the data to validate our hypothesis. We analyzed the data under the limelight of FGLS model through the panel data. GLS model is the generalized least square model, where we are computing the relation using some fixed autocorrelation value.

FGLS model is an estimator with a vector with parameter θ.

Function is $V = V\left(\theta\right)$

Let, the estimator of V be $V' = V\left(\theta'\right)$

Then FGLS estimator is:

$$\beta'_{FGLS} = \left(X^T V'^{-1} X\right)^{-1} X^T V'^{-1} Y$$

The fitted values are

$$Y'_{FGLS} = X\beta'_{FGLS}$$

The resulted value is computed as:

$$Y' = \beta + \sum_{k=2}^{k} \beta_K X + e_t$$

Here,

Table 1. Showing P-value result

Country	P>lzl
India	0.001

Source: Computed by Author

Y' = Natural Logarithm of Cybercrime number for years
β = Parameter Estimation
X = Exogenous variable, unemployment.

After fitting the model P_{value} is being computed as:

We can reject our null hypothesis only when the computed $P_{value} < 0.05$, and in this case it's far behind. Thus we can safely reject our null hypothesis and can conclude that our alternate hypothesis stands true.

From the result, we can say that there is increase in cybercrime with the increase of unemployment in India.

Now, connecting it with inflation rate, we can say that there is a pretty good rise in the goods price and there is also a increase in the unemployment rate considering the long run Phillips curve, where we neglect the trade-off factor. People with no job and increasing good price can't afford and they are choosing the path of crime. (Lau et al., 2014)

GLOBAL ECONOMIC CONDITION, WITH A LIMELIGHT TO INDIAN ECONOMY

The internet usage throughout India is growing rapidly. Online crime metamorphosed into a money spinner business that backs down millions of dollars. One of the major potential reason giving sparks in cyber crime is the increase of e-business, ecommerce. Online crime has become one of the major economic crimes now-a-days. Economic crime means well –organized and high funded criminal activity and

Figure 9. Showing that cyber crime is one of the major economic crime
Source: PWC report on economic survey

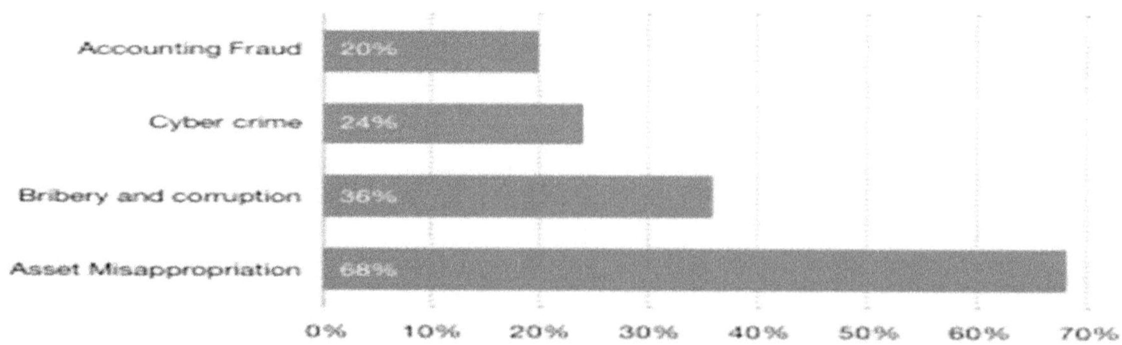

breakdown the framework and here cyberspace is the medium that paved the way to fraud. Respondents have reported it to be one of the major economic crimes, internet banking.

Now-a-days maximum bank frauds and financial crimes are the results of online fraud. The rate of growth of online crime per year is near about 50% in India which is a big reason to worry.India ranks fifth among the countries victimized by online crime.The cost of carrying-out a business in this era is to secure IT systems and over 50% respondents feels that IT industry is at high risk. These days attackers requires very nominal amount of skills because price of IT and exploit tools are going down as well as the usage of is becoming easy and more sophisticated. According to survey report of Cyber Savvy CEO and CSRI it is evident that there are mainly 5 types of attack where one of a major motive is monetary value which impact in downfall of economy. They are:

- *Economic crime*
- *Espionage*
- *Activism*
- *Cyber-Terrorism*
- *Cyber-Warfare*

In fact, asset misappropriation by digital means is prevailing in our country mainly from 2008 and stealing of entity resource mislead to billing scheme, cheque and data tampering and rise in organization give rise to situation. As prevention is better than cure so it is better to take control measures which significantly impact in *'fraud controls paradox'* .Apart from quantitative loss there is also qualitative loss. It has been reported that online frauds impacts financial condition by more than 35% directly and also results in collateral damage. Some of the major collateral damage is:

- Unemployment rate: The rate is increasing and in 2015 it reached to pathetic condition.

- Decline in employee morale
- Damage of business relations
- Damage of reputation of brand globally

Figure 10. 10 years of unemployment rate
Source: CEICDATA

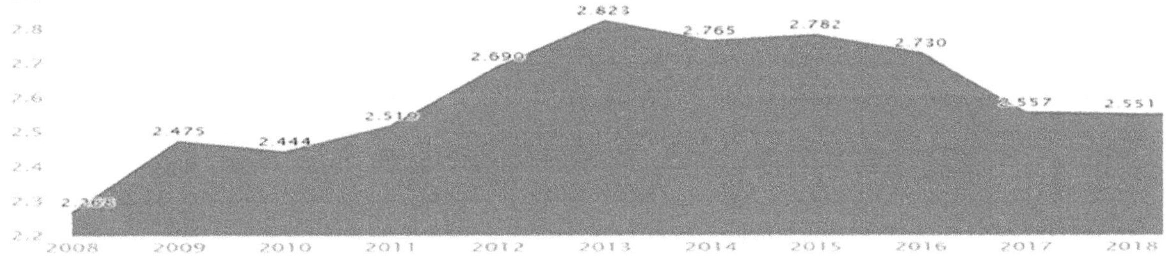

Figure 11. Showing the characteristic of fraudsters
Source: Economic survey by PWC

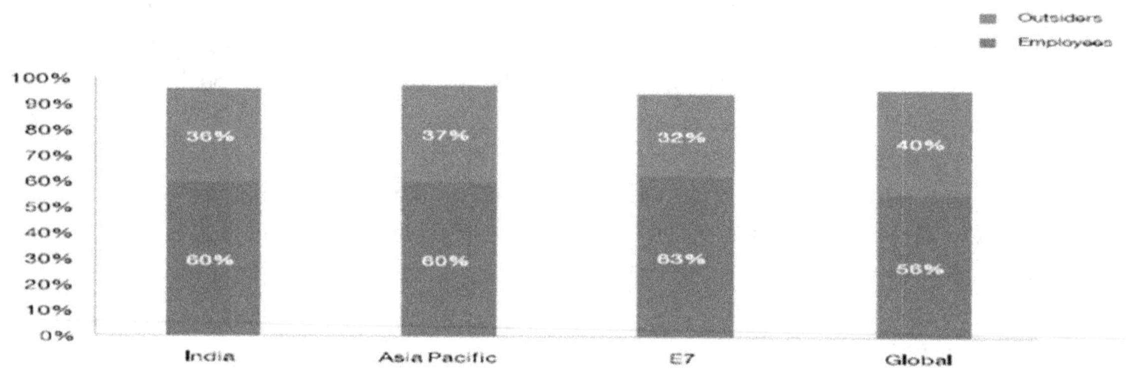

So to optimize this we need to analyze the parameters and different indicators. The starting approach is to extract information of the executioner and detect the flaws. Following the steps it was found that a major portion of perpetrator was among them only. (Orphanides et al., 2016)

Also it is evident from the report that tendency of committing the crime for the internal staff is maximum (up to 65%) for an employee working in an organization for 0-5 years. It is evident that breaking into the cyberspace creates seismic shift in the cyber security landscape. Maximum of the organizations still now uses proactive approach reducing their susceptibility.(Prathap and Ramesha, 2018)

Approach to Minimize Risk

So to optimize the problems it is important for CEO of the company to understand and take actions on risk management. One of the key prevention may be KYC (Know your Customer) and get sufficient knowledge about the associates and executives and it a robust risk analysis and reduction of potential risk should be performed. It is important to develop corporate culture and controls over different electronic or automated generated reports. For the internal staffs, internal audit process helps to detect frauds and from the different international and national reports it is evident that it is decreasing due to budget constraint because number of employee (n) is less and the volume of work (v)is more and thus we get an inverse relation which is:

$$v \propto \frac{1}{n}$$

But in India risk assessment is not carried out due to lack of knowledge, cost, apprehend lack of value etc.

GLOBAL ECONOMY IMPACT DUE TO CYBERCRIME

Technology sets off a powerful and sensational weapon for criminals to captivate different online crimes and illegal bank and other financial transactions. In the context of digital crime, the criminals or attackers associate with each other providing a competitive service and in this era the main target is to make a safer cyber world. Cybercrime has reached a top rank in terms of Security Strategy in many EU states .Internet economies are tempting for a range of cyberspace related crimes that uses information communication technology.

Figure 12. showing the cost of cyber-crime
Source: Research gate

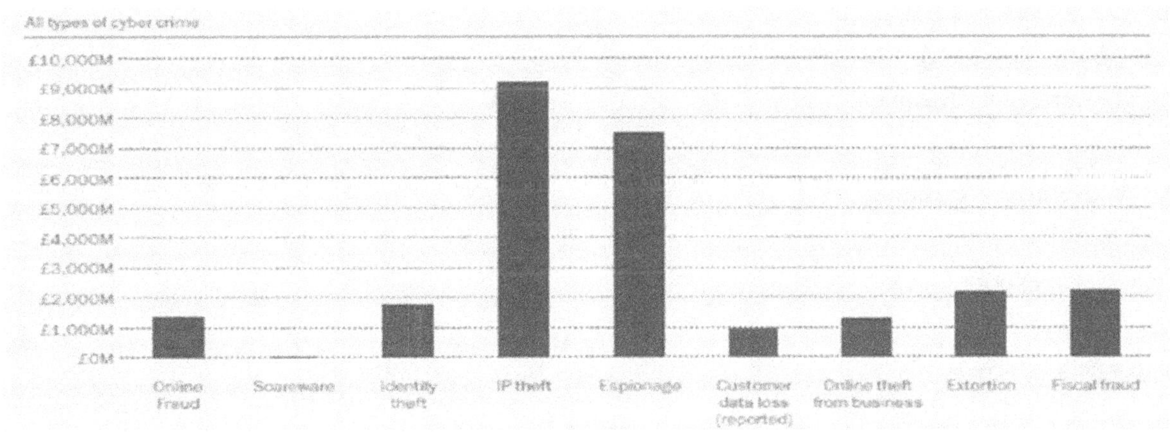

Also around 70% cases were unreported and thus results in loss of around \$600bn and this results in downfall of economy of a country. (Revathi & Suriakala, 2018)

Concept of GDP

Gross Domestic Product (GDP) is the monetary value of different products and services during a particular period of time. It gives a representation of size and rate of change of economy of a country. Basically it is a quantitative measure of economy of country and it is roughly calculated as:

$$GDP = cn + T_{inv} + gs + \left(exp - imp\right)$$

Where cn =consumption done by private companies or by individuals for business purpose

T_{inv} =Total Investment

gs =Spending and expenses of government

Figure 13. Showing the loss of GDP due to cybercrime in terms of percentage
Source: Statista

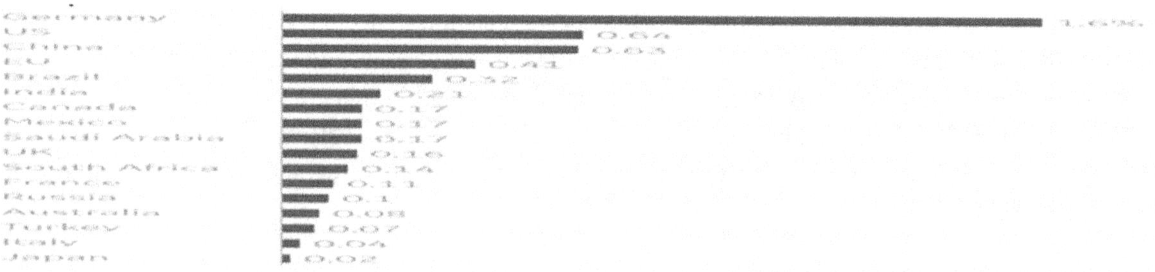

$$gs = G_i + G_s$$

G_i =Government investment

G_s =Government spending

exp =Total export

imp =net import

There are many types of GDP and Currently US has highest nominal GDP.

GDP affected by cybercrime:

The negative impact of cybercrime affects GDP results in downfall of economy of the country. Europe has put up with the highest economic negative impact due to online crime, which is around at 0.84% of the regional GDP, in comparison with US which is around 0.78%, which is a lot in terms of GDP.

Some of the European countries giving penalty of cybercrime:

Poland: over € 390million /annum

Table 2. Some of Cryptocurrency of the market

Name	Market Capital
Bitcoin	$166.28 Bn
Ethereum	$19.40 Bn
Litecoin	$9 Bn
Bitcoin cash	$6.35 Bn
EON	$3.65 Bn

Source: Created by author

Germany: over€ 3.5billion /annum

Cryptocurrency(CC)-Is it a blessing ormalediction of digital world to the world?

Now-a-days, crypto currency and bit-coin is becoming very popular. It is basically a digital asset which is used for digital transaction and uses cryptographic logic to ensure and endure different financial transaction. CC represent the intangible objects applied with the help of digital logic in different applications and networks especially in P2P network. (Sengupta & Mukherjee, 2018)

Bitcoin uses block chain technology and the anatomy is illustrated in the Figure 14.

Figure 14. Showing the steps involved in transaction in crypto currency with the help of block chain
Source: PWC US

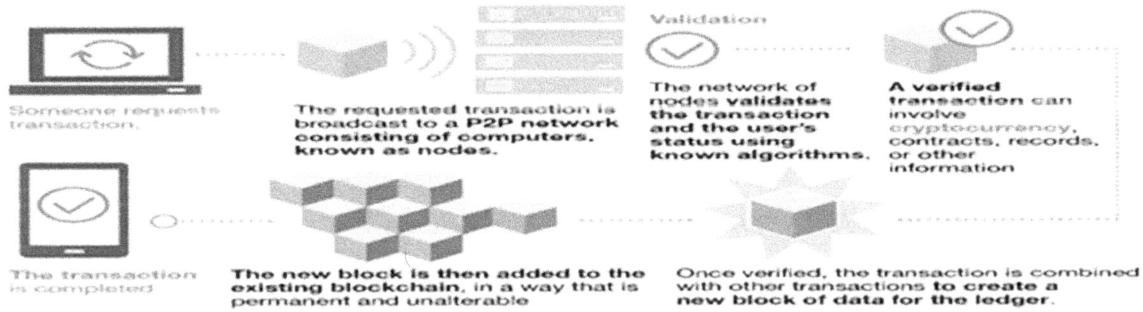

There are over 1600 Crypto currencies available over the internet and the total value is $282 billion and the value is increasing day-by-day. The question lies about their safety and security. From the reports it is evident that a total of $4.2 billion have been stolen in Cryptocurrency. In last year it turned out to be one of the major scam. Some of the major assets which are theft prone are bitcoin, bitcoin cash,XRP,Litecoin,Dogecoin etc. Recently, from one of the largest cryptocurrency exchange of the world Binance, around $40 million dollar worth bitcoin is hacked.

Apart from the security threats there are also other impacts of cryptocurrency some of which is stated below which may hamper the economy of a country

- Unlimited virtual currency is issued without the calculation and concern of demand and supply which may lead to inflation and even falling down of the system
- Affecting the real monetary system and Fluctuation of its value
- Money Laundering is also one of the negative impact

The hacking of the cryptocurrency leads to black market and also in this system an existing user can create more than one account leading to illegal transaction and hacking which paved the way in downfall of economy. (Choudhary et al., 2019)

Figure 15. Showing increasing trend of internet usage
Source: Statista

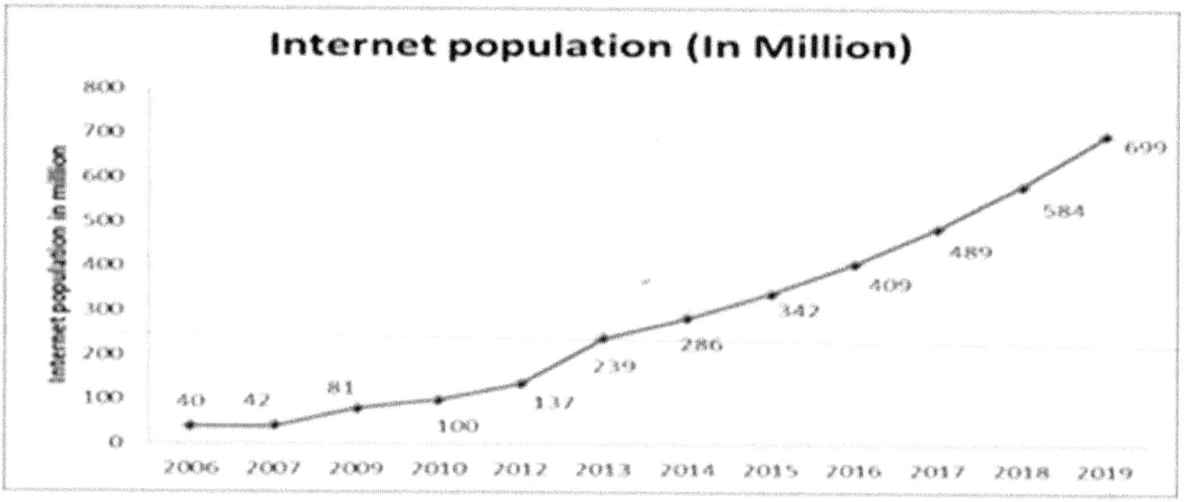

INTRODUCTION TO CYBER CRIME

Online and Cybercrime Overview

In Tech-savvy world one of the essential necessities is internet. In the digital era of 21st Century, the growth of Internet has taken a significant rise.

As we know that there are 2 sides of coin and so the one of the pitfalls is online crime or cyber-crime. 'Cyber' means it is related to computer, networks as well as internet which is rising sun of the era and crime means some illegal activities or unlawful act. So cyber crime is defined as crime or illegal digital activity committed on internet or network system by unlawful access to the devices and targeted to an individual or groups or organization in order to harm or cause a lot of damage. First recorded online crime took place in 1820. Experts say it is one of the top economic crimes caused in India. In 2013, India is one of the top countries in online crimes and is an increasing trend and the amount involved in

Figure 16. Showing the amount involved in online crime is increasing day-by-day
Source: World economic forum

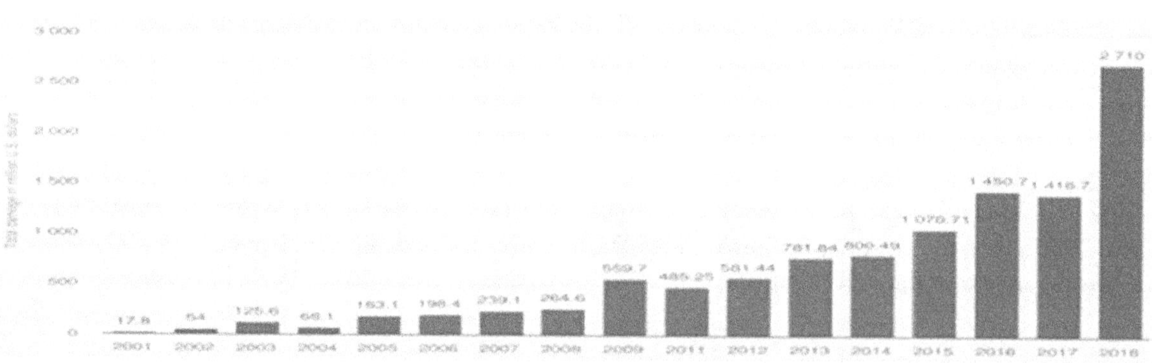

Business community is a major victim of these crimes. Based on Device from which the attack has to be done, it can be categorized into 2 parts:

1. Using Device(computer/laptop) as target
2. Using Device as a weapon or tool

Based on target it can be mainly classified into 3 parts:

1. Individual
2. Group and property
3. Organisation /community

There are many types of online crime like cyber-bullying, banks frauds, hacking, cyber terrorism and Software piracy and which is increasing day by day creating a panic and phobia within citizens and they are unable to fully enjoy the beauty and blessing of the digitalization.

Necessity to Map and Reduce Cybercrime

India is 3rd most attacked country for cyber attacks and the online crime is increasing day-by-day and man is coming under potential threat of using internet as a resource medium for fund transfer etc. At the same time everyone is getting addicted to different social media which results in cyber bullying,

Social Media Trolling etc. So we have to reduce the exploitation.

Cyber Security refers the enactment of protecting systems as well as devices, networks, and different programs from attacks. It draws in people, processes as well as technologies working as a composite block or unit known as system. This reduces vulanabirities and understands risk. (Yadav et al., 2017)

Figure 17. Prediction which shows the increasing trend of social media
Source: Statista

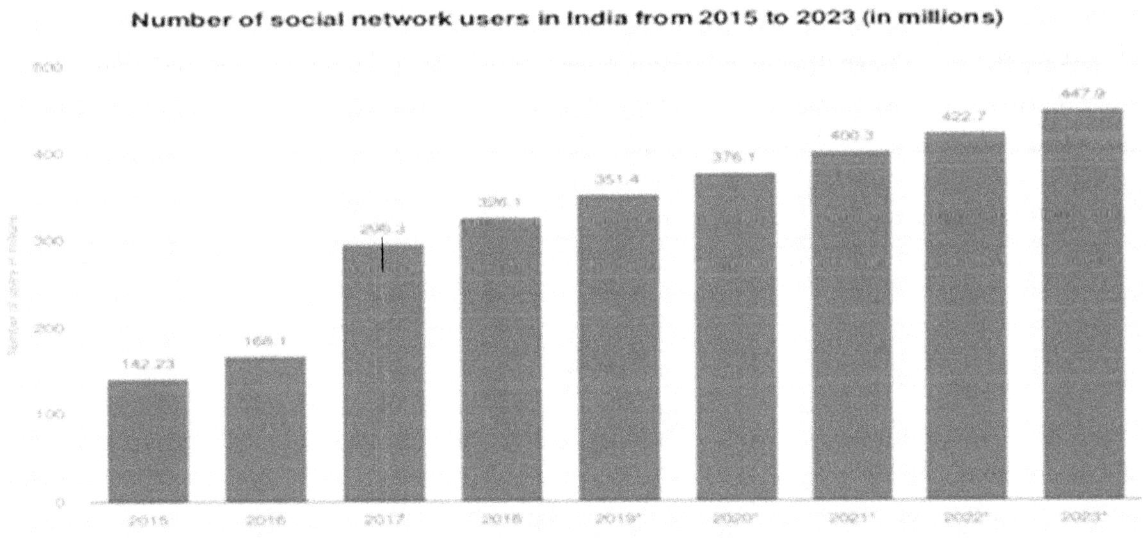

Figure 18. Cost of data breach in our country is increasing day by day
Source: IBM

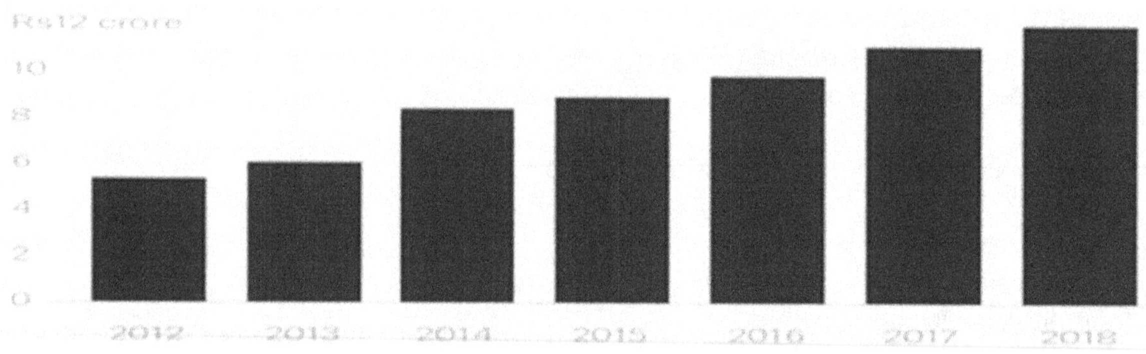

From Figure 2 and Figure 3 it is evident that amount involved in crime is increasing day by day and the use of internet is on increasing edge and from current technical scenario if online crime is not minimized then there will be massive loss and so it is essential to reduce it.

BACKGROUND AND MOTIVATION BEHIND ONLINE CRIME

The term was first coined by William Gibson and virtual environment or '*cyberspace*' is the basic background and starting point of cyber crime. To prevent rather minimize we have to reduce vulnerability or incompetence of a unit to endure effects of the disastrous environment and to minimize the risk ie damage control due to vulnerability. Also Data breaches are one of the major reasons.

There are many types and motives behind committing. Some of them are stated below:

1. **State Actors:** Steal important information, documents and intellectual property of a country/big organisations/community .Main motive is document extraction
2. **Organized Cyber Criminals:** Engaged in frauds, scam. Target is only money
3. **Hacktivist:** Revenge or political revenge, social media troll
4. **Terrorist:** Theyexploit new technology to push their agenda forward. Main motive is political, financial, military information and main target is cyber-terrorism
5. **Espionage Specialists:** Purloin intellectual property from competitors
6. Other motives

DIFFERENT TYPES OF ONLINE CRIME

There are many types of cyber crime prevailing but each time some new features is updated and improvement strategies is one of the key factors how the escape the eyes of cyber security analyst. But the classical sequence remains same. It is mainly of 5 steps described in the Figure 21

Figure 19. Different motives behind online crime

1. Money
2. Beat competitors
3. Steal documents and intellectual properties
4. Cyber terrorism and cyber war
5. Revenge by different means like troll, bullying
6. Human trafficking and other offensive crimes
7. Software piracy and attacking system
8. E-Commerce
9. Law-enforcement
10. Others

Some common web application attacks are buffer overflow, DOS, phishing etc. There are many types of online crimes some of which are stated below:

- **Cyber-Bullying:** The excessive use of internet and widespread affects human life by playing important role in every aspect and it is not limited to adults but also to children and teenagers which leads to bullying or rather now-a-days cyber bullying because it has changed from physical level to virtual or digital level. It has mainly 2 formats overt and coverts .It is mainly dominant among teenagers and youths. From the different reports it has been noted that mainly this is carried out by teenagers and adolescents. Maximum victims cannot express themselves to their guardians

Figure 20. Different sectors affected by online crime

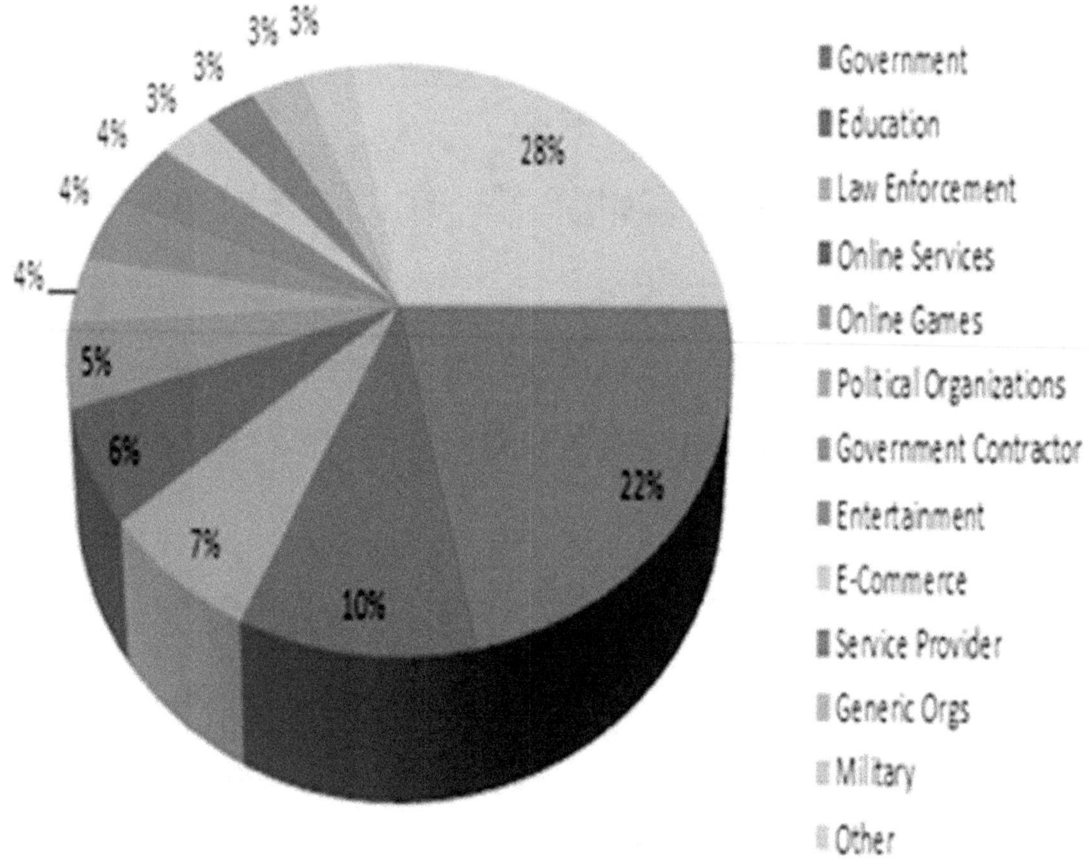

or to anyone which results in certain psychological problem like depression, anxiety and also some disorders. The main platform is social media and different chat rooms. The bullying may be gossips,"mean or unfriendly treatment", spreading rumours, doxing, threatening etc. From the

Figure 21. Showing steps of classical steps of an attack
Source: Cisco

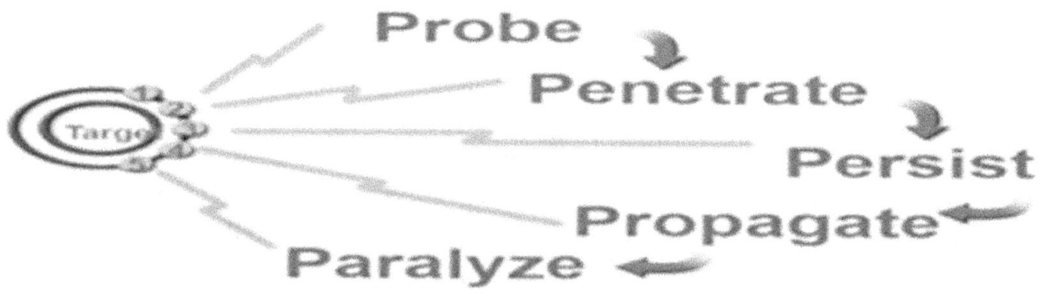

Figure 22. Showing that India is one of the leading countries affected by cyber bullying
Source: Ispos

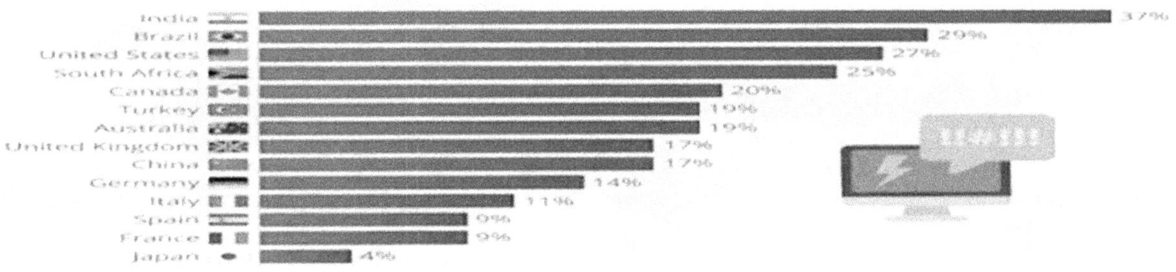

reports it is evident that India is one of the leading countries victimized by this. Females between the ages 17-25 are the major victims. (Afroz et al., 2013)

- **Cyber Stalking:** It is a part of cyber bullying where by repeated use of electronic device and by medium as internet victim has to face harassment, insult, and defamationfalse acquisition. Here main motive is revenge, anger, personal issues. Here, attackers use self-publishing media outlet and mainly uses flaming techniques. Here also tendency of victimized is more for females. Corporate stalking is done for the harassment of Competitor Company.
- **Cyber/Internet Troll:** It is also a type cyber-bullying and online harassment. Troll is depicted as a giant. Here group is more targeted rather than individual. Here interests are political, corporate revenge and also some sponsored troll is there and so sometime monetary transaction is involved.
- **Frauds:** In terms of laws, fraud means bereave a victim from his/her own legal rights. Widespread of internet and rise in online banking has increased the frauds and consequently the financial loss gives a spark. In the year of 2010 it took a sharp edge and hit a percentage in 90% .The robustness of banking sector is one of the key factors determine the economy and productivity and economy

Figure 23. Showing increase of mobile banking
Source: Newspaper report

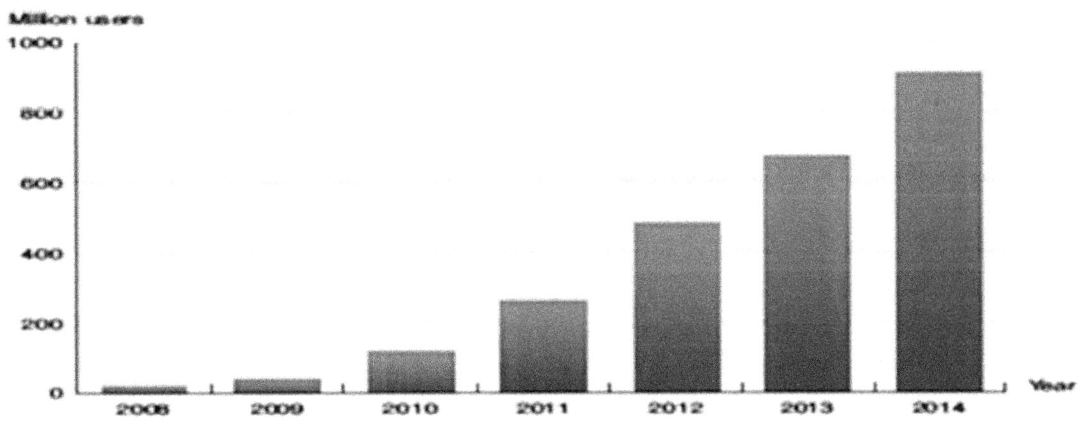

Figure 24. Showing the increasing trend of hacking and data breaches in different fields
Source: Jefferies, identity theft resource centre

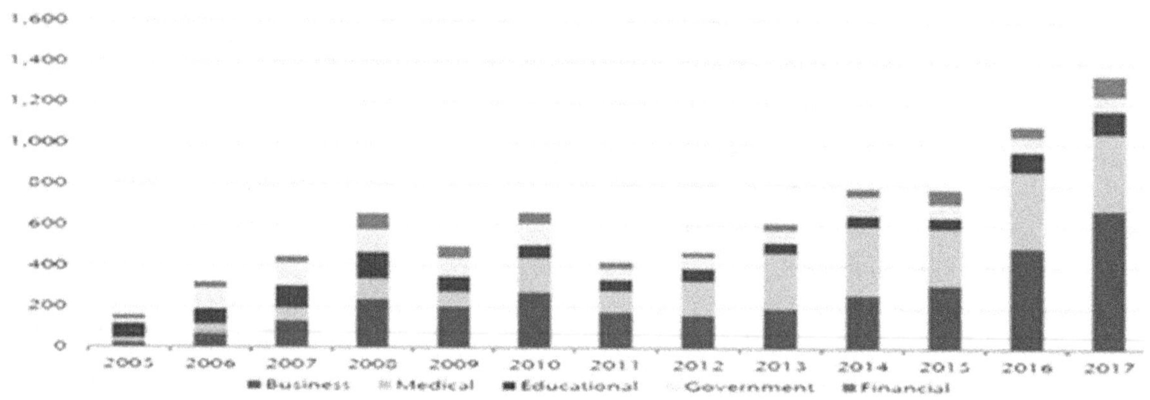

of a country. The increase in usage of smart phone mobile banking also gets popularized. This mainly took place from 2011-12 and number of frauds also intensified rapidly.

Fraud detection is difficult because of its dynamic behavior and uneven distribution over the different profiles and highly imbalanced and unstructured dataset. Due to black box approach process of manual detection becomes slower.

- **Hacking:** It is an approach or some strategy and aptitude of breaking the framework of device without authorisation and authentication. There are mainly 3 types of hackers: white hat hackers who are the paid hackers or ethical hackers engaged in infiltration testing, black hat hackers where they mainly bypass the protocols just for personal reasons like financial gain, personal reasons, revenges etc. Grey hat hackers are the blend of both the types. Some of the tools used by them are exploit by using different kit(shortcoming), root kit, sniffing, vulnerability scanner. (Al-Suwaidi et al., 2018)

Some of common types of attack are stated below:

- **DOS:**Denial-Of-Service (DOS) attack engulf a system's measures so that it is unable respond to requests.It basically provides an approach to reduce in scope of a network without increasing as well as inward access.DOS assaults work by flooding the gateway switches .It can be used in business purpose.
- **DDOS:**It basically disrupts normal traffic in the network of target which is similar to creating a traffic jam in highway. It uses numerous, changing, source IP addressesand is initiated from a large number of other host machines that are already vitiated by malicious software remotely controlled by attacker. Anatomy of the attack is illustrated below in the Figure

Figure 25. Showing the anatomy of the DDOS attack
Source: Research gate

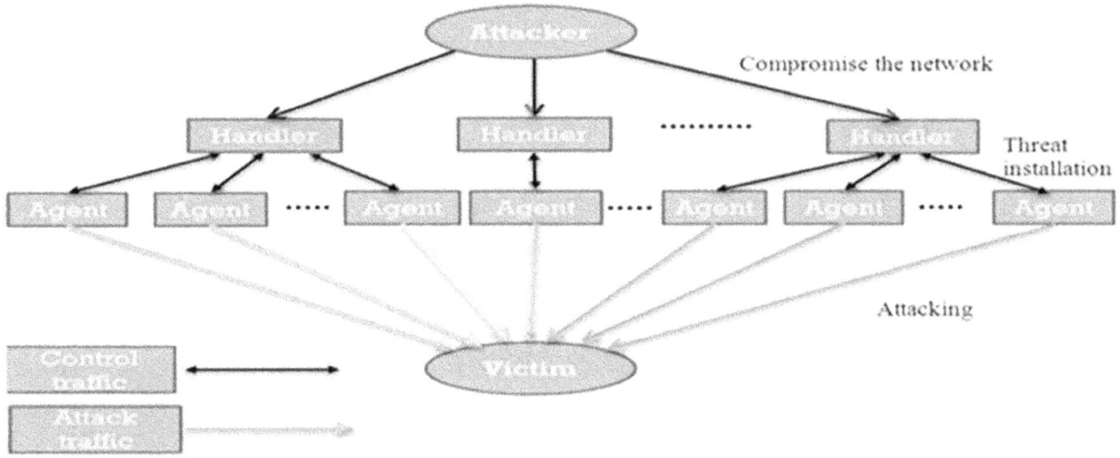

Some of the examples of DDOS and DOS are TCP SYN flood attack, teardrop attack, and ping-of-death attack. Botnets are the systems victimised with malware under control for the execution of DDOS attacks. (Armin et al., 2015)

- **Man-in-the-Middle Attack:**A MitM attack takes place when attacker put itself in between the networks and hampers the communications of a client and a server. Some common examples are Session hijacking, IP Spoofing etc.

- **Phising:** It is an online scam where attacker imitateslegitimate entities in order to get the access. From the name itself it is ensured that putting a teaser and waiting for the return hand-shake is the

Figure 26. Showing Man-in Middle attack
Source: Cisco

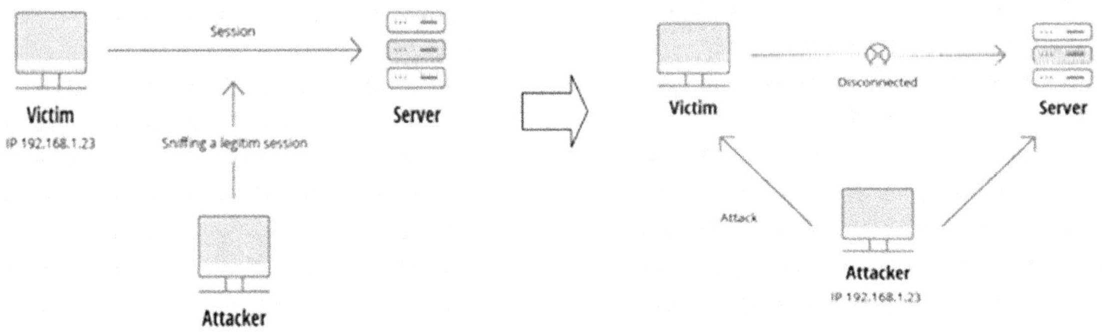

Figure 27. Showing phishing cycle
Source: Mailjet

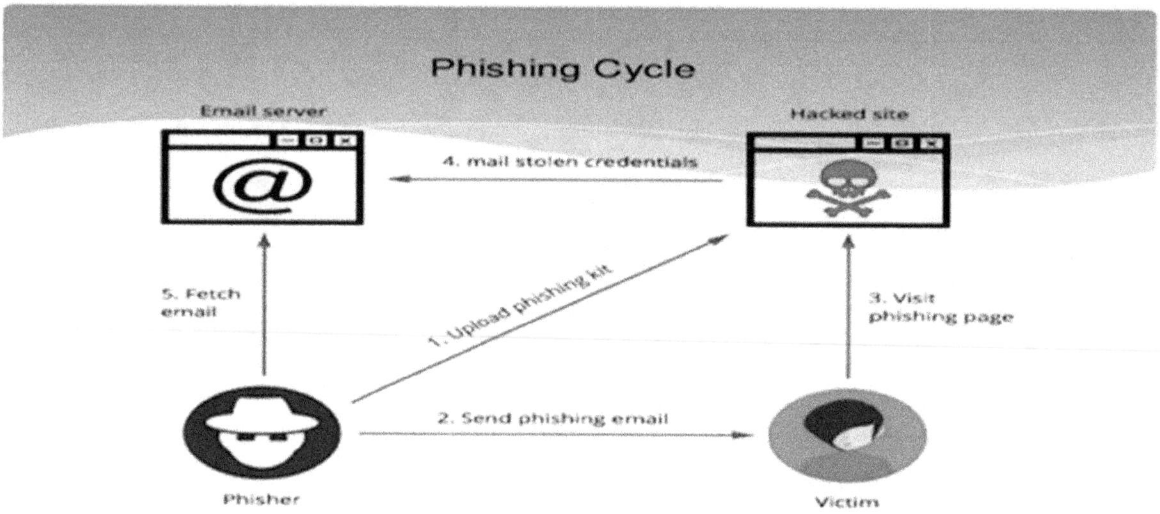

approach for the attack. Here mainly sending emails or digital signal or electronic messages that seem to be from trusted source is the one of the popular approach and attachment or link in the email contain certain malware which attacks the target.

Some of the other major attacks are SQL injection, XSS attack, Malware attack etc .Also cyber terrorism and the cyber war now-a-days is big reason to worry because the attacker infiltrate through the security barriers of the different military or government intellectual data leading to destruction of

Figure 28. Showing different countries prepared to resist cyber war
Source: World economic forum

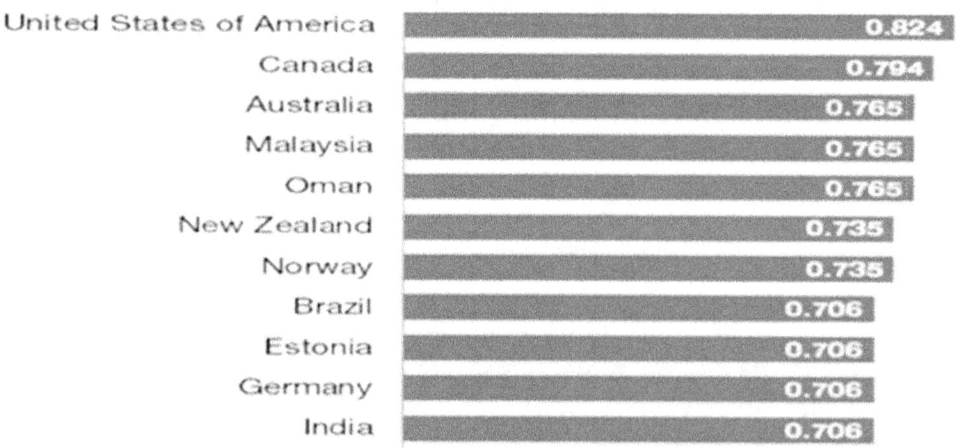

different properties and assets of a country. This type of destruction leads to cyber-war and the military network is hampered. (Kumar, 2016, pp. 256-251)

From the statistics it is evident that India is not enough prepared to resists against cyber war with respect to other European and American countries.

So it is very much essential to combat against cyber-crimes paving a way to create a safe & secure digital India.

CRIME PATTERN DETECTION AND ANALYSIS

During building of a model or product it is very much essential to apply diminution strategy from perceptible potential threats to its security for reducing the risk from threats to do countermeasures and protect the system before it get victimised. (Nallaperumal, 2018, pp. 1-4)

Some of the possible strategies that can be implemented are:

- Conisation of surface
- Defensive coding and review,
- Platform Independent coding
- Security patches

To understand the structure of attack we need to understand the chain or kill chain in terms of cyber world. The concept actually comes from the military concept of understanding the structure of attack strategies of the enemies. Cyber kill chain is an approach to understand the chronology of events and

Figure 29. Showing cyber kill chain
Source: Created by author

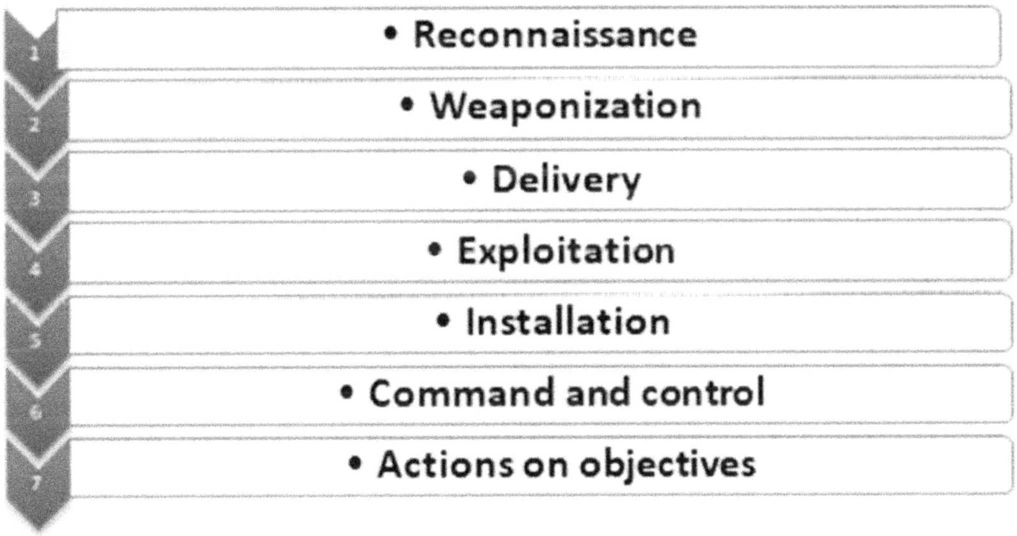

Figure 30. Describing the pyramidical structure

steps involved in an attack. Designing and monitoring it helps to study it because by following some cycle or pattern attackers infiltrate into network as well as extract information from it.

They are mainly 7 steps involved which are illustrated in the Figure 29

For basic protection and fundamental information security we have to follow the CIA triad which will protect the different parts and by following the different rules we can give a protection layer to it. It forms a pyramidical structure .One of the major use is in the organisational security.

They are

- CONFIDENTIALITY
  - Main Techniques is encryption or creating certain pattern
- INTEGRITY
  - Main Techniques is hashing(like SHA-1,SHA-2)
- AVAILABILITY
  - Main Technique is Load balance, RAID

Another fundamental information security technique is Authorisation, authenticity and non-repudiation. If the system is checked with proper authorisation then it the access to operate is given. For authenticity some security questions and verification if done it is better. (Singh et al., 2017)

RISE OF BIG DATA IN CLOUD

The outbreak of internet and users led to creation of huge amount of data in every time space. This data can't be handled by database methodologies. The endless growth in volume is due to data which are being captured by organizations, social media sharing, and use of smart products i.e. IoT enabled devices has resulted in a massive data outbreak.

Figure 31. Big data market value trend
Source: (Chatterjee, 2021)

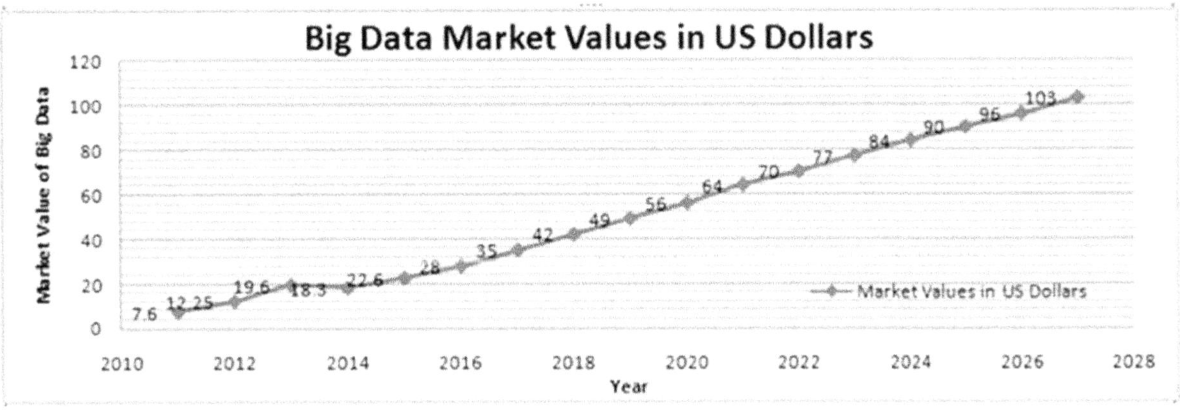

The Cloud computing delivers various service models to its user as service-based services in pay you go form. It delivers infrastructure, platform, and software as services. These services are namely, Infrastructure as a Service (IaaS), Platform as a Service (PaaS), and Software as a Service (SaaS). [1]*

Figure 32. Service Model of Cloud Computing
Source: (Chandrika & Dalwal, 2019)

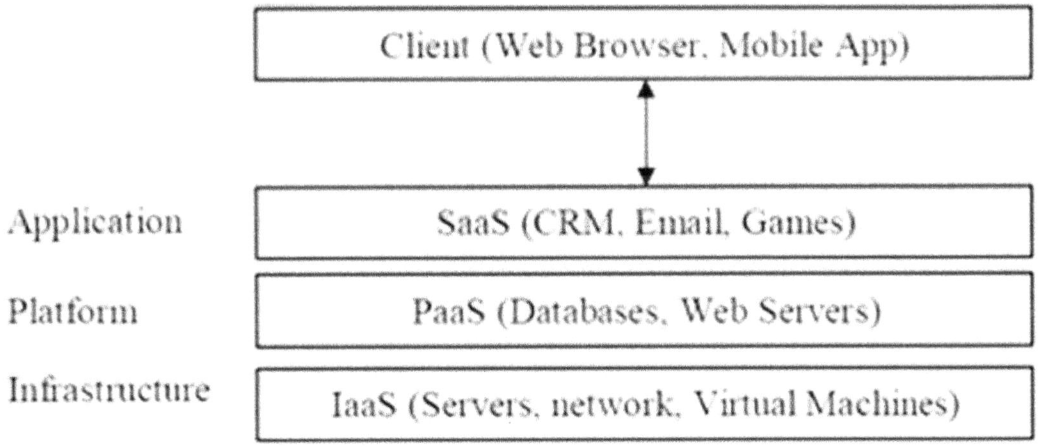

Table 3. Comparison of various big data cloud provider

	Google Cloud	Microsoft Cloud	Amazon Cloud
Big Data Storage	Google Cloud Service	Azure	S3
Big Data Analytics	BigQuery	Hadoop on Azure	Elastic Map Reduce (Hadoop)
Map Reduce	App Engine	Hadoop on Azure	Elastic Map Reduce

Source: (Chatterjee, 2021)

Big Data and Cloud Computing goes hand by hand. Big Data allows people to compute through distributed queries across multiple dataset and return results accurately. On the other hand cloud computing provides Hadoop, a class of distributed data-processing platforms. The data sources are being stored in a distributed fault-tolerance database and proceed through a programming model for large dataset with a parallel distributed algorithm. (Kumar et al., 2015)

Big Data utilizes the distributed storage technology service of Cloud Computing rather than the storages installed as hard drives in the computers. In the following table some big data cloud providers are being listed.

Security Challenges in Cloud Computing

Cloud Computing provides many facilities and advantages, but as everything coin has two sides it also has its own challenges and problems. Security issues are there, protection of users data is one of the most difficult challenges to be face and to be won. Various security issues are there as it merges a lot of technologies such as database, OS (Operating System), virtualization, resource scheduling, memory management, and concurrency problem.

- **Privacy of Data:** One of the biggest challenge, as everyone's personal data is very important. And consumers nowadays feel free to upload and save their data in a cloud platform. But they are not aware of where there data is going and how they can be bypassed by unauthorized persons to exploit the data. Thus throwing is as a big challenge.
- **Confidentiality of Data:** Another challenge of the modern cloud community, data confidentiality means that only authorized person can have the access to the owner's data. Encryption comes very handy here to ensure security.
- **Integrity of Data:** Data Integrity means that the data which has been stored in the cloud can be stored safely from being modified from any unauthorized access. Digital Signature and RAID technique are often adopted to solve this challenge.
- **Data Remanenece:** In order to maintain the cycle, it's recommended that the data from cloud should be deleted after a particular time cycle. The reformatting or deletion doesn't ensure complete deletion, those data can be accessed later by somehow means. Thus creating a challenge for the cyber security personals.
- **Transmission of Data:** Transmission of data means that the data when transferred from user to cloud it goes via a path, a client path, then the data is returned to cloud from the client server. While transmitting this data it is advised that the data should be encrypted so that it can be transferred safe. But as encryption-decryption process takes a lot of time this step is often ne-

glected thus while transmitting the data, it remains unsafe and unauthorized access becomes easy. (Kakkad et al., 2019)

Cloud Data-Center and Architecture

The current scenario is such we need to provide securities to the challenges seen in the contemporary market. Marking and refining the connecting links in the cloud network increases the security and reliability in the system. When one datacenter is down or not in use state, in that state it's called an open space and from there vulnerability can arise. That point becomes the open port through which the attack could be carried out. Underperformance of any datacenter or taking up too much time to respond generally indicates any vulnerability attack. In similar condition if the attacker can overuse or misuse the data assignments in the datacenters it can lead to massive carbon footprints and heating in the system, and too much heating can lead to poor performance even malfunction of device. Figure3 suggests the cloud datacenter architecture.

Figure 33. Cloud Data-Center Architecture
Source: (Hosseini & Vakili, 2018)

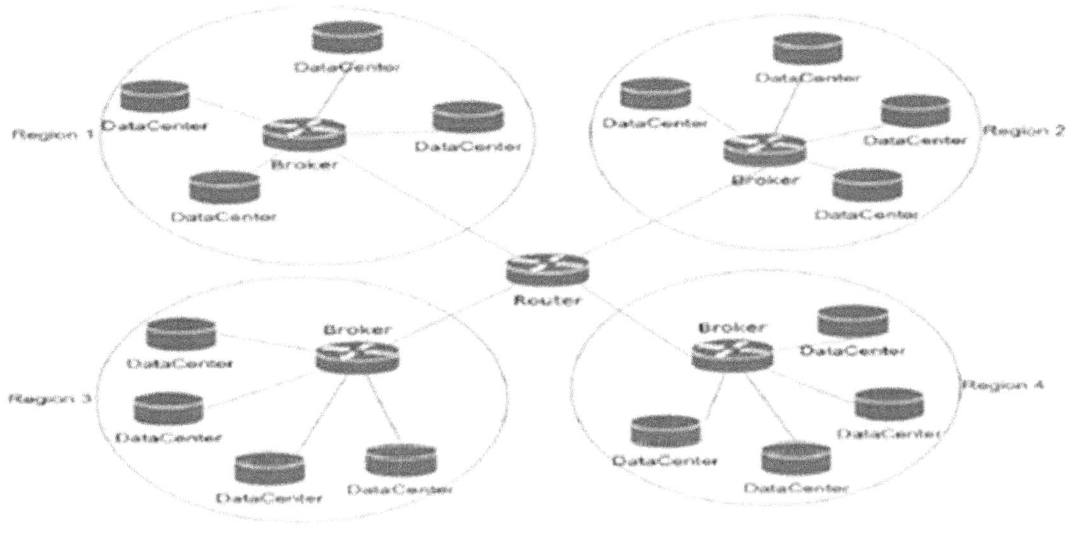

In this section we frame the attacks in the cloud architecture using Bell theory and optimization is done in the parameter of data center instead of nodes. The cost measurement function which will operate the cost for returning value is framed using fuzzy logic system and Game Theory. (Carminati et al., 2014)

In game theory is basically examines the strategic decision makings, here users or actors are taken and an optimal solution is being created based on the inputs from the users. In this case these actors are user and a tester. User opts for low cost datacenter with as high as possible security and the job of tester here to maximize the response cost function. This game is being tested and played in Nash Equilibrium strategy, where the selection probabilities is optimal for user and failing to breach is optimal for tester. The vulnerability test for data centre is being done using Method of Successive Weighted Averages (MSWA).

GAME THEORY

Game theory is a branch of mathematics which in general deals with the strategic implementations with human dilemma taken in and as parameters. This game can be defined as interaction between a set of players with some set of rules implied, here the output or the result not only depend on the input of the players but the influence of others also play a significant role. The motive of a player is to make a decision which can yield them maximum pay-off. (Rao et al., 2014)

Types of Game Model

- **Co-Operative Game Model:** In this game the players form a group or set to get off a common goal of each. Players maybe people, companies etc.
- **Non-Cooperative Game Model:** This model of strategy implies such strategic planning or actions taken through which a player can make their dominance and achieve their own motives, Nash Equilibrium is the solution in such cases, in our model we have proposed the game through this model.
- **Static Game Model:** Players make their decision in concurrent manner with knowledge of strategies of other players.
    - **Static Prisoner's Problem:** In this type of problems the caught prisoner have opportunity to either help or defy the statements of authority, but the sentence of punishment is totally different for two cases if caught lying.
    - **Static Zero Sum Game:** Players interest are complemented in this case i.e. if player 1 gets best outcome then player 2 will get worst outcome.

MATHEMATICAL MODELLING

Pair of alternative solutions represented by:

a_1 = Cheap SAAS model for cloud user
b_1 = Expensive cost function by cloud user
a_2 = non-protected cloud SAAS model
b_2 =protected cloud system through hacking

In this case we offer 4 kinds of outcome/results for each state of utility.

Let u represents the cheap and protected states
v represents non-protective & cheap
w represents protective and costly, then

$$\begin{bmatrix} u & u-w \\ u+v & u+v-w \end{bmatrix}$$

The above cases form certain conditions from which certain conditional probabilities can be formed. They are:

$$\left(1-y\right)=P\left({}^{A_2}\!/_{A_1}\right)$$

$$\left(1-y-z\right)=P\left({}^{A_2}\!/_{B_1}\right)$$

$$y=P\left({}^{B_2}\!/_{A_1}\right)$$

$$y+z=P\left({}^{B_2}\!/_{B_1}\right)$$

E_{a1} & E_{b1} are the expected values of choices $a1$ and $b1$ respectively represented as

$$E_{a1}=u\left(1-y\right)+\left(u-w\right)y \tag{1}$$

$$E_{b1}=\left(u+v\right)\left(1-y-z\right)+\left(u+v-w\right)\left(y+z\right) \tag{2}$$

A factor x ,is taken as a person of conflict choices.
Multiplying equation 1 with 2 we get:

$$E=\left(1-x\right)E_{a1}+xE_{b1}=u+vx-wy-wxz \tag{3}$$

Where E is assumed to be expected value of the entire game but the index of choice of cloud data centres.

From the equation 3 we get:

$$E_{b1}-E_{a1}=v-wz \tag{4}$$

$x\rightarrow 0$; When $v-wz$ is large

$x \rightarrow 1$; When $v - wz$ is zero

$$\therefore x = \left(1 + k^{wz.v}\right)^{-1} \tag{5}$$

The parameter k must exceed the unity. The partial dependence of x on the product wz indicates that changes in action. Three functions summarize relationship among 3 variables

$$x = F_1\left(v, w, z\right) \tag{6}$$

$$u = F_2\left(v, w, x, y, z\right) \tag{7}$$

$$E = F_3\left(u, v, w, x, y, z\right) \tag{8}$$

Maximized of the E with respect to x, y, z equal to zero; partial derivatives are taken for maximizing the required function:

$$\frac{\partial E}{\partial w} = \frac{\partial F_3}{\partial w} + \frac{\partial F_3}{\partial x} \cdot \frac{\partial x}{\partial w} + \frac{\partial F_3}{\partial u} \cdot \frac{\partial u}{\partial w} = 0 \tag{9}$$

$$\frac{\partial E}{\partial y} = \frac{\partial F_3}{\partial y} + \frac{\partial F_3}{\partial y} \cdot \frac{\partial u}{\partial y} = 0 \tag{10}$$

$$\frac{\partial E}{\partial z} = \frac{\partial F_3}{\partial z} + \frac{\partial F_3}{\partial x} \cdot \frac{\partial x}{\partial z} + \frac{\partial F_3}{\partial u} \cdot \frac{\partial u}{\partial z} = 0 \tag{11}$$

From the equation we can see that E is not directly affected through w & z but they are creating an indirect impact on the role of change. Isolation of change in u with the help of equation of 9 and 11

$$\frac{\partial F_3}{\partial u} \cdot \frac{\partial u}{\partial w} = \frac{\partial F_3}{\partial F_2} \cdot \frac{\partial F_2}{\partial w} + \frac{\partial F_3}{\partial F_2} \cdot \frac{\partial F_2}{\partial x} \cdot \frac{\partial x}{\partial w} \tag{12}$$

$$\frac{\partial F_3}{\partial u}.\frac{\partial u}{\partial z}=\frac{\partial F_3}{\partial F_2}.\frac{\partial F_2}{\partial z}+\frac{\partial F_3}{\partial F_2}.\frac{\partial F_2}{\partial x}.\frac{\partial x}{\partial w} \tag{13}$$

Response of u in impact of $w\,\&\,z$ on x:

$$\frac{\partial F_3}{\partial u}=1$$

$$\frac{\partial F_3}{\partial w}=-y-xz$$

$$\frac{\partial F_3}{\partial x}=v-wz$$

$$\frac{\partial F_3}{\partial y}=-w$$

$$\frac{\partial F_3}{\partial z}=-wr$$

Nash equilibrium strategy:

In normal form, the game is:

1. 1. Finite no of players:

$$M=\left\{a_1,a_2,\ldots\ldots a_n\right\}$$

2. 2. Strategy S_i assigned to each other

$$i\in M$$

The strategy of all combination can be written as:

$$S = \prod_{i \in M} S_i$$

3. 3. Pay-off function:

$$U_i : S \to \mathbb{R}$$

$$\forall s \in S : u_i(s) \in \mathbb{R}$$

A combination of the different strategies $S^* \in s$ is called *Nash Equilibrium*iff

$$\forall i \in M \forall s_i \in S_i :$$

$$u_i S^* = u_i\left(s_i^*, s_{-i}^*\right) \geq u : \left(s_i^*, s_{-i}^*\right)$$

Definition:

A strategy $\widehat{s_i} \in S_i$ is called best combination $s_i \in S_{-i}$ iff conditions

$$\forall s_i \in S_i : u_i(\widehat{s_i, s_{-i}}) \geq u_i\left(s_i, s_{-i}\right)$$

$$u_i(\widehat{s_i, s_{-i}}) = \max\left\{u_i\left(s_i, s_{-i}\right); s_i \in S_{-i}\right\}$$

Mixed Strategies in Nash Equilibrium:

A mixed strategy of player i then it is a probability distribution on S_i .

Vector $p_i = \left(p_i^1, \ldots. p_i^{m_i}\right) \in R^{m_i}$ with $p_i^\alpha \geq 0 \forall \alpha$ $\& \sum_\alpha p_i^\alpha = 1$.

The support of such a mixes strategy is defines as a set of those α with $p_i^\alpha > 0$. After normalisation we get,

$$\sum_i = \left\{ p_i \in \mathbb{R}_+^{m_i} : \sum_{\alpha}^{m_i} p_i^{\alpha} = 1 \right\}.$$

So the mixed strategy profile becomes a pair:

$$p = (p_1, p_2) \in \sum_1 * \sum_2 =: \Sigma.$$

$$p(s) = p_1^{s_1} p_2^{s_2}$$

The expected value of the pay-off function is:

$$\prod_i (p) \Delta \sum_{s \in S} p(s) \prod_i (s)$$

$\prod_i (p)$ is a continuous function of p where:

$$\prod_i (p) = \sum_{\alpha=1}^{m_j} p_j^{\alpha} \prod_i (e_j^{\alpha}, p_{-j}) \, for \, (j = 1, 2..)$$

Weighted sum of the pay-off within strategy α

$$\prod_i (p) = \prod_{\alpha=1}^{m_1} \cdot \prod_{\beta=1}^{m_2} p_1^{\alpha} \prod_i (\alpha, \beta) \, p_2^{\beta}$$ n matrix notation:

$$\prod_1 (p) = p_1 A p_2 \, , \prod_2 (p) = p_1 B p_2 = p_2 B^t p_1$$

Under the assumption there exist is Nash *equilibrium:*

$$\prod_i \left(x_1^*, \ldots x_n^*\right) = \max \prod_1 \left(x_1^*, \ldots, x_{i-1}^* \, , x^i, .. x_n^*\right); s_i \in k_i$$

For $x, y \in X : k_1 .. k_n$

$$f(x, y) = \sum_{i=1}^{n} (\prod_i (y_1 \ldots y_n) - \prod_i (y)$$

f is concave and continuous in 2nd argument $f\left(x, x^*\right) \leq 0$ for all $x \in X$

This means when applying it:

$$x = \left(x_1^*, \ldots, x_{i-1}^*, x^i, \ldots x_n^*\right)$$

$$\prod_i \left(x_1^*, \ldots, x_n^*\right) \geq \prod_i \left(x_1^*, \ldots, x_{i-1}^*, x^i, \ldots x_n^*\right) \forall i$$

LITERATURE REVIEW

Bell proposed a game theory approach to identify the intrusion detection, through failure of nodes which will apparently affect the network performances. Here it's a two player game router and virtual network tester. The aim for network tester is to maximize the trip cost whereas the same for router is to find least cost path. (Ruangnapakul, 2019, pp. 104-111)

Viera et al proposed intrusion detection system based on network nodes in cloud computing environment. This method used grid type detection but the major drawback in this methodology is it cannot detect any kind of new attack.

Han et al proposed a method, in which it will detect the potential attacks and it will cluster into 3 different types of sub sections namely high, moderate, and low risk afterward it will be 2 player optimization strategy play in which the attackers are forced to behave as common element and total cost of attacker gets increased in a significant manner, but the major drawback in this case is it's basically designed for independent datacenters it cannot support multiple datacenter. (Kumar & Agarwal, 2018)

Ferdousi et al proposed a method in which the datacenters are placed in a disaster management system and the protection policy was dynamic in nature.

PROPOSED METHODOLOGY

In this methodology we assume there is a hypervisor H where n users are supported to run the VM setups. In this scenario we assume the motive of the attacker is to start and run as many VM setups as he can to slow down and heat up the environment which will subsequently affect the performance and increase to cost function. (Kshetri, 2018, pp. 83-87)

Game Model:

In this scenario we consider 2 players, one is the attacker and one is the defender (cloud system provider). The players are considered to be rational and all the moves they count is rational in nature. The aim of attacker is to maximize the damage done, and $\max\left(VM\ Workload\right)$ and the aim of defender is to $\min\left(VM\ Workload\right)$ both the cases considered when the VM are in ideal situation not with the instruction processing.

The attacker denoted by A_i have targets of VM of user denoted by U_i, the function defined in the equation –

$$VM\left(U_i,t\right) = \left\{VM_{i1}, VM_{i2}, VM_{i3} \ldots, VM_{in}\right\}$$

$$Tgt\left(A_i\right) = VM\left(U_i,t\right)$$

The approach of attacking the VM for attacker can be any, it can be by breaching, DDoS, executing attack payload function, backdoor attack etc. The attacker holds the key to activate the number of VM at time, and the security protection level will define the cost function. (Rubin-Delanchy et al., 2016)

The goal of the defender is to now find and imply a VM allocation policy through which the attacker have to co-reside in a VM. There are mainly 4 types of server choosing strategy namely, least number of VM, maximum numbers of VM, random selection of VM, and round robin rule VM selection. The game is played in Nash Equilibrium, mixed strategy method. A cost function is introduced for checking and marking the vulnerability of the cloud computing system. The target of the game is finding out the datacenter which failure will impact the most in the system.

As we know that data center is a pool of server, where consumer jobs can be executed on virtual machines. The sever computation can be measured by no of cores and each can able to handle single virtual machine.

$$Server = Freq * AvailableResources * No.ofcore$$

Power consumption of a server can be divided into two types: Static and dynamic power consumption.

$$P_S = P_{Static} + P_{Dynamic}$$

It's assumed that whenever a datacenter fails each failure scenario is related to a failed data center. The selection of datacenter i with the probability as:

Pseudo code of MSWA-I

Initialisation:

Set $\gamma_0 = 0$, $k = 1$ the real number $d \geq 0$ and the stop criteria $\varepsilon > 0$

Calculate initial point x^1 & $y^1 = F\left(x^1\right)$

Main iteration:

While $\left\|x^k - y^k\right\| \geq \varepsilon$ do

$$\beta_k = k^d, \gamma_k = \gamma_{k-1} + \beta_k, \alpha_k = \beta_k / \gamma_k;$$

Figure 34. Convergence of probabilities of datacenter
Source: Computed by Author

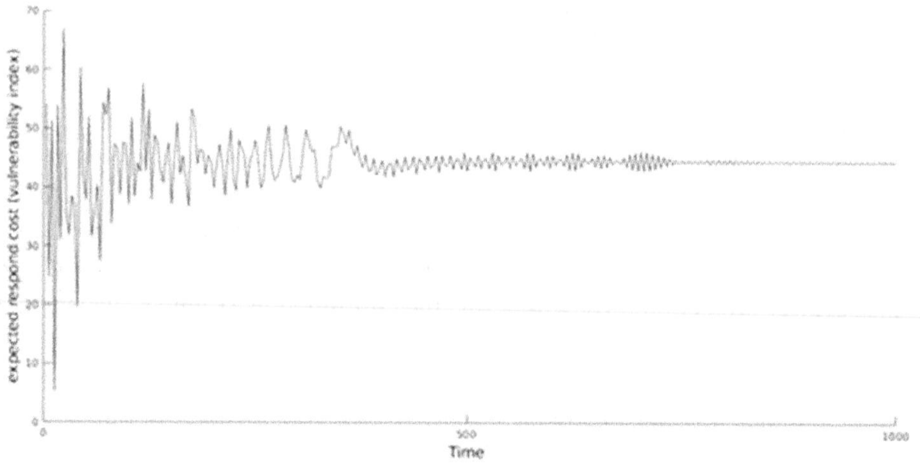

$$x^{k+1} = x^k + \alpha_k \cdot \left(y^k - x^k\right);$$

$$y^{k+1} = F\left(x^{k+1}\right);$$

$$k = k + 1;$$

End

Output: x^k

ANALYSIS ON THE BACKDROP OF THE RESULT

One of the most important jobs is to secure the data, privacy of the data. It's a two player Nash equilibrium based modeling; the vulnerability is being measured through the probabilistic heuristic model based on Method of Successive Weighted Averages (MSWA) algorithm. Around 200 tasks being generated; condition is all the tasks are generated in random order with the range in between 2000 to 3000 million instruction per second. (Sattar et al., 2018)

The user gets the probability for the datacenter with minimum cost function and the attacker is the probability of delaying the response of datacenter through which he can exploit the vulnerability. The time taken response in convergence of the probabilities of datacenter is considered as the best case scenario in this case.

Figure 35. Probabilistic Curve for each datacenter
Source: Computed by Author

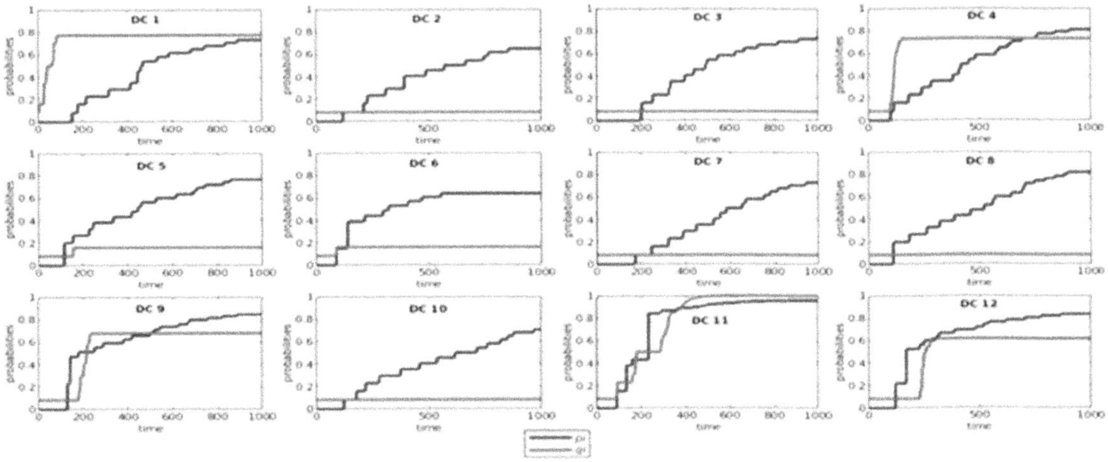

CONCLUSION

In this paper we investigated the prime cause behind the online crime; we have taken an econometric ride to examine the results. Through experiment we can conclude that economic condition of a country plays a pivotal role in the increment of cybercrime across the country. We saw that poor unemployment rate give rise to increased crime rate and which is increasing hand in hand with inflation. A brief overview on impact of cybercrime on economics also given. (Jani, 2018)

In the analytics part firstly we showed the background of cybercrime, and cybercrime pattern analysis also being done. A proposed method is given by which we can identify the vulnerability of a cloud data center using the concept of game theory. We have proposed the game strategy as non-cooperative with the solution being Nash Equilibrium strategy. The results we achieved are quite satisfactory in nature that is by which we can consider our algorithm to be satisfied.

REFERENCES

Abdelhamid, N., Thabtah, F., & Abdel-jaber, H. (2017, July). Phishing detection: A recent intelligent machine learning comparison based on models content and features. In 2017 IEEE international conference on intelligence and security informatics (ISI) (pp. 72-77). IEEE. doi:10.1109/ISI.2017.8004877

Afroz, S., Garg, V., McCoy, D., & Greenstadt, R. (2013, September). *Honor among thieves: A common's analysis of cybercrime economies. In 2013 APWG eCrime Researchers Summit.* IEEE.

Al-Suwaidi, N., Nobanee, H., & Jabeen, F. (2018). Estimating Causes of Cyber Crime: Evidence from Panel Data FGLS Estimator. *International Journal of Cyber Criminology*, *12*(2), 392–407.

An, J., & Kim, H. W. (2018). A data analytics approach to the cybercrime underground economy. *IEEE Access: Practical Innovations, Open Solutions*, *6*, 26636–26652. doi:10.1109/ACCESS.2018.2831667

Armin, J., Thompson, B., Ariu, D., Giacinto, G., Roli, F., & Kijewski, P. (2015, August). 2020 cybercrime economic costs: No measure no solution. In *2015 10th International Conference on Availability, Reliability and Security* (pp. 701-710). IEEE.

Basu, A. (2014). Social network analysis: A methodology for studying terrorism. In *Social networking* (pp. 215–242). Springer. doi:10.1007/978-3-319-05164-2_9

Bhatt, U., Iyyani, D., Jani, K., & Mali, S. (2018, April). Troll-Detection Systems Limitations of Troll Detection Systems and AI/ML Anti-Trolling Solution. In *2018 3rd International Conference for Convergence in Technology (I2CT)* (pp. 1-6). IEEE.

Bhatt, U., Iyyani, D., Jani, K., & Mali, S. (2018, April). Troll-Detection Systems Limitations of Troll Detection Systems and AI/ML Anti-Trolling Solution. In *2018 3rd International Conference for Convergence in Technology (I2CT)* (pp. 1-6). IEEE.

Carminati, M., Caron, R., Maggi, F., Epifani, I., & Zanero, S. (2014, June). BankSealer: An online banking fraud analysis and decision support system. In *IFIP International Information Security Conference* (pp. 380-394). Springer. 10.1007/978-3-642-55415-5_32

Choudhary, P., Singh, U., Dalal, S., & Bisen, D. (2019, February). Social Network Analysis (SNA): A Vision for Counter-Terrorism Approach in Modern Indian Defence Sector. *Proceedings of 2nd International Conference on Advanced Computing and Software Engineering (ICACSE)*. 10.2139srn.3349028

Das, P., & Das, A. K. (2017, January). Behavioural analysis of crime against women using a graph based clustering approach. In *2017 International Conference on Computer Communication and Informatics (ICCCI)* (pp. 1-6). IEEE. 10.1109/ICCCI.2017.8117714

Jani, S. (2018). *The Growth of Crypto currency in India: Its Challenges& Potential Impacts on Legislation.* Research Gate Publication.

Kakkad, V., Shah, H., Patel, R., & Doshi, N. (2019). A Comparative study of applications of Game Theory in Cyber Security and Cloud Computing. *Procedia Computer Science*, *155*, 680–685. doi:10.1016/j.procs.2019.08.097

Kshetri, N. (2018). Introducing the IT Economics Department. *IT Professional*, *20*(1), 83–87. doi:10.1109/MITP.2018.011311501

Kumar, P. V. (2016, March). Growing cyber crimes in India: A survey. In *2016 International Conference on Data Mining and Advanced Computing (SAPIENCE)* (pp. 246-251). IEEE. 10.1109/SAPIENCE.2016.7684146

Kumar, S., & Agarwal, D. (2018). Hacking attacks, methods, techniques and their protection measures. *International Journal of Advance Research in Computer Science and Management*, *4*(4), 2253–2257.

Kumar, S., Koley, S., & Kuamr, U. (2015). Present Scenrio of Cyber Crime in INDIA and its Preventions. *IJSER*, *6*(4), 1972–1976.

Lau, R. Y., Xia, Y., & Ye, Y. (2014). A probabilistic generative model for mining cybercriminal networks from online social media. *IEEE Computational Intelligence Magazine*, *9*(1), 31–43. doi:10.1109/MCI.2013.2291689

Nallaperumal, K. (2018, December). CyberSecurity Analytics to Combat Cyber Crimes. In *2018 IEEE International Conference on Computational Intelligence and Computing Research (ICCIC)* (pp. 1-4). IEEE.

Orphanides, C., Akhgar, B., & Bayerl, P. S. (2016, August). Discovering knowledge in online drug transactions using conceptual graphs and formal concept analysis. In *2016 European Intelligence and Security Informatics Conference (EISIC)* (pp. 100-103). IEEE. 10.1109/EISIC.2016.026

Prathap, B. R., & Ramesha, K. (2018, February). Twitter sentiment for analysing different types of crimes. In *2018 International Conference on Communication, Computing and Internet of Things (IC3IoT)* (pp. 483-488). IEEE. 10.1109/IC3IoT.2018.8668140

Rao, Y. S., Saini, H., & Panda, T. C. (2014). *Effect of Cyber Crime Indian Economy*. Academic Press.

Revathi, S., & Suriakala, M. (2018, February). An Intelligent and Novel Algorithm for Securing Vulnerable Users of Online Social Network. In *2018 Second International Conference on Computing Methodologies and Communication (ICCMC)* (pp. 214-219). IEEE. 10.1109/ICCMC.2018.8487760

Ruangnapakul, N., Salam, Y. D., & Shawkat, A. R. (2019). A Systematic Analysis of Cyber bullying in Southeast Asia Countries. *International Journal of Innovative Technology and Exploring Engineering*, *8*(8S), 104–111.

Rubin-Delanchy, P., Lawson, D. J., & Heard, N. A. (2016). Anomaly detection for cyber security applications. In Dynamic Networks and Cyber-Security (pp. 137-156). doi:10.1142/9781786340757_0006

Sattar, Z., Riaz, S., & Mian, A. U. (2018, November). Challenges of Cybercrimes to Implementation of Legal Framework. In *2018 14th International Conference on Emerging Technologies (ICET)* (pp. 1-5). IEEE. 10.1109/ICET.2018.8603645

Sengupta, R., & Mukherjee, S. (2018). Crime, Deprivation and Social Sustainability—Evidence across States in India. *Indian Journal of Human Development*, *12*(3), 354–377. doi:10.1177/0973703018811173

Singh, J., Singh, G., & Singh, R. (2017). Optimization of sentiment analysis using machine learning classifiers. *Human-centric Computing and information. The Sciences*, *7*(1), 32.

Yadav, S., Timbadia, M., Yadav, A., Vishwakarma, R., & Yadav, N. (2017, April). Crime pattern detection, analysis & prediction. In *2017 International conference of Electronics, Communication and Aerospace Technology (ICECA)* (Vol. 1, pp. 225-230). IEEE. 10.1109/ICECA.2017.8203676

Chapter 6
A Decadal Walk on BCI Technology:
A Walkthrough

Ahan Chatterjee
https://orcid.org/0000-0001-5217-4457
The Neotia University, India

Aniruddha Mandal
https://orcid.org/0000-0003-0721-3108
The Neotia University, India

Swagatam Roy
https://orcid.org/0000-0002-8012-5529
The Neotia University, India

Shruti Sinha
https://orcid.org/0000-0001-6611-3831
The Neotia University, India

Aditi Priya
https://orcid.org/0000-0003-1430-5684
The Neotia University, India

Yash Gupta
The Neotia University, India

ABSTRACT

In this chapter, the authors take a walkthrough in BCI technology. At first, they took a closer look into the kind of waves that are being generated by our brain (i.e., the EEG and ECoG waves). In the next section, they have discussed about patients affected by CLIS and ALS-CLIS and how they can be treated or be benefitted using BCI technology. Visually evoked potential-based BCI technology has also been

DOI: 10.4018/978-1-7998-4706-9.ch006

thoroughly discussed in this chapter. The application of machine learning and deep learning in this field are also being discussed with the need for feature engineering in this paradigm also been said. In the final section, they have done a thorough literature survey on various research-related to this field with proposed methodology and results.

INTRODUCTION:

Communication is the basic need for a human being to co-exist in a society. Man lives within technology and in the context of current technical scenario artificial intelligence seems to have a greater edge. In the Tech savvy and modern digital era, Artificial intelligence, robotics and automation industry is the rising sun but it is very difficult to understand the human brain. BCI is bridging the pathway between the brain and other external devices, and it is giving promising applications in the field of neuroprosthesis and neurorehabilitation as it is controlling the external devices straight from the patient's intention. With the help of BCI we can understand as well as interpret the different perceptive of human brain. Hearing-impaired or rather differently abled people often find difficulty to connect with their peers, thus they use sign languages to communicate with other people around them, but most of the people find it difficult to understand those things as they are not trained to recognize them (Wolfpaw et al., 2002). This leads to create a frustration among the challenged people who unable to express their feeling and a communication gap is also created among them. Thus it would be very helpful for the differently abled of our society meeting up the communication gap between them and the world.

ELECTROENCEPHALOGRAPHY WAVES

Electroencephalography waves, is the futuristic concept in the field of Brain-Computer Interface technique with a medical image processing background. It's defined as an electrical activity captured from the surface of the scalp with the help of metal electrodes and conducting medium. The analysis of EEG waves is giving new paths to in the field of neuroprosthesis and neurorehabilitation.

EEG is measured directly from the cortical surface (electrocardiogram) by using depth probes called electrograms. When there is a transfer of data from neuron to neuron a local current is being developed due to synaptic excitation of dendrites and that is being measured by EEG. All the recordable electrical activities is being generated in the human scalp region, the local current penetrates the skin, skull and other layers to be measured (Larsen, 2011). As they travel all the layers the current becomes weak to be recorded so it has to be amplified, and digitalized and then stored in the memory.

Brain Waves:

Brain waves are defined as some oscillating electrical voltages in the brain. The general amplitude of brain wave is around a millionth of a volt. Different regions of brain emit different frequencies and we need to put those electrodes in that manner. Majorly there are 5 types of brain waves as described in table -1, and figure 1 representing the waves.

Table 1. Different Brain waves and their characteristics

Frequency Band	Frequency (Hz)	Brain State
Gamma (γ)	>35 Hz	Concentration
Beta (β)	12 – 35 Hz	Anxiety Dominant, Active, External Attention, Relaxed
Alpha (α)	8 – 12 Hz	Very Relaxed
Theta (θ)	4 – 8 Hz	Deeply Relaxed, inward focused
Delta (δ)	0.5 – 4 Hz	Sleep

Source: Created by the Author, based on information

BCI Helmet Electrode Placement 10/20

The method of EEG wave capture is carried out through a BCI helmet and there is a proper fixed procedure to carry on the experiments. The Encephalographic measurements use a system which compromises namely,

- Electrode and Conductive medium
- Amplifier with Filters
- Analog-to-Digital (A/D) Converter
- Recording Device

The placement setups of the electrodes are being recommended by American EEG Society. The rule is to place odd number of electrode on the left scalp and the even number of electrode on the right side of the scalp. The distance between 2 adjacent electrodes should be either 10% or 20% distance of the scalp. Table 2 represents zones of brain and figure 2 visualizes the zones with electrodes used.

In case of a more detailed EEG wave result we can put some extra electrodes in the spaces covering the 10/20 system. A new nomenclature named as Modified Combinatorial Nomenclature (MCN) is there. It uses 1,3,5,7,9 in left hemisphere (% of the inion-to-nasion distance) and 2,4,6,8,10 in right hemisphere.

Table 3 represents different electrodes used for different activities capture.

Figure 1. Different Brain waves and their characteristics

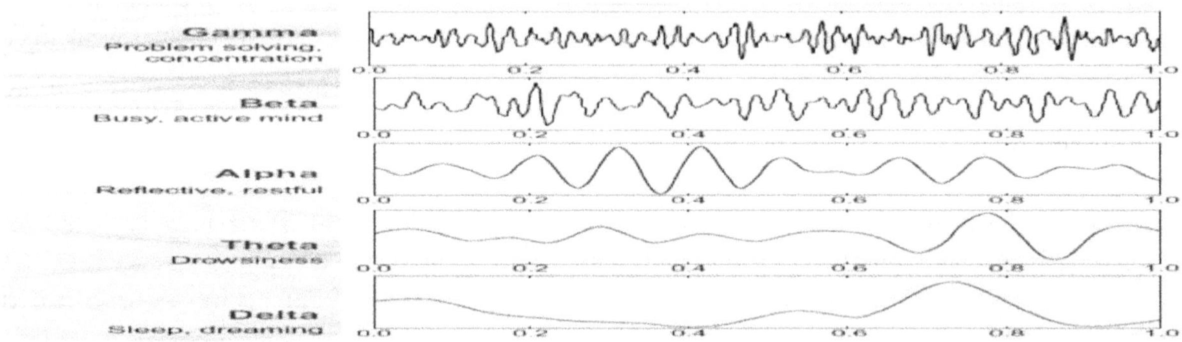

Table 2. Letter representation for different brain lobes/zones

Representation Letter	Zone
F	Frontal Region
C	Central Region
P	Parietal Region
T	Temporal Region
A	Anterior Region

Source: Created by the Author, based on information

Figure 2. Different Electrode Position in Brain

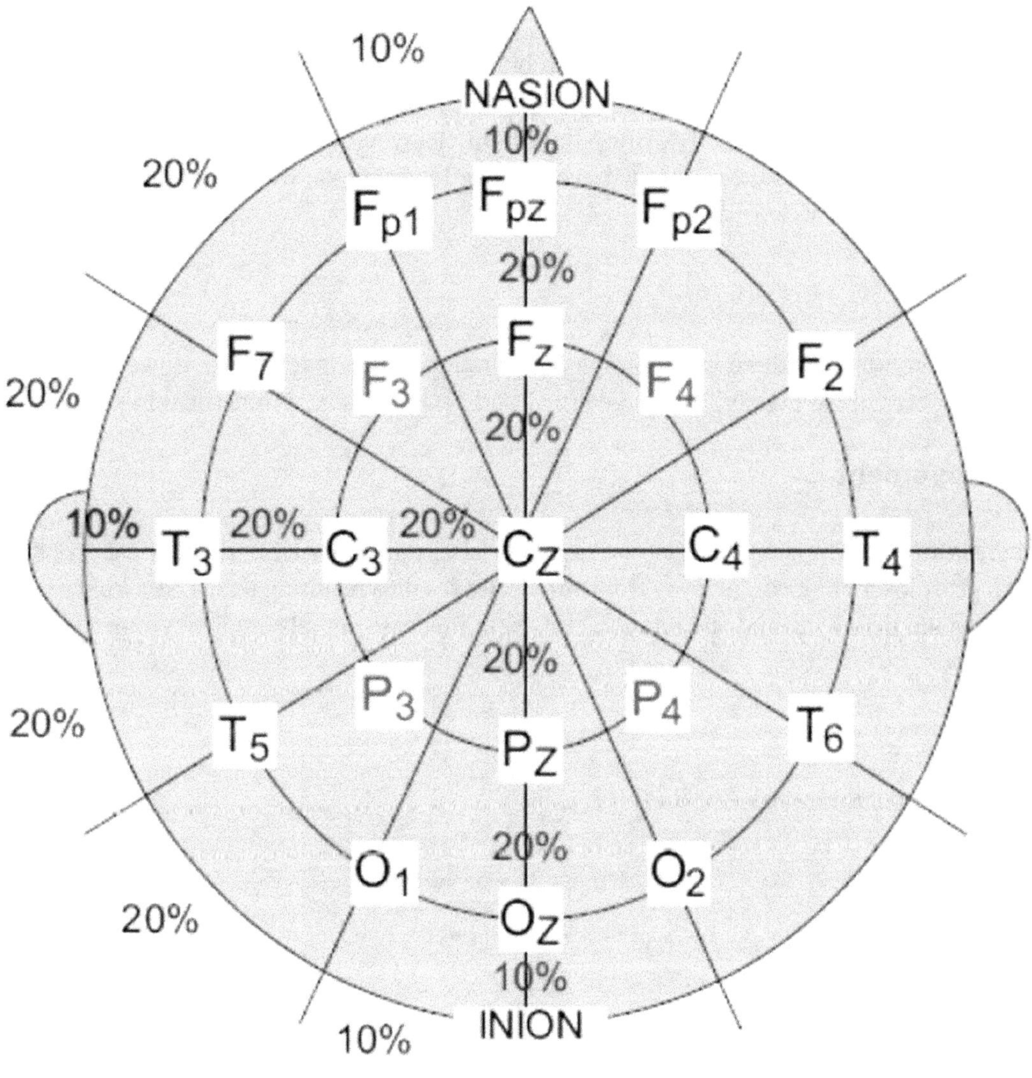

Table 3. Electrode and their Function abilities

Electrode	Zone	Function Captured
F7	Near Center	Rational Activities
Fz	Center	Intention and Motivational
F8	Center	Emotional Impulse
C3, C4, Cz	Cortex	Sensory and Motor
P3, P4, Pz	Posterior	Perception and Differentiation
T5, T6	-	Memory Functions
O1, O2	-	Primary Visual Areas

Source: Created by the Author, based on information

Artifacts

Artifacts are those signals which have no cerebral origin of the waves. The main origins of artifacts are namely, ocular, muscular, and mechanical. The subject related artifacts are those unwanted physiological signals that may significantly disturb the EEG (Zhang et al., 2019). Technical side of these can be eliminated by decreasing the electrode impedance or lowering the wire length. Some of the artifacts are being discussed below.

Eye Blink

In case of an eye blink there is a low signal generation which is relatively slower (<4 Hz). The signal mainly appears in the mainly frontal area (Fp1 and Fp2) and it is symmetrical in nature.

Eye Movement

If there is an eye movement then also we see a low EEG signal (<4 Hz). It's caused as eye makes a dipole which gets closer of moves far away from our electrodes thus resulting this. These kind of signals mainly appear in the mainly frontal and temporal lobes. In this case signals are not symmetrical.

Figure 3. Artifacts through Eye Blink

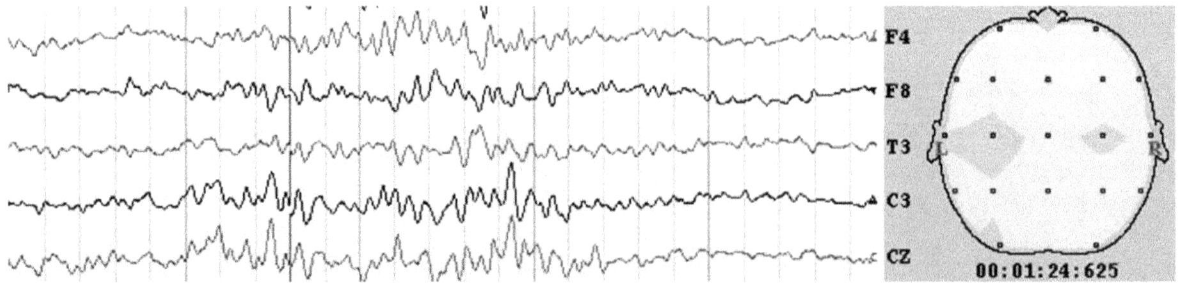

Muscular Artifacts

EEG signals related to muscular movements causes a high frequency signals (>13 Hz), it is generally much higher than the cerebral signals. The jaw muscle and frontal muscles appear in these cases (Nurse, 2015).

Speech Acquisition Using EEG Waves

Speech processing is basically the study of speech signals through processing. The speech are being transferred through digital representation so it can considered as the special case of digital signal processing.

Speech Processing would be easier if there were some simple linear relationships between articulation and acoustics, and in between acoustics and perception (Akram et al., 2013).

Uses of BCI

- Continuous monitoring of alertness, coma.
- Locate damage of head injury
- Monitor Cognitive
- Control Anesthesia depth
- Test Epilepsy drug effects

Mathematical Background Behind Brain Wave Signal

The spatially coherent EEG phenomena like the alpha rhythm wave can be generated using small number of isolated generators.

The spatial-temporal variation of head surface potential can be represented using Fourier series, Mathematical background of BCIthe spatial-temporal variation of heat surface potentials may be represented by the Fourier series

$$\varnothing\left(x,y,t\right)=\sum_{n=0}^{\infty}An\left(x,y\right)\sin\left(\omega nt+Qn\right) \quad (1)$$

Where,

x=x-coordinate of head surface
y=y-coordinate of head surface
$\Lambda n^2/\Delta\omega$ |wave =8-12 cycles/sec

The spatial distribution of potential over the head is equally as important as the temporal variation due to problem during calling reading the Fourier series can be expanded in a double Fourier series-

$$An\left(x,y\right)=\sum_{i=0}^{\infty}\sum_{n=0}^{\infty}Blmn\sin\left(K_{xl}+Q_{xl}\right)\sin\left(K_{ym}Y+Q_{ym}\right) \quad (2)$$

Frequency component of the EEG is made up of a number of waves of different spatial wavelengths

X- 2 π /Kxl
Y- 2 π /Kym

- Brain wave equation:

The instantaneousgross potential in the position r of our cordialsurface can be represented as-

$$\phi(r,t) \sim \phi_1 \eta_1 g(r,t) + \phi_0^+ + h^+(r,t) - \phi_0^- h^-(r,t) \tag{3}$$

Where,

ϕ_1 =local surface potential generated by firing of one cortical cell in dv
$\phi_0^{\pm}$ =local surface potential generated by single EPSP in dv
η_1 =count of density of cortical cells
$\eta_0^{\pm}$ =count of density of excitatory connection of part of cells located inside dv.
$g(r,t)$=the fraction of cells in dv that are firing at time t.
$h^{\pm}$ (r, t)= count of density of inhibitory synaptic firings.

$$\mathrm{I}\, I \int_{corticalsurface} d^2 r^1 \int_0^{\infty} f(e, | r - r^1) dc = \eta_0^{\pm} \tag{4}$$

This relation holds for both excitatory and inhibitory states
L - Volume elements

$$dv = 1 d^2 \mathrm{r}$$

The equation establishes relation between h (r, t) and g(r, t)
In some the cases long-length EEG phenomenon are consistent with linearity

$$B\eta_1 g(r,t) = h^+(r,t) - h^-(r,t) \tag{5}$$

From these we can conclude one statement,

$$No.ofcells\ firing\ at\ a\ giventime \propto Instantaneous\ difference\ between\ thenumberofexcitatory\ firings\ and\ inhibitory\ synaptic \tag{6}$$

$$h(r,t) \equiv h + (r,t) - h - (r,t) \tag{7}$$

To obtain the linearized of equation (5) we combine equation (4), (5), (7).

$$h(r,t) = h_0(r,t) + \frac{1}{B\Delta_1 g}\int_0^{\infty} .de \int_{corticalsurface} h\left(r', t - \frac{|r - r'|}{c}\right) . f_+\left(c, |r - r'|\right) - f_-\left(c, |r - r'|\right) d^2 r' \quad (8)$$

Equation 8 is defined as an inhomogeneous Fredholm equation of space-time variation of the synaptic firing density.

$$Retardedtime = t - \left(|r - r'| / c\right)$$

Equation (9) is formed on evidence of basal dendrites

$$f_{\pm}\left(c, |r - r'|\right) = \left(\frac{\Delta_0 \lambda_{\pm}^2}{2\pi t}\right) e^{\left(-\lambda_{\pm}|r-r'|F_{\pm}(c)\right)} \quad (9)$$

The coefficient is made/ chosen of normalization of equation (4).

The integral equation for the synaptic firing density of cortex, is given by the equation (10).

Fourier laplace transformation done for our feasibility

$$h\left(K_x, K_y, p\right) = \int_{-\infty}^{+\infty} dx \int_{-\infty}^{+\infty} dy \int_{-\infty}^{+\infty} dt.h\left(x, y, t\right) e^{\left(ik_x x + ik_y y - pt\right)} \quad (11)$$

Neglecting boundary value limitations,

$$CorticalSurfacearea \rightarrow \infty$$

$$h(k,t) = \frac{1}{2\pi i}\int_{D_r} h_0(k,p) H(k,p) e^{(pt)} dp \quad (12)$$

Where,

$h_0(k,p) \rightarrow$ Fourier-laplace transform

$h(k,\beta) \rightarrow$ Depends only on the properties of cortex.

$$h\left(K_x, K_y, p\right) = \sum_i R_i e^{\left[p_i(k)t\right]} \quad (13)$$

Here;

R_i is defined as residue of ith poles of $H(k, p)$.

$$p_i(k) = r_i(k) + i\omega_i(k) \tag{14}$$

$H(k,t)$ can move to some finite maximum at which non-linear effects invalidate.

$$mod(\omega_j) \gg mod(r_j) \tag{15}$$

Substituting the equation (12) in (11)

$$k_x = k$$

$$k_y = 0$$

$$\therefore k_x - k_y = k$$

$$\therefore \int_0^{\alpha} e^{(ikr\cos\theta)} d\theta = 2\pi J_0(kr) \tag{16}$$

$$\int_0^{\alpha} rJ_0(ar)e^{-br} dr = \frac{b}{(a^2+b^2)^{3/2}} R_e(b) > 0 \tag{17}$$

$$\left[H(k,p)\right]^{-1} = 1 - \frac{\eta_0^+}{\eta_1 . B} \int_0^{\alpha} \left(\lambda_+^3 c^3 + p\lambda_+^2 c^2\right) \times \left[\left(k^2+\lambda_+^2\right).c^2 + 2\lambda + pc + p^2\right]^{-\frac{3}{2}} .F_+(c)\,dc \; +$$
$$\frac{\eta_0^+}{\eta_1 . B} \int_0^{\alpha} (\lambda_-^3 c^3 + p\lambda_-^2 c^2) \times \left[\left(k^2+\lambda_-^2\right).c^2 + 2\lambda + pc + p^2\right]^{-\frac{3}{2}} .F_-(c)\,dc \tag{18}$$

The computational wave equation with few cortical wave equations is given as:

$$H(k,p)]^{-1} = 1 - \frac{\eta_0^+}{\eta_1 . B} \int_0^{\alpha} (\lambda_+^2 c^2 + p\lambda + c) \times \left[\left(k^2+\lambda_+^2\right).c^2 + 2\lambda + pc + p^2\right]^{-1} .F_+(c)\,dc + C \tag{19}$$

$F_{+}\left(c\right) \approx \delta\left(c - c_{0}^{\pm}\right)$ is taken valid for a cortical wave propagating is narrowly distributed about the characteristic velocities $c_{0}^{\pm}$

- Excitatory and inhibitory circuits:

$$\lambda c_{0} \equiv \lambda_{+} c_{0}^{+} \approx \lambda_{-} c_{0}^{-}$$

$$H\left(k, p\right) = \left[p^{2} + 2\lambda c_{0} p + \left(k^{2} + \lambda^{2}\right) c_{0}^{2}\right] \times \{p^{2} + 2\lambda c_{0}\left(1 - D\right) p + \left(k^{2} + \lambda^{2}\left(1 - 2\beta\right)\right) c_{0}^{2})\}^{-1}$$

Electrocorticography Waves

Electrocorticography (ECoG) is a kind of electrophysiological monitoring in which electrodes are placed directly on the exposed surface of the brain that is either above or below of the Dura mater but not within brain parenchyma, to record electrical activity from the cerebral cortex. ECoG has been recognized as a favorable platform for Brain Computer Interface (BCI) research and applications (Birbaumer, 2006).

Signal Acquisition

The electrodes are placed in the cortex of the brain but before that the surgeon performs craniotomy, removing a part of skull or opening a window to access the brain surface. The electrodes are placed guided by the results of preoperative EEG and magnetic resonance imaging (MRI). Muller, 2016)

These electrodes are of platinum, clinically having a diameter of 4mm (2.3 exposed). These electrodes are generally configured in form of grid or strips with an interelectrode distance of 10mm and implanted for a period of few days to 1-2 weeks. (Nam et al., 2013)

The subjects on whom the operation has been performed have had major neurosurgical procedure. The ability of the subject to cooperate in the study may differ from time to time and fluctuations and pain seizures may occur so the research team needs to be always ready to gather data when the subject is stable and willing to participate. (Muller et al., 2017)

Protocol Design of BCI

The protocol design of BCI depends on ECoG feature to be used for BCI control is chosen and the feature is used for BCI controlled cursor movement or required output (Botrel & Kübler, 2018).

In first protocol we need to choose the ECoG signal feature on which we want to focus for BCI control. This is similar to that of EEG or other single unit readings. For every single neuron recordings we get signal of ECoG which involves the production of frequency bands and locations. This shows a great difference in feature of ECoG when it is in motion or in rest. Most studies of ECoG had been focused on hand gesture or word articulations which it had been encoded. We have broad coverage of the ECoG feature due to the ready access to specific aspects of cognitive, motor and language function. Different

Figure 4. Brain Parts

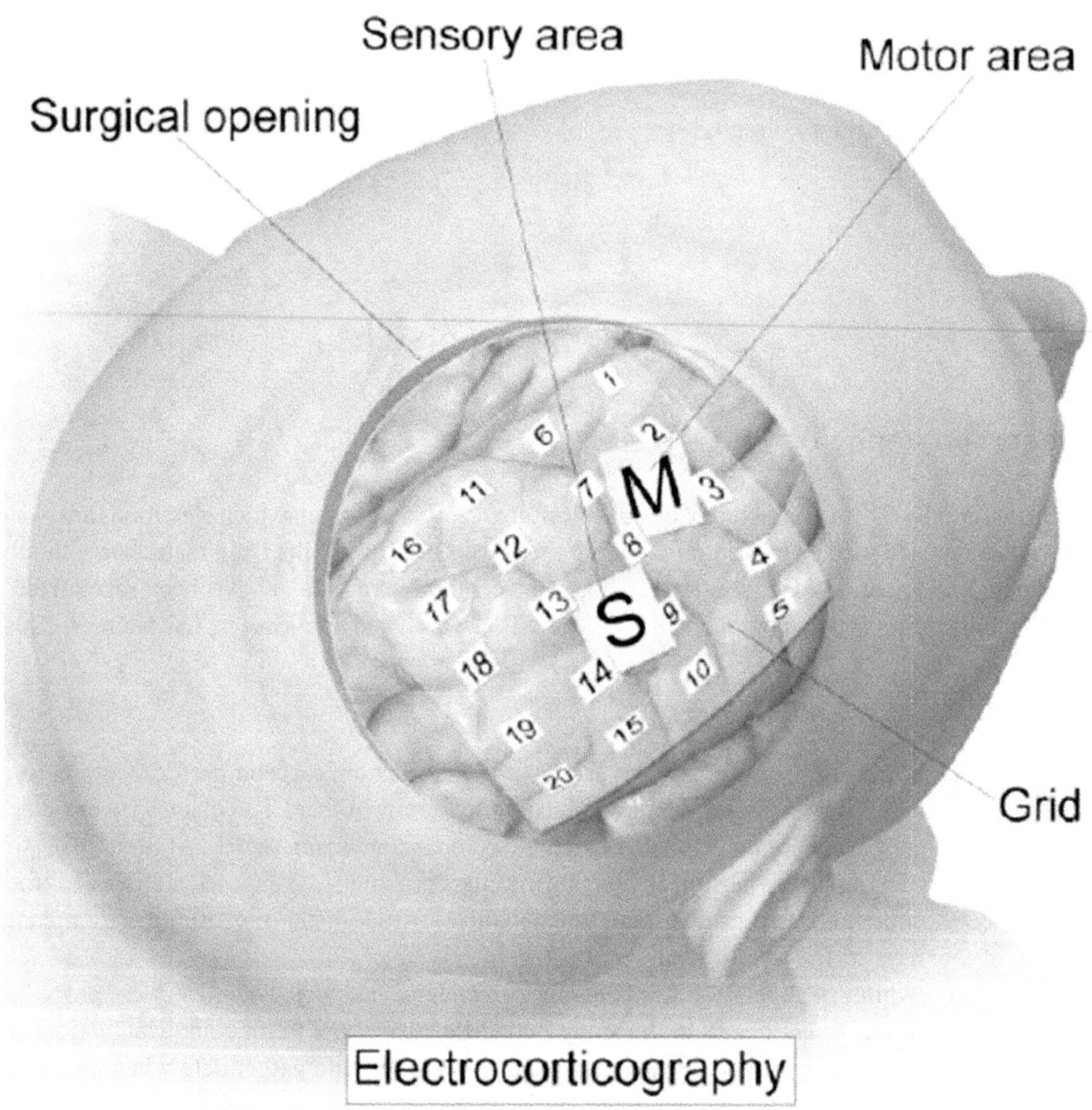

brain functions are related to different task performed and these produce different feature of ECoG which can be combined to control BCI. (Kaongoen & Jo, 2017)

In the second part of protocol the subject is trained to operate the BCI by feature of ECoG to perform a specific task as in cursor movement. In most of the work till date linear combination of one or more of the active feature controlled a particular task. The parameter of the linear transformation can be set or changed depending upon the data of the chosen feature.

Figure 5. ECoG control of vertical cursor movement using imagination of specific motor

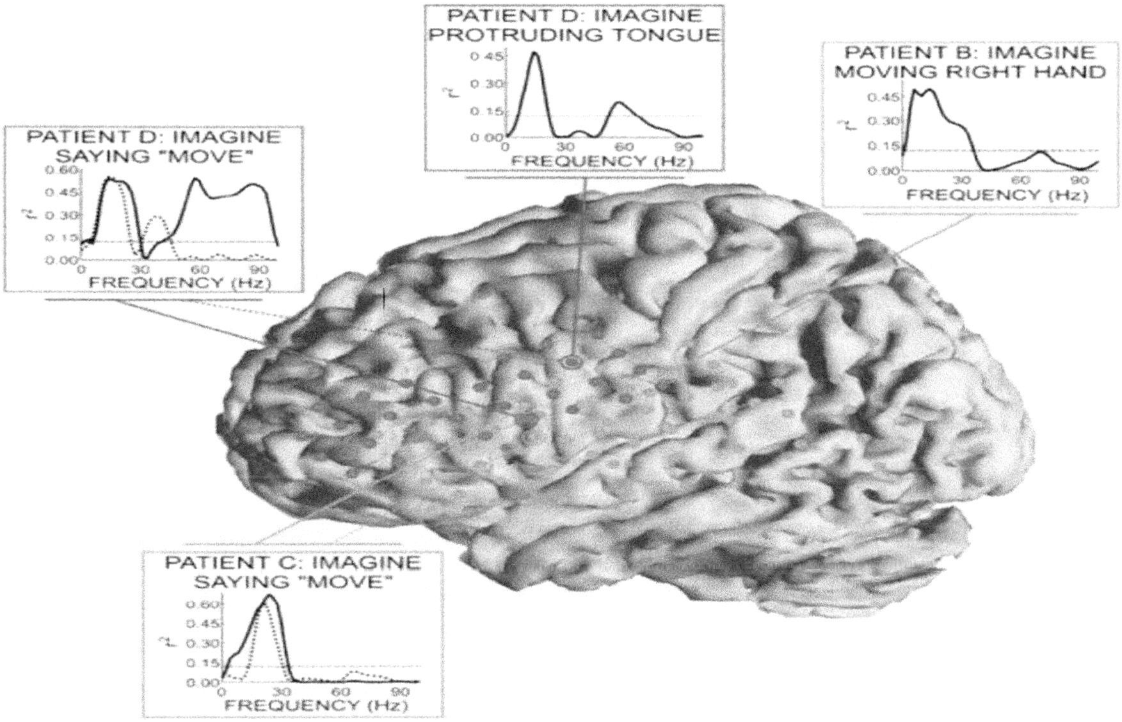

ECoG Based Multimodal BCI Control

The data are recorded for the ECoG feature change depending upon the task performed. The same task are performed by different subjects it produces different ECoG feature (Clerc et al., 2016). Depending upon the fig 3 over a training of 3-24 min the 4 subjects attain an online success rate of 75-100%. And correlation of two dimensional joystick movement the following accuracy was found depending upon the subject as shown in fig 4.

Visual Evoked Potential (VEP)

The fundamental unit of the nervous system of humans is neuron, which is a series of cells that can communicate with electrical signals. On stimulation of the human body, the neurons get excited and signals are sent to the brain. The stimulation can be of the following types:

- Visual
- Auditory
- Somatosensory

An example of visual stimulation can be when light bounces off an object and enter the eye, it excites the receptors in the eyes, which in turn sends signal to the brain. This pattern of electric potential of the

Figure 6. Learning curves for ECoG control of vertical cursor movement using motor imagery

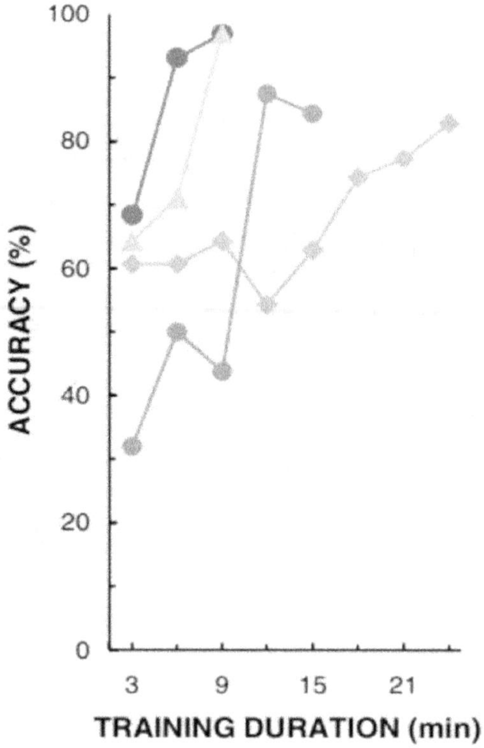

nervous system under stimulation is called Evoked Potential and under visual stimulus it is known as Visual Evoked Potential (VEP)

VEP based on BCI can be an useful tool to analyze the target on which the user is visually fixated with the help of concerted recording of EEG waves (Chun et al., 2016). A unique pattern is evoked on every unique stimulus sequence. VEPs collected from different stimulus sequence should be orthogonal or

Near-orthogonal to each other in some transform domain for reliable and better analysis of the target. The stimulus sequence poses a problem for analyzing the VEP based BCI. From the different stimulus sequence modulation approach used, the VEPs can be classified into:

- Time Modulated VEP (t-VEP).
- Frequency Modulated VEP (f-VEP).
- Pseudorandom Code Modulated VEP (c-VEP).

Figure 7. A graph of VEP in a human brain

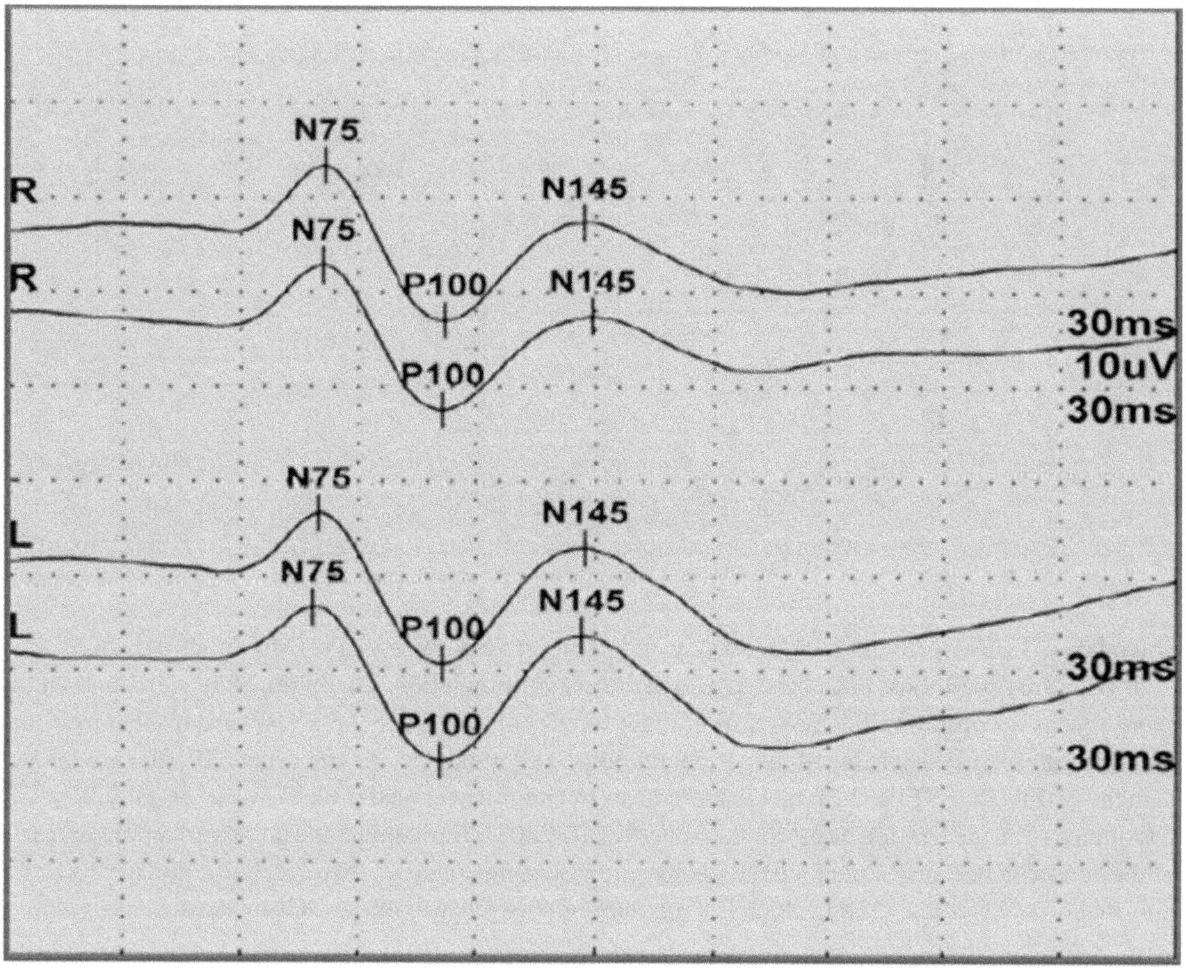

t-VEP Based BCI

The flash sequences of the different targets are mutually independent, which is achieved by randomization of the duration of ON and OFF states of each target's flash sequence. This evokes flash Visual Evoked

Figure 8. Diagram of VEP based BCI

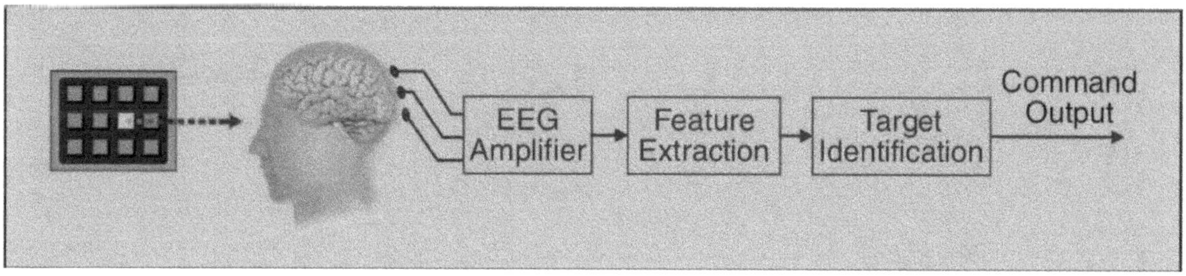

Figure 9. (a) Stimulus sequences of targets of a t-VEP based BCI, flashing mutually independently. (b) Evoked response to a single stimulus.

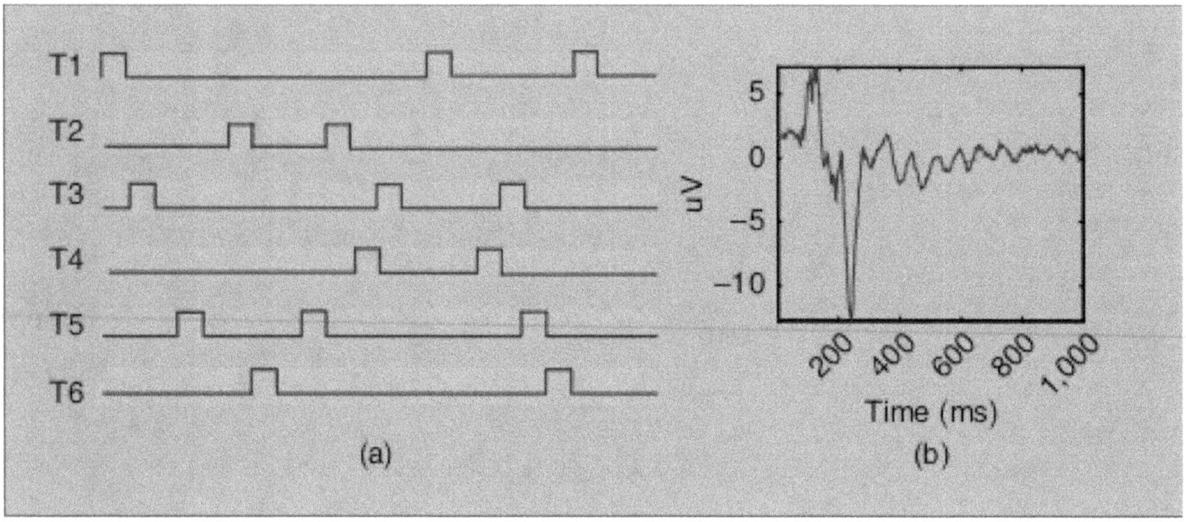

Potentials (FVEP) which has shorter latency and duration than other VEPs. (Muller, 2014)

FVEPs are time-locked and phase-locked to the visual stimulation. Therefore, after averaging out the short epochs according to flash onset of a target fixed will maximize the FVEPs from that target and minimize the FVEPs contributed by peripheral non-fixation targets. As the central FVEPs are greater than the peripheral FVEPs, the target that produces the largest averaged peak-to-valley amplitude can be considered to be the fixation target (Franceschetti et al., 2007). To maintain the non-overlapping scenario of the consecutive FVEPs, the t-VEP BCIs have low stimulus rate and therefore they also have a lower information transfer rate (ITR), which is a measure of performance of the BCI systems.

Figure 10. Diagram of t-VEP based BCI

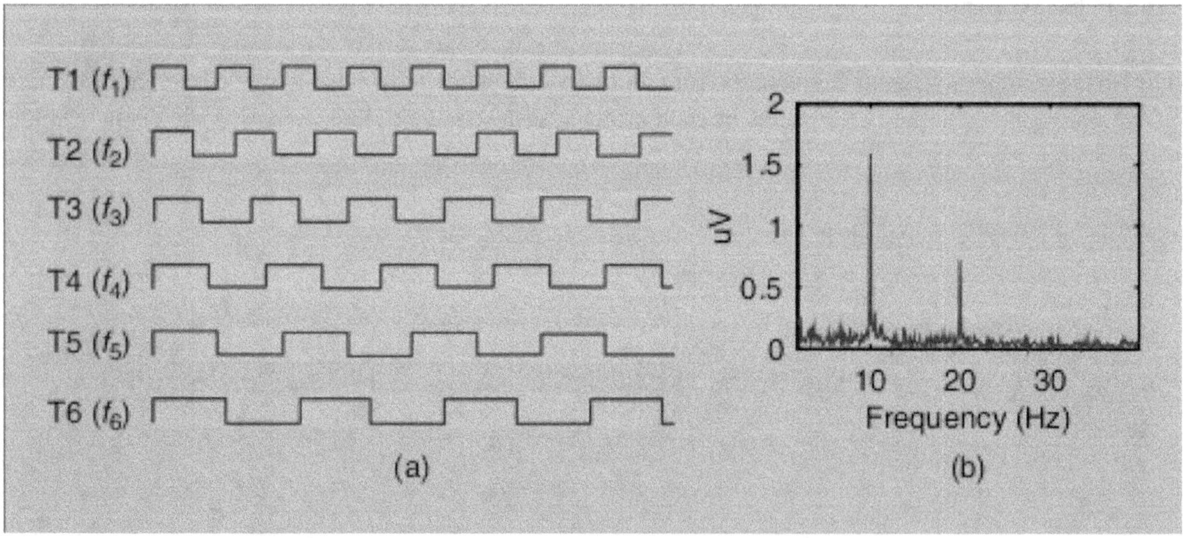

f-VEP Based BCI

In f-VEP based BCI, each target is flashed at the different frequencies and generating response which have same fundamental harmonics as that of flickered stimulus. Figure 3 shows the stimulus sequence of f-VEP based BCI and the power spectrum of an evoked response. (Mattia et al., 2016)

For target analysis of f-VEP based BCI power spectral analysis is generally used. From the EEG data collected, x obtained from n-target f-VEP BCI with flicker frequencies f_1, f2, f3, fn respectively, target analysis may be implemented through the following steps:

1. The EEG data (x) is collected and the power spectrum P(f) is calculated using Fast Fourier Transform (FFT) technique or other spectral analysis technique.
2. Calculation of signal-to-noise ratio (SNR), S_k which is the ratio of P(f_n) to the mean value of the corresponding frequency points. (Punaswal et al., 2017)
3. Selection of the fixation target K, corresponding to the maximum S_k.

Since the flicker frequency of the f-VEP BCI are usually >6Hz, the consecutive response from the target gets overlapped with each other, thus generating a periodic sequence of VEPs, known as Steady-state visual evoked potential (SSVEP), which is frequency-locked. Therefore, f-VEP BCIs are generally called SSVEP BCIs (Gao et al., 2014). A common problem in this system is the high user variation, which can be reduced with careful selection of channel location, stimulus frequency and speed of selection. From the reference of the results made out in *A Practical VEP-Based Brain-Computer Interface, Yijun Wang, Ruiping Wang, Xiaorong Gao, Bo Hong, Shangkai Gao, 2006,* the average ITR recorded to be about 43 bits/min. The results indicate that >90% of the population can be applied upon with a high ITR in living environments. Since ITR depends on the number of selections, accuracy and speed of selection, lower speed enhances the SNRs of SSVEPs, thus higher accuracy. An expectation of these two parameters must be considered to gain the highest intra-user ITR. A way to improve the SNR is optimization of the stimulus parameters such as flicker modulation depth, modulation wave and alternation color. Increasing the number of selections is also an approach to enhance the ITR. The amplitude of the SSVEP varies in a complex manner with the frequency of modulation. The amplitude peaks in the three regions known as subsystems- low-frequency, medium-frequency and high-frequency region. The factors that control the amplitude variation are namely- electrode position, luminance and flicker modulation depth. (Kim & Kim, 2018)

In Figure 4, the variations of amplitude vs. frequency response for three SSVEP subsystems are shown. The larger amplitude response falls in the lower frequency region.

A variation of lower amplitude response against background noise is also there as SNR decreases if background noise remains unchanged. But in higher frequency regions the noise decreases. In Figure 4(b), we can see that SNRs of the three subsystems are almost identical. SNR here, is the ratio of $y(f)$to the mean value of n adjacent points. (Lim & Ku, 2019)

Where, y is the amplitude spectrum calculated by a 1024-point FFT and f is the stimulus frequency. When the stimulation frequencies are in the medium-frequency (31-35 Hz) and high-frequency (39-45 Hz) region the detection accuracy is >95%.

The potential and efficiency of f-VEP BCIs has been demonstrated in many laboratories and clinics. The factors behind the robustness of f-VEP BCI are simple system configuration, no prerequisite knowledge of user, and high Information Transfer Rate (ITR). (Ramsey, 2014)

Figure 11. Stimulation Frequency Graph

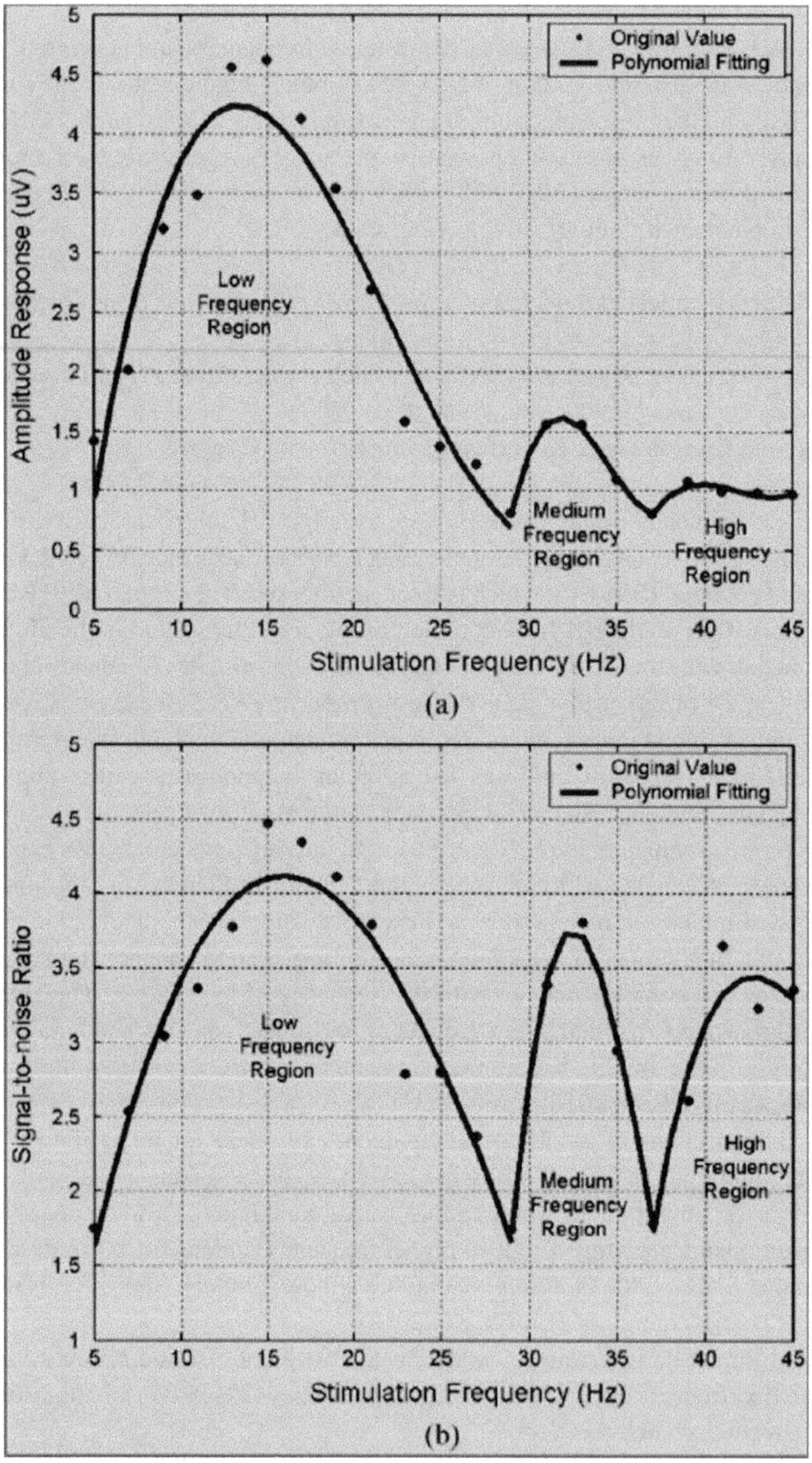

Figure 12. (a) Stimulus sequence for targets in a c-VEP BCI. (b) A waveform of the evoked response. (c) Power spectrum of the evoked response from a target. (d) Auto-correlation of the evoked response.

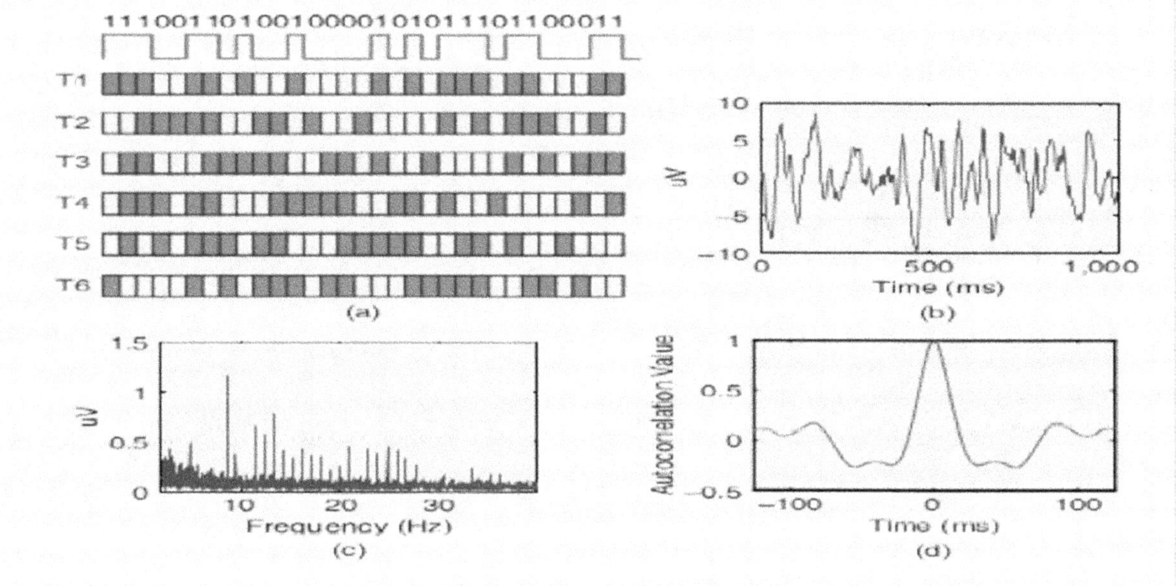

c-VEP Based BCI

In c-VEP BCI, pseudorandom sequences are used. The m-sequence is mostly used in this system. The binary m-sequence is created using maximal linear feedback shift registers which make them an invaluable tool in linear and nonlinear systems both (Grosse-Wentrup, 2019). The autocorrelation function of the m-sequence is very close to an unit impulse and nearly orthogonal to its time lag sequence. Figure 5 shows the stimulus sequence of the targets. (Lim et al., 2013)

Auditory Evoked Potential (AEP)

Auditory Evoked Potential or **AEP** directly translates to a substantial stimulus or electrical signal which is generated by the brain when proved by auditory signals. The AEP is basically represented in a time-locked manner when the stimulus is made and is one of the major components in the study of *Brain Computer Interface* or BCI.AEP consist of all audio related signals for example: amplitudes, reproducible negative or positive peaks. An auditory evoked potential's signal is comparatively smaller, amplitude wise than an EEP signals. The signals of an AEP can be demarcated into two parts known as (1) *transient state* and (2) *steady state*. The signals of AEP are radiated while preceding/receiving at a very slow pace to avoid overlapping of the stimulated signals is known as (1) transient state AEP. On the other hand, the process in which AEP signals are radiated while receiving at a very high speed which involves of the stimulated signals is known as (2) steady state AEP. (Sadouanov, 2018)

An individual's listening abilities level can be reflected by using signals of AEP. By a simple stimulation of a clicking sound into the ears of the personal being treated. These clicks having different concentration levels of specific frequencies example *20dB, 25dB, 30dB, 40dB, 50dB and 70dB*. By

Figure 13. AEP system based on hearing perception level detection. (HPL stands for Hearing perception level)

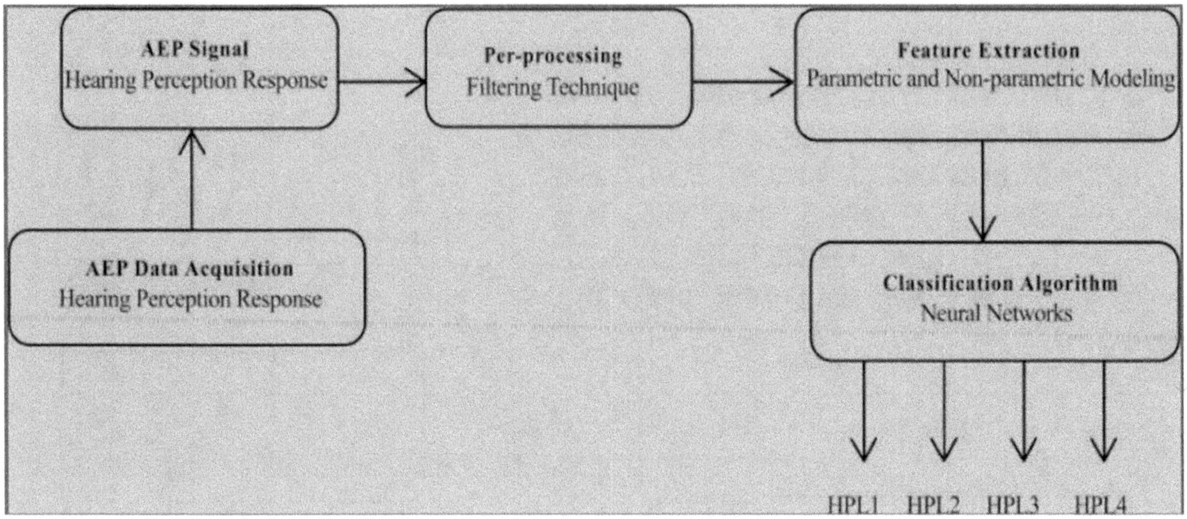

doing so we can establish a perception level of a person's hearing capabilities (Guan et al., 2017). On performing further research on the AEP signal stimulation, we can determine different listening insights of different personals both efficiently and with accuracy. The followed block dia. represents the system of determination of hearing:

To determine the early segments of an AEP the use of **ABR** is composed. ABRs are made up of several peaks or waves. These waves or peaks are usually described by using roman numerals from **I-VII.I**; **III** and **IV** are generally clinically more momentous. The ABR is mainly used for infant hearing screening as its diagnosis is done by testing the auditory functions and treating pathologies affecting brainstem pathways (Gwak et al., 2013).

The portions from 200-400 millisecond AEP signals are called **Mismatch Negative** (MMN). This MMN is provoked when two similar stimulated sounds followed by an aberrant sound passed. This recorder over the scalp and its response is one of the earliest receiving in AEPs. MMN is used by bursting number of feedbacks of responses used by BCI in understanding the processing of brain, formingbiological interface of auditory system and other forms of auditory related memories. Rita ceponiene *et al.*explored the MMN constituent of potential newborns and their focus towards different frequencies and sound durations. Neonates possess ordinary contrivances to observe and sense auditory prompt frequency and different duration contrivances or mechanisms as indexed by this process of MMN.

Clinical Importance of Peak V of ABR: Roberston *el at.* evaluated alternate ways for reckoning the range of ABR signals. The frequencies in mid and high ranges are recognized as more important for determination of underdevelopment potentials and threshold estimation. Rushaidin *et al.*assessed the peak of rapid energy of peak **V** of the ABR, differentiating normal and unusual hearing persons on the basis of the respected individual threshold values.

Figure 14. Normal ABR signal (presence of peak V visible)

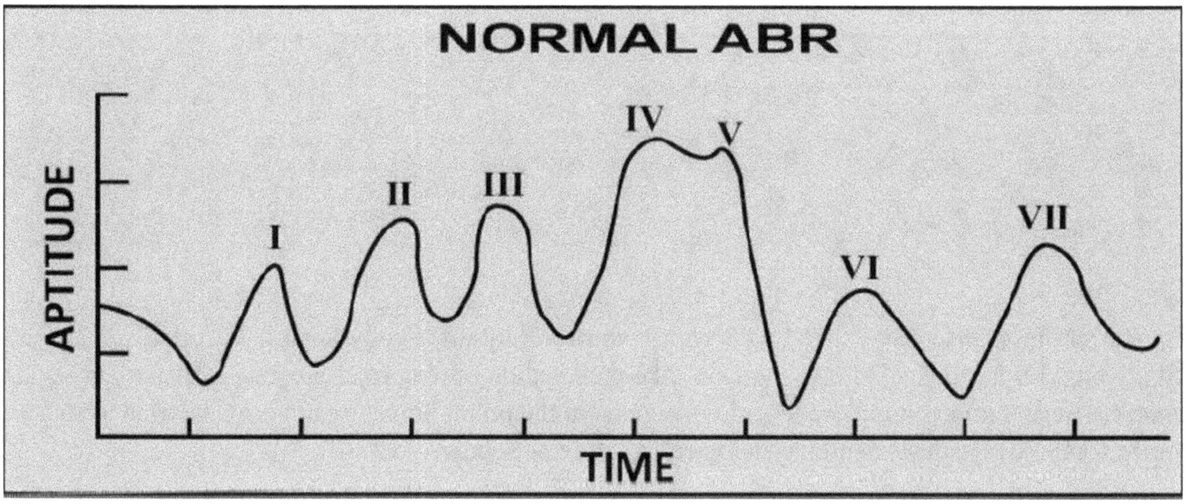

CLIS – ALS

One of the most important motives of BCI community is to communicate with the patients suffering from CLIS-ALS.

ALS – Amyotrophic Lateral Sclerosis

ALS, this disease can result in locked in syndrome (LIS) in which a patient is aware but cannot communicate due to complete paralysis of nearly all voluntary muscles in the body except for vertical eye movements and blinking. In CLIS eye muscles also get paralyzed (Han et al., 2019).

BCI wants to develop a system that does not instruct patients to perform any activity to indicate their intention but rather directly responses yes or no for that question. Using functional near-infrared spectroscopy (fNIS) as measurement modality a successful communication over several weeks with two CLIS patients was reported.

It can be said that the feasibility of the communication concept based on different basic assumptions. The first one is that mental decisions and reactions is examined, in a dimension that both exceeds and complements noticeable behavior, from the array of observable bioelectric signals, second assumption is that all meaningful EEG phenomenon should be viewed as complex structure of elementary wavelets, and create continuous flow of neurological messages. The third assumption is the operant conditioning procedure can increase the reliability and stability of these time signatures.

NEUROPHYSIOLOGIAL AND COGNITIVE CHANGES IN CLIS-ALS

Early, ALS was believed to be a disorder of nervous system that cause abnormal and involuntary movements but does not affect psychological processes.

Figure 15. Represents target of BCI
Source: Fliuidi

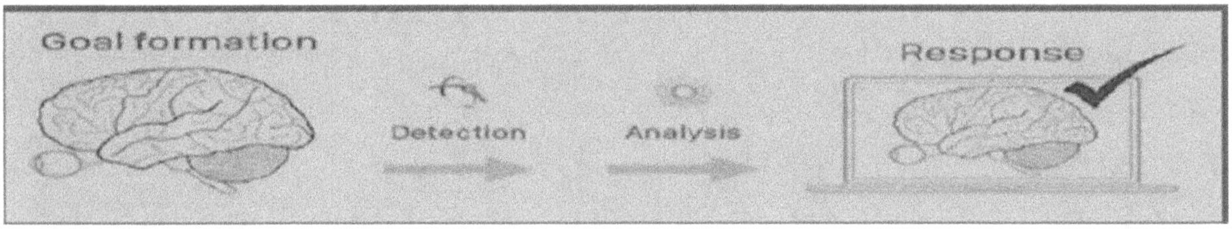

Current researches have linked ALS with both neurophysiologically as well as cognitive changes. Study on ALS has linked ALS to a prison- like propagation of abnormal proteins that starts in sensor motor areas but affects high level cognitive regions of the brain. Someresearchers have tried to characterize CLIS-ALS patients neurophysiologically. (Song & Sepulveda, 2017)

FUTURE DIRECTIONS FOR BCI IN CLIS-ALS

The hypothesis on the extinction of goal-directed thinking in CLIS-ALS is often considered binary - either CLIS-ALS patients hold the capacity for goal-directed thinking, or they do not. It is possible that CLIS-ALS patients drift in and out of awareness in a non-controllable fashion. If this assumption is correct, BCI-based communication may only be possible in short time windows of a few minutes at a time. To test this hypothesis, we need to construct a BCI system that continually monitors a patient and starts a communication attempt whenever it detects an activity pattern that indicates the patient is ready to communicate. (Song & Lee, 2018)

APPLICATION OF MACHINE LEARNING AND DEEP LEARNING IN BCI

Our target is to understand the electrophysiological signals and so with the help of *EEG* and *ECoG* waves, *fMRI, fNIRS,* measurement of the brain activity will take place.

There are mainly 2 methods Invasive and non-invasive. When the neurons communicate, exchange and passage of signals between them takes place and with the help of pair of electrode configuration of the BCI helmet is done and from the scalp of human brain the signals are noted by calculating the *action*

Figure 16. Taking input from scalp by following both Invasive and non-invasive methods

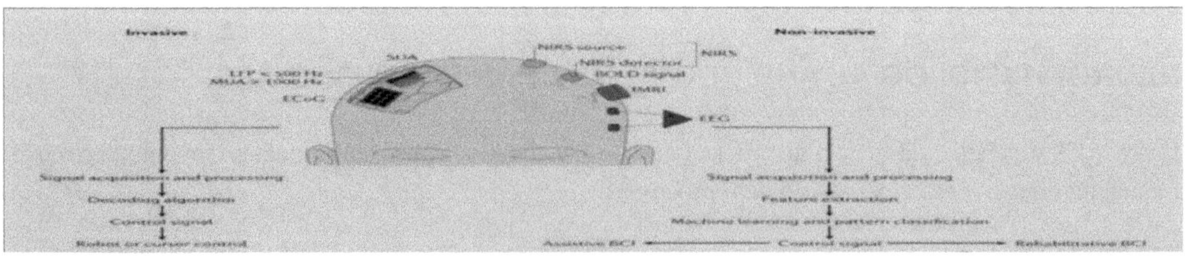

Figure 17. Reprents the pre-processing steps of BCI system
Source: Wolpawa et al., 2002

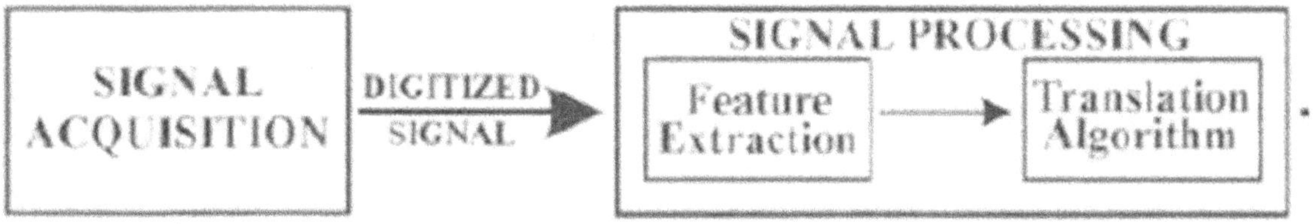

potentials and noting the week voltage gap(in μV) and then it is amplified for our research purpose (Kang, 2016).

For understanding the brain-activity by finding insights and creating pattern we take the help of different waves ($\alpha, \beta, \gamma, \delta, \theta$) which are having different frequency range. Also *neurofeedback* is a n important factor in determination of different factors

ML AND DL APPROACH

The embarkation of any model starts with data acquisition and pre-processing and feature extractions play an important role in building a perfect model.

With the different emotions and different actions signals having different frequencies are generated. By correlation of intentions of the user we can rank in the order of signal vector features. Since the different signals has different frequency so if we can give the frequency within a range we can apply regression models. But for regression we need continuous signal but as the signals varies so classification algorithm is suited here. We can apply Nearest neighbours, nonlinear Bayesian classifiers, SVM algorithms to categorise into different classes and giving a discrete output. But there is a problem:

Can this category of signal categorise the different actions of brain?

So for that we need a comparison of study of required signals and noise. So we have to calculate the *signal-to-noise ratio (SNR)* which can be written as:

$$SNR = \frac{P_{SIGNAL}}{P_{NOISE}} = \left(\frac{A_{SIGNAL}}{A_{NOISE}} \right)^2$$

Where $P = power\, of\, signal$ and $A = amplitude\, of\, signal$

So after considering the SNR factor it is observed that classification accuracy is low and also the generalisation ability of real world is not up to the mark. So we have to take the help of deep learning

Table 4. All the Literature Review

Author	Proposed Methodology	Results
Faraz Akram, Mohamed K. Metwally, Hee-Sok Han, Hyun-Jae Jeon, Tae-Seong Kim	A P300 Based BCI speller system by integrating a word suggestion mechanism.	Proposed methodology reduces typing time significantly, when tested produces an average words typing time of 1.66 minute whereas conventional took 2.9 minutes for the same words.
Loic Botrel, Andrea Kübler	Identification of reliable predictors of BCI performance	Suggested a relation between predictor and criterion based on correlation and regression analysis.
Jinsung Chun, Byeonguk Bae and Sungho Jo	A BCI based technique to control 3D objects in a three dimensional VR.	The proposed interface shows that the interface has similar or better performance than others and can be used as an alternative interface of VR.
Cuntai Guan, Neethu Robinson, Vikram Shenoy Handiru, Vinod A Prasad	Detection and tracking of multiple direction movements in EEG based BCI system	Classified the multiple movement directions, and achieved an average accuracy of 80.24% +/- 9.41 in discriminating the directions.
Kiuk Gwak, Robert Leeb, Jose del R. Millan, Dae-Shik Kim	A tactile stimulation system for BCI feedback by employing tactile illusion of movement to produce a continuous movement within six coin motors.	The results show that there are no identified artifacts in the EEG signal and no performance degradation be identified in case of online BCI experiments.
Xu Han, Shangen Zhang, Xiaorong Gao	A novel method to reduce the training time by replacing the traditional sinusoidal template or signal template with a dynamic SSVEP model and conducting a sampling training strategy.	The results shows the dynamic model based template outstripped the sinusoidal template and for signal template, the proposed method reduces the training time significantly while keeping the performance degradation within an insignificant range.
Eun Song Kang, Bum-Chae Kim, Heung-Il Suk	Empirical suggestion of resolving calibration session in Motor Imagery-based BCIs via collaborative learning.	Results suggest utilizing categorical information of training samples and sample of generic patterns for generalizing a BCI model.
Netiwit Kaongoen, Sungho Jo	Investigation of the effect of selective attention on the amplitude of Auditory Steady State Response (ASSR) in a binary-class BCI system.	The results show that the amplitude of ASSR is significantly increased by an approximate of 20% when the subject is selectively attended to the target stimulus.
Minju Kim, Sung-Phil Kim	The effect of artifact removal by different methods on the ERP waveform as well as BCI classification accuracy was investigated.	The results demonstrated that the ERP waveforms through ICA showed a less across-trial variability in P300 amplitudes compared to other methods as well as higher BCI classification accuracy.
Jeong-Hwan Lim, Han-Jeong Hwang, Chang-Hwan Im	A new paradigm for steady-state visual evoked potential (SSVEP) based BCI was proposed which can be used for disabled individuals with impaired oculomotor function. The proposed system allows users to express their binary intentions without the need to open their eyes.	The results showed an average ITR of the online experiments, reaching 10.83 bits/min, which demonstrated the feasibility of the BCI paradigm.
Chang Liu, Songyun Xie, Xinzhou Xie, Xu Duan, Wei Wang, Klaus Obermayer	A video feedback car control system based on SSVEP was designed and a study on an improved multiple signal classification (MUSIC) method was applied to classify the SSVEP signals to improve the performance of frequency-domain analysis.	An average online accuracy for four directions is about 87.5% and subjects could control the smart car by adjusting their distribution of the attention and drive the car through an obstacle fluently.
Batyrkhan Saduanov / Tohid Alizadeh, Jinung An, Berdakh Abibullaev	Perceiving and interaction with the world through the help of a robot controlled by a BCI system via telepresence for paralyzed people.	The real-time accuracy of P300-BCI was above 78% on average.
Minryung R. Song, Sang Wan Lee	Hippocampus-Striatum Network Inspired Architecture towards a flexible BCI.	The proposed methodology would afford us the leverage to measure latent states of the hippocampus-striatum functional network, thereby enabling robust recognition of users' intent in context-changing scenarios.

approaches. As the feature vectors have high dimensions so ANN ie Artificial Neural Network and due to high dimension, multilayer perceptron can be used. In terms of mathematics, due to *spatial frequency* obtained so study of sinusoidal waves can be done and Fourier Transformation can be applied for detailed calculation (Kaongoen & Jo, 2017).

So with the help of BCI we can at least somehow meet up the communication gap leading a friendly environment for differently abled people gives them different privilege and establishment in the society creating a heavenly world.

LITERATURE REVIEW

CONCLUSION

In this paper we have discussed the major pillars of BCI technology varying from EEG waves, ECoG waves to CLIS Patient treatment. We have also discussed little research work which have been done on this field.

REFERENCES

Akram, F., Metwally, M. K., Han, H. S., Jeon, H. J., & Kim, T. S. (2013, February). A novel P300-based BCI system for words typing. In *2013 International Winter Workshop on Brain-Computer Interface (BCI)* (pp. 24-25). IEEE.

Birbaumer, N. (2006). Breaking the silence: Brain–computer interfaces (BCI) for communication and motor control. *Psychophysiology*, *43*(6), 517–532. doi:10.1111/j.1469-8986.2006.00456.x PMID:17076808

Botrel, L., & Kübler, A. (2018, January). Reliable predictors of SMR BCI performance—Do they exist? In *2018 6th International Conference on Brain-Computer Interface (BCI)* (pp. 1-3). IEEE.

Chun, J., Bae, B., & Jo, S. (2016, February). BCI based hybrid interface for 3D object control in virtual reality. In *2016 4th International Winter Conference on Brain-Computer Interface (BCI)* (pp. 1-4). IEEE.

Clerc, M., Bougrain, L., & Lotte, F. (Eds.). (2016). Brain-computer interfaces. Academic Press.

Franceschetti, M., Dousse, O., David, N. C., & Thiran, P. (2007). Closing the gap in the capacity of wireless networks via percolation theory. *IEEE Transactions on Information Theory*, *53*(3), 1009–1018. doi:10.1109/TIT.2006.890791

Gao, S., Wang, Y., Gao, X., & Hong, B. (2014). Visual and auditory brain–computer interfaces. *IEEE Transactions on Biomedical Engineering*, *61*(5), 1436–1447. doi:10.1109/TBME.2014.2300164 PMID:24759277

Grosse-Wentrup, M. (2019, February). The elusive goal of BCI-based communication with CLIS-ALS patients. In *2019 7th International Winter Conference on Brain-Computer Interface (BCI)* (pp. 1-2). IEEE.

Guan, C., Robinson, N., Handiru, V. S., & Prasad, V. A. (2017, January). Detecting and tracking multiple directional movements in EEG based BCI. In *2017 5th International Winter Conference on Brain-Computer Interface (BCI)* (pp. 44-45). IEEE. 10.1109/IWW-BCI.2017.7858154

Gwak, K., Leeb, R., Millán, J. D. R., & Kim, D. S. (2013, February). A novel tactile stimulation system for BCI feedback. In *2013 International Winter Workshop on Brain-Computer Interface (BCI)* (pp. 29-31). IEEE. 10.1109/IWW-BCI.2013.6506619

Han, X., Zhang, S., & Gao, X. (2019, February). A study on reducing training time of BCI system based on an SSVEP dynamic model. In *2019 7th International Winter Conference on Brain-Computer Interface (BCI)* (pp. 1-2). IEEE. 10.1109/IWW-BCI.2019.8737318

Kang, E. S., Kim, B. C., & Suk, H. I. (2016, February). An empirical suggestion for collaborative learning in motor imagery-based BCIs. In *2016 4th International Winter Conference on Brain-Computer Interface (BCI)* (pp. 1-3). IEEE. 10.1109/IWW-BCI.2016.7457450

Kaongoen, N., & Jo, S. (2017, January). The effect of selective attention on multiple ASSRs for future BCI application. In *2017 5th International Winter Conference on Brain-Computer Interface (BCI)* (pp. 9-12). IEEE. 10.1109/IWW-BCI.2017.7858144

Kim, M., & Kim, S. P. (2018, January). A comparsion of artifact rejection methods for a BCI using event related potentials. In *2018 6th International Conference on Brain-Computer Interface (BCI)* (pp. 1-4). IEEE. 10.1109/IWW-BCI.2018.8311530

Kobayashi, N., & Nakagawa, M. (2018). BCI-based control of electric wheelchair using fractal characteristics of EEG. *IEEJ Transactions on Electrical and Electronic Engineering*, *13*(12), 1795–1803. doi:10.1002/tee.22742

Larsen, E. A. (2011). *Classification of EEG signals in a brain-computer interface system* (Master's thesis). Institutt for datateknikkoginformasjonsvitenskap.

Lim, H., & Ku, J. (2019, February). High engagement in BCI action observation game by relevant character's movement. In *2019 7th International Winter Conference on Brain-Computer Interface (BCI)* (pp. 1-3). IEEE. 10.1109/IWW-BCI.2019.8737252

Lim, J. H., Hwang, H. J., & Im, C. H. (2013, February). "Eyes-closed" SSVEP-based BCI for binary communication of individuals with impaired oculomotor function. In *2013 International Winter Workshop on Brain-Computer Interface (BCI)* (pp. 79-80). IEEE. 10.1109/IWW-BCI.2013.6506637

Liu, C., Xie, S., Xie, X., Duan, X., Wang, W., & Obermayer, K. (2018, January). Design of a video feedback SSVEP-BCI system for car control based on improved MUSIC method. In *2018 6th International Conference on Brain-Computer Interface (BCI)* (pp. 1-4). IEEE. 10.1109/IWW-BCI.2018.8311499

Mattia, D., Astolfi, L., Toppi, J., Petti, M., Pichiorri, F., & Cincotti, F. (2016, February). Interfacing brain and computer in neurorehabilitation. In *2016 4th International Winter Conference on Brain-Computer Interface (BCI)* (pp. 1-2). IEEE. 10.1109/IWW-BCI.2016.7457446

Müller, K. R. (2014, February). Multimodal imaging, non-stationarity and BCI. In *2014 International Winter Workshop on Brain-Computer Interface (BCI)* (pp. 1-1). IEEE.

Müller, K. R. (2016, February). Machine learning for BCI: towards analysing cognition. In *2016 4th International Winter Conference on Brain-Computer Interface (BCI)* (pp. 1-2). IEEE. 10.1109/IWW-BCI.2016.7457453

Müller-Putz, G. R., Ofner, P., Schwarz, A., Pereira, J., Pinegger, A., Dias, C. L., . . . Sburlea, A. I. (2017, January). Towards non-invasive EEG-based arm/hand-control in users with spinal cord injury. In *2017 5th International Winter Conference on Brain-Computer Interface (BCI)* (pp. 63-65). IEEE.

Nam, S., Kim, K. H., & Kim, D. S. (2013, February). Motor trajectory decoding based on fMRI-based BCI—A simulation study. In *2013 International Winter Workshop on Brain-Computer Interface (BCI)* (pp. 89-91). IEEE. 10.1109/IWW-BCI.2013.6506641

Nurse, E. S., Karoly, P. J., Grayden, D. B., & Freestone, D. R. (2015). A generalizable brain-computer interface (BCI) using machine learning for feature discovery. *PLoS One*, *10*(6), e0131328. doi:10.1371/journal.pone.0131328 PMID:26114954

Punsawad, Y., & Wongsawat, Y. (2017). Multi-command SSAEP-based BCI system with training sessions for SSVEP during an eye fatigue state. *IEEJ Transactions on Electrical and Electronic Engineering*, *12*, S72–S78. doi:10.1002/tee.22441

Ramsey, N. F. (2014, February). Exploration of the brain for optimal placement of BCI implants in paralyzed people. In *2014 International Winter Workshop on Brain-Computer Interface (BCI)* (pp. 1-3). IEEE. 10.1109/iww-BCI.2014.6782543

Saduanov, B., Alizadeh, T., An, J., & Abibullaev, B. (2018, January). Trained by demonstration humanoid robot controlled via a BCI system for telepresence. In *2018 6th International Conference on Brain-Computer Interface (BCI)* (pp. 1-4). IEEE. 10.1109/IWW-BCI.2018.8311508

Song, M. R., & Lee, S. W. (2018, January). Meta BCI: Hippocampus-striatum network inspired architecture towards flexible BCI. In *2018 6th International Conference on Brain-Computer Interface (BCI)* (pp. 1-3). IEEE.

Song, Y., & Sepulveda, F. (2017, January). An online self-paced brain-computer interface onset detection based on sound-production imagery applied to real-life scenarios. In *2017 5th International Winter Conference on Brain-Computer Interface (BCI)* (pp. 46-49). IEEE. 10.1109/IWW-BCI.2017.7858155

Wolpaw, J. R., Birbaumer, N., McFarland, D. J., Pfurtscheller, G., & Vaughan, T. M. (2002). Brain–computer interfaces for communication and control. *Clinical Neurophysiology*, *113*(6), 767–791. doi:10.1016/S1388-2457(02)00057-3 PMID:12048038

Zhang, X., Yao, L., Wang, X., Monaghan, J., Mcalpine, D., & Zhang, Y. (2019). *A survey on deep learning based brain computer interface: Recent advances and new frontiers*. arXiv preprint arXiv:1905.04149

Chapter 7
A Fusion-Based Approach to Generate and Classify Synthetic Cancer Cell Image Using DCGAN and CNN Architecture

Ahan Chatterjee
https://orcid.org/0000-0001-5217-4457
The Neotia University, India

Swagatam Roy
https://orcid.org/0000-0002-8012-5529
The Neotia University, India

ABSTRACT

The most talked about disease of our era, cancer, has taken many lives, and most of them are due to late prognosis. Statistical data shows around 10 million people lose their lives per year due to cancer globally. With every passing year, the malignant cancer cells are evolving at a rapid pace. The cancer cells are mutating with time, and it's becoming much more dangerous than before. In the chapter, the authors propose a DCGAN-based neural net architecture that will generate synthetic blood cancer cell images from fed data. The images, which will be generated, don't exist but can be formed in the near future due to constant mutation of the virus. Afterwards, the synthetic image is passes through a CNN net architecture which will predict the output class of the synthetic image. The novelty in this chapter is that it will generate some cancer cell images that can be generated after mutation, and it will predict the class of the image, whether it's malignant or benign through the proposed CNN architecture.

DOI: 10.4018/978-1-7998-4706-9.ch007

INTRODUCTION

One of the most discussed diseases of the 21st Century is none other than Cancer. As per statistics per year around 10 million people pass away globally due to cancer. (Chatterjee et al. 2020) Detection of cancer cell in the early stage is plays as one of the key factor in increased survival rate.(Yang et al.,2017) Application of Deep Learning is brining paradigm shift in medical image prognosis. (Rubin et al., 2019) There is a shortage of labeled data in the field of bio-medical, thus the importance data augmentation increases. (Perez at al. 2017) Along with the generated image data, the classification of those data correctly is another important task as with the early detection of malignant cell we can act quickly and better in the prognosis to save one's life. The arrival of Deep Neural Network (DNN) has brought dynamic changes in the field by improving algorithms by leaps and bounds. (Xie et al., 2015)

One of the most promising approaches of image synthesis is Generative Adversarial Networks (GAN). Data Augmentation becomes easier with the applicability of GAN as it's capable of creating high quality realistic image from the training dataset. Along, with the data generation we need to classify those generated images into malignant and benign class for prognosis, and for this we use Convolutional Neural Network (CNN). CNN is highly capable to classify image based on the patterns and features present in an image. (Kitrungrotsakul et al., 2019) In this paper, our study focuses on generating new cancer cell using GAN, which doesn't exist currently but it may appear in human body as a malignant cell due to constant mutation of cancer cells. (Chen et al. 2014) As the cancer cell is constantly mutating it's becoming more dangerous than before, thus detection at early stage becomes more important. (Zhang et al. 2016). Then the generated image is being tested through a CNN Architecture, to classify those cells into malignant and benign. In this way we create a data library for new cancer cells and classify them into classes for future prognosis.

The paper is structured as; Sect II. contains literature review, Sect III. contains our proposed methodology and model architecture, Sect IV. contains results, and analysis on that backdrop, and Sect V. as concluding remark with future scope of study.

1. Literature Review

Shin et al. used Image to Image Condition in GAN to generate the synthesized data to classify T1 brain tumor class on ADNI dataset. They showed it can increase accuracy of classifier if trained in GAN images rather than original dataset. Iqbal et al. proposes an innovative method of medical imaging using GAN (MI-GAN) to generate retinal images. The results show the newly formed image contains the structure from original image. Senaras et al. proposed a conditional GAN technique (cGAN) to synthesize. Another work by Mahapatra et al. shows to generate a high resolution image from a low resolution image through proposed model of P-GAN. Another work suggested implementing GAN for hyper spectral images, authored by Zhu et al. Fosto Kamga guy et al. proposed to use transfer learning algorithms such as VGG16, Alexnet, InceptionV3 to create a fusion-schema model to generate images. Shang et al. computed the missing data problem in a dataset using Cyclic-GAN technique. Gurumurthy et al. proposed a DeLiGAN method to solve problems arise due to limited dataset for training. Premchand and Dutt proposed a methodology through which GAN can be implemented for speech denoising. One of the major challenges in implementing CNN model is the requirement of cleaned required data, thus to solve synthetic images are taken.

PROPOSED METHODOLOGY

In current date also there is an acute shortage of medical image data through which we can carry out proper classification of cancer cells. (Pan et al., 2015) In this section we propose a CNN based architecture which can classify synthetic image which has been formed using GAN architecture. The primary research findings of our proposed algorithm are as follows:

1. Generation of synthetic Blood Cancer Cell Image using proposed GAN Architecture
2. Synthesize and validate the missing modalities in the synthetic generated images.
3. Classifying the synthetic images using CNN Architecture.

Generative Adversarial Network (GAN)

The introduction of Deep Neural Network led to a paradigm shift in the applicability of AI in various fields. There is a massive advancement in the algorithm design due to increased accuracy of neural networks. Generative Adversarial Network is such one classic example of this. This algorithm is capable of generating synthetic images which simply doesn't exist but looks totally realistic. The GAN is based on the Game Theory approach where there are 2 players here, 2 neural networks which will try to optimize each other's result until the synthetic images has similar features with the original training data. GAN works on Zero-Sum principle, and has 2 blocks to generate the synthetic image namely generator and discriminator. (Thuy et al. 2019)

- **Generator:** Generator or the Generative neural network is mainly responsible for creating synthetic image with the goal to get undetected. It generates the image without training the features of the image of the input dataset, i.e. without learning semantics of the input image data.
- **Discriminator:** The discriminator neural network, learns to classify that the given sample is from the same data distribution or not. The major goal of a discriminator network is to detect the fake content in the set. It's basically a classifier network which classifies whether the image is real or not.(Chaudhari et al. 2019)

The GAN model is based on two separate convo neural net architecture. In our proposed architecture we have used Deep Convolutional GAN (DCGAN). The deep generative models are equipped to act against the backdrop of the difficulty in approximated computing, which is generated in Maximum Likelihood estimation.(Rubin et al., 2019) Thus training the both neural networks give us an upper hand to generate better results. The basic architecture of GAN is showed in figure 1.

The maximization of D is represented in equation 1. The function should be maximized as it will give decision over real data.(Han et al. 2018)

$$E_{x \sim P_r(x)}\left[\log D\left(x\right)\right] \tag{1}$$

e fake sample function of $G\left(z\right)$ is represented in Eq. 2.

Figure 1. Architecture of GAN

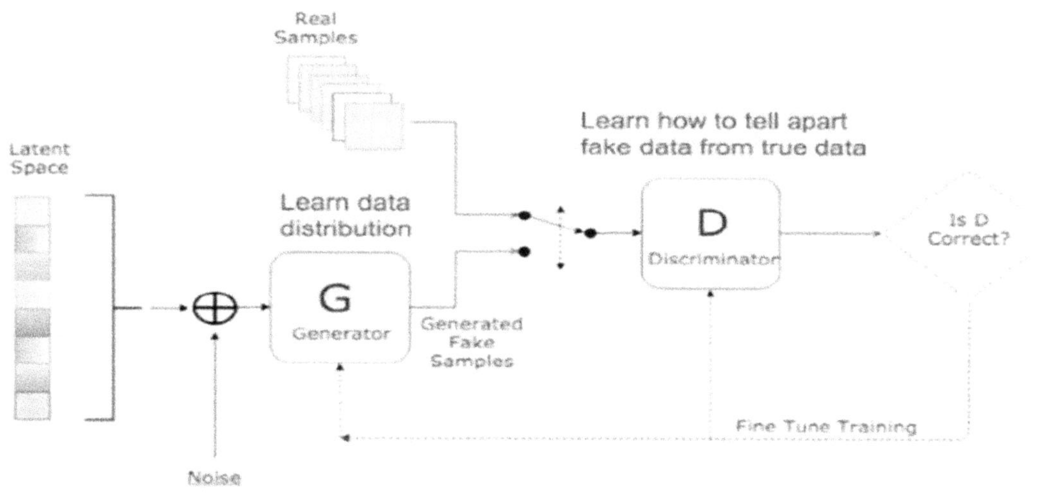

$$G(z) = z \sim p_z(z) \tag{2}$$

D will return an output with a probability $D(G(z))$ of close to 0 in order to maximize the function given in Eq. 3.

$$E_{z \sim p_z(z)} [\log(1 - D(G(z)))] \tag{3}$$

Table 1. Denotes meaning of variable

Symbol	Meaning
P_z	Data distribution over noise z
P_g	The generator distribution over data x
P_r	Data distribution over real sample x
G.	Generator
D	Discriminator

Source: Created by Author

Similarly on the other hand we have architecture of G where the function should be minimized in order to get the subsequent result.(Mani et al. 2001) The minimization formula is represented in Eq. 4.

$$E_{z\sim p_z(z)}[\log(1-D(G(z)))] \tag{4}$$

Thus combining the both parameters together e design the GAN network architecture as whole. (Rashid et al. 2019) The both parameters play a minmax game in which we have to optimize the function to get optimal result shown in Eq. 5

$$min_G\, max_D\, L(D,G)=E_{x\sim P_r(x)}[\log D(x)]+E_{z\sim p_z(z)}[\log(1-D(G(z)))] \tag{5}$$

Solving the equation:

$$min_G\, max_D\, L(D,G)==E_{x\sim P_r(x)}[\log D(x)]+E_{x\sim P_g(x)}[\log(1-D(x))] \tag{6}$$

The term $E_{x\sim P_r(x)}[\log D(x)]$ don't have any impact on the result of G during gradient descent. (Salehinejad, et al 2018)

The loss function should be optimized in order to get the best results. Thus optimizing the loss function for D.

$$L(G,D)=\int_x (P_r(x)\log(D(x))+P_g(x)\log(1-D(x)))\,dx \tag{7}$$

We aim to optimize the value of $D(x)$ thus maximizing $L(G,D)$

$$\hat{x}=D(x),\ A=P_r(x),\quad B=P_g(x)$$

Substituting and calculating the values in the integral

$$f(\hat{x})=Alog\hat{x}+Blog(1-\hat{x}) \tag{8}$$

$$\frac{df(\hat{x})}{d\hat{x}}=A\frac{1}{\ln 10}-B\frac{1}{\ln 10}\frac{1}{1-\hat{x}} \tag{9}$$

$$\frac{df\left(\hat{x}\right)}{d\hat{x}}=\frac{1}{\ln 10}\left(\frac{A}{\hat{x}}-\frac{B}{1-\hat{x}}\right) \tag{10}$$

$$\frac{df\left(\hat{x}\right)}{d\hat{x}}=\frac{A-\left(A+B\right)\hat{x}}{\hat{x}\left(1-\hat{x}\right)} \tag{11}$$

Now, equating $\frac{df\left(\hat{x}\right)}{d\hat{x}}=0$ we will get the optimized value for D

$$D_o\left(x\right)=\frac{P_r\left(x\right)}{P_r\left(x\right)+P_g\left(x\right)}\in\left[0,1\right] \tag{12}$$

When both of our architecture are optimized we reach a state where $P_r=P_g$ and $D_o\left(x\right)=1/2$ The new loss function is defined in the Eq. 15

$$L\left(G,D_o\left(x\right)\right)=\int_x (P_r\left(x\right)\log\left(D_o\left(x\right)\right)+P_g\left(x\right)\log\left(1-D_o\left(x\right)\right))\,dx \tag{13}$$

$$L\left(G,D_o\left(x\right)\right)=\log\frac{1}{2}\int_x P_r\left(x\right)dx+\log\frac{1}{2}\int_x P_g\left(x\right)dx \tag{14}$$

$$L\left(G,D_o\left(x\right)\right)=-2log2 \tag{15}$$

is is the optimized function of our GAN model, the model is arranged with Deep Layers which has been elaborately discussed in section 3.2, the model architecture which we have implemented along with the our proposed algorithm.

Convolutional Neural Network (CNN)

The Convolutional Neural Network or CNN is a neural network architecture generally used to classify images. In this paper we have used CNN architecture after the last layer of our DCGAN model to classify the synthetic images which have been created by the generator. (Lee et al. 2001)

CNN is often regarded as ConvNet, is equipped with deep feed forward architecture in it. Using that feature it makes it more able detect better than other fully connected layers. It's based on the concept of weight sharing. One of the major benefit in CNN is it takes very less amount of parameters thus it's safe from over fitting of the model. (Bouvrie, et al. 2006)

Figure 2. Basic Constituent layer of CNN

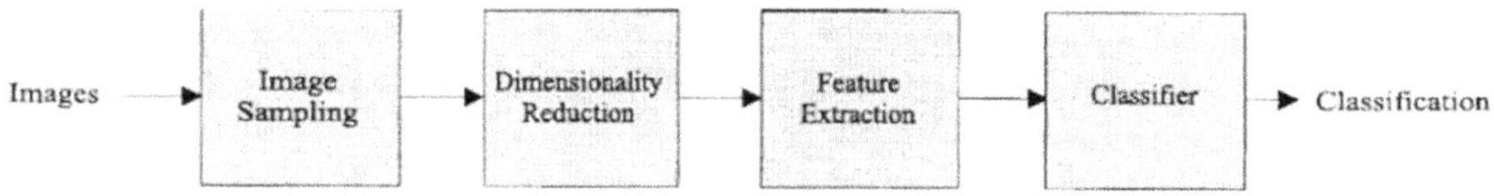

General Model of CNN

In general ANN model takes one single input and output layer along with multiple hidden layers which act as the processing unit for the computation of the result. The vector x produces an output layer of Y, performing any particular function as represented in Eq. 16.

$$F\left(x,W\right)=Y \tag{16}$$

Here W denotes the weight vector which represents the interconnectivity strength of adjacent layer neurons. The general model of CNN consists of 4 layers namely, (a) Convolutional Layer, (b) Pooling Layer, (c) Activation Layer, and (d) Fully Connected Layer.

Convolutional Network

An image is constituted of pixels. The image is being classified by the feature extracted from the input layer, and based on the feature set the output class is being determined. The adjacent layer neurons are connected among each other and they extract features from the receptive field to form the weight vector. (Indolia et al. 2018)

The weight vector is alternatively known as filter, or kernel, which slides over the input vector field to form the filter map. This method of sliding over the input vector either horizontally or vertically is called Convolutional operation. This creates N feature map with the N features which have been extracted. The output which will subsequently act as input for the next layer is given in Eq. 17.

$$a_{ij}=\sigma\left(\left(Wx_{ij}+b\right)\right) \tag{17}$$

Where,

x is the input vector
W is the kernel which slides over an image
b is the bias
σ is the non-linearity introduced in the network architecture.

Figure 3. Weight Vector Initialization

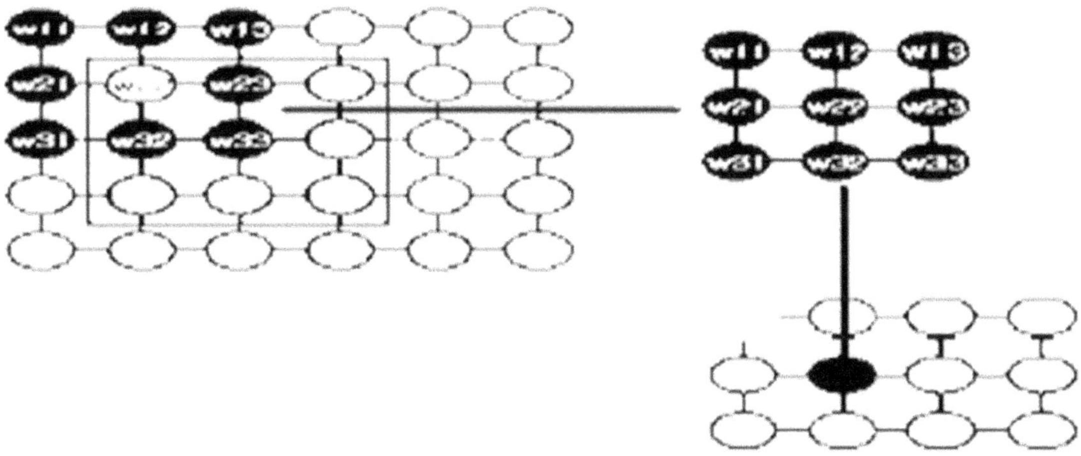

Pooling Layer

The primary aim to introduce pooling layer is to reduce the number of trainable parameters by introducing translation invariance. To perform this operation a window is being created and the input elements are passed through it a pooling function as shown in Fig. 4. The maximum figure is taken from a matrix thus this method is max-pooling and the layer doing this is known as max-pooling layer. (Bouvrie, et al. 2006)

Fully Connected Layer

The fully connected layer is being fed by the output which has been generated from the last 2 layers. The loss function, gradient descent reduces the cost function here over the training dataset, with constantly updating the weights in the layer with every passing epochs, where an epoch is defined as the journey of traversing the whole network.

Figure 4. Pooling Window

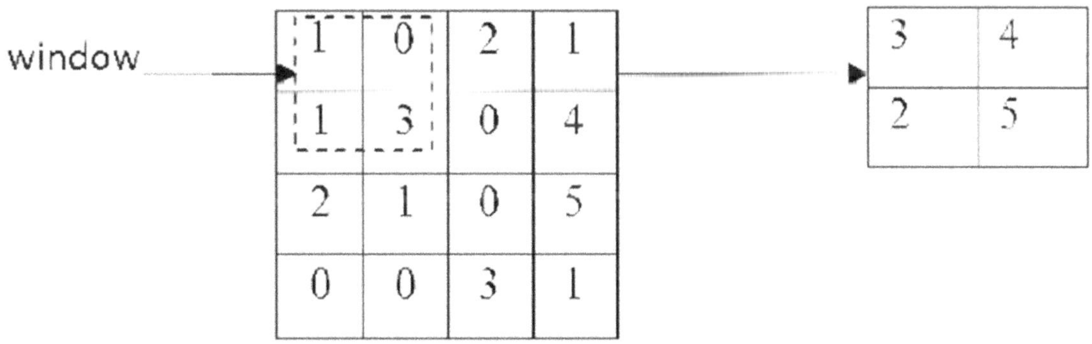

Activation Function

The introduction on non-linearity is needed in the architecture, and in order to introduce that particular non linearity we generally use Rectified Linear Unit (ReLu) activation function. It's chosen based on the advantages it gives like, doesn't allows the gradient to disappear which gives a major upper hand in loss calculation in CNN architecture. Sometimes we opt for leaky ReLu if a large gradient is being passing through the network, as at that time ReLu fades away.

Calculation of Gradient Descent in CNN Architecture

In the time of training through the filters, we use a technique named backpropagation to change and update the pre-initialized weights that are being allotted. The overall network is at first feed forwarded, and then the computing at every layer starts and total error component is introduced in the last layer. In order to compute an optimized network the computed gradients are being backpropagated the calculation is as follows.

After passing the input vector through Eq. 18 the next steps are being calculated.

$$C_q^l = \left(\sum_{p=1}^{n} S_p^{l-1} * K_{p,q}^l + b_q^l \right) \tag{18}$$

$$C_q^l = \left(\sum_{p=1}^{n} \sum_{u=-x}^{x} \sum_{v=-x}^{x} S_p^{l-1}\left(i-u, j-v\right).K_{p,q}^l\left(u,v\right) + b_q^l \right) \tag{19}$$

Where,

$$n = Number\ of\ feature\ map\ in\ last\ layer$$

$$p,q = Layer\ map\ of\ last\ layer\ and\ previous\ layer$$

$$\varphi = Activation\ Function\left(ReLu\ or\ Sigmoid\right)$$

$$* = Convolutional\ Operation$$

$$b = bias$$

$$x = Size\ of\ Filter$$

$$S_p^0 = Input\ Image\ where\ first\ ConvNet\ will\ be\ performed$$

$$z = Pool\,Size\,Window$$

After Conv. Layer pooling is applied as represented in Eq. 20

$$S_q^l\left(i, j\right) = \frac{1}{4}\sum_{u=0}^{z}\sum_{v=0}^{z} C_q^l\left(2i - u, 2j - v\right) \tag{20}$$

Afterwards it's being passed through fully connected layers for output class prediction.

$$\widehat{output} = \sigma\left(w \times f + b\right) \tag{21}$$

$$Where, f\ is\ the\ final\ output\ vector$$

Then the result is passed through the softmax activation function, which is represented in Eq. 22.

$$y'\left(i\right) = \frac{e^{\widehat{output}}}{\sum_{1}^{labels} e^{\widehat{output}}} \tag{22}$$

The loss function is computed through Eq. 23

$$L = \frac{1}{2}\sum_{i=1}^{no.of\ training\ pattern}\left(y'\left(i\right) - y\left(i\right)\right)^2 \tag{23}$$

Where,

$$y'\left(i\right) - Target\ Output$$

$$y\left(i\right) = Predicted\ Output$$

The Error is now needed to be back-propagated

$$\Delta W\left(i, j\right) = \frac{\partial L}{\partial W\left(i, j\right)} \tag{24}$$

$$\Delta W\left(i,j\right)=\frac{\partial L}{\partial wy'}\cdot\frac{\partial y'}{\partial W\left(i,j\right)} \tag{25}$$

$$\Delta W\left(i,j\right)=\partial(\frac{1}{2}\sum_{i=1}^{P*}\frac{\left(\left(y'\left(i\right)-y\left(i\right)\right)^{2}\right)}{\partial y'}\cdot\frac{\partial y'}{\partial W\left(i,j\right)} \tag{26}$$

$$\Delta W\left(i,j\right)=\left(y'\left(i\right)-y\left(i\right)\right)\times\frac{\partial}{\partial W\left(i,j\right)}(\sigma(\sum_{j=1}^{no.of\ nodes}W\left(i,j\right)\times f\left(j\right)+b\left(i\right))\ ) \tag{27}$$

$$\Delta W\left(i,j\right)=\Delta y'\left(i\right)\times f\left(j\right) \tag{28}$$

1.1. Proposed Algorithm

In this section we propose our novel algorithm through which we can generate and then classify the blood cancer cells.

Input :

a.Discriminator : *Original Datset*

b.Generator : *Noise with Partial Range of Original Datset*

Output : ***Predicted Class*** $\left(\textbf{\textit{Malignant}}\text{or } \textbf{\textit{Benign}}\right)$ ***of Synthetic Blood Cancer Cell Image***

Start

G :

$$Sy_{Smpl}=Gn\left(GUns+Data\left(Lb,UB\right)\right)$$

Do

$\{$

$\boldsymbol{D}:$

$Diff\left(Sy_{Smpl}\right)$

$\boldsymbol{D}\,Diff\ \ N_{smpl}\ \&\ Sy_{Smpl}\,using\,MLE$

$if\left(MLE = High\right)$

$Accept\,Sy_{Smpl}$

$Else$

$\boldsymbol{G}\left(Fdbk\right)$

$\boldsymbol{G}:$

$$Sy_{Smpl} = Gn\left(GUns + Data\left(Lb,UB\right)\right) + Fdbk$$

$G\,clts\,Error\,betweeen\,Sy_{Smpl}\ \&\ N_{smpl}$

$\}\ while\left(Sy_{Smpl} = Accepted\right)$

$\boldsymbol{CNet}\left(Sy_{Smpl}\right):$

$for\left(i = 0:3\right)$

$\{$

$$ConvoNN\left(St = 16\left(Sy_{Smpl}\right)\right)\ \&\ fv\,Generate$$

$Mplng\left(fv\right)$

$Dpl(0.2)$

$St++;$

}

$FlCl(Mplng)$

$Opl(FlCl)$

End

The meaning of the variables used in the proposed algorithm is explained in Table 2.

Model Architecture

The classification model of CNN is based after the GAN where the synthetic images of blood cancer cell are being generated. The generator network takes an input vector of size 100, of which are taken as sample at random from a normal distribution. The input is being propagated through the fully connected internal layer architecture and as a result it creates a RGB image with constituent's $128 * 128 * 3$ pixels.

The model is equipped with a fully connected layer along with 4 deconvolutional with stride 1 and a kernel size of $4 * 4$. All the layers except the output layers uses Leaky ReLu as its activation function, and the last output layer uses tanh activation function.

The discriminator is designed with a classic architecture. The input network is a n RGB image with the pixel size of $128 * 128 * 3$. Similar architecture has been followed here with 4 Convolutional networks and keeping other layers same as the generator architecture. Except the output layer we have used Leaky ReLu activation function with a slope of 0.1, and the last layer is fitted with sigmoid function which will eventually return the probability of the image being fake or real.

After passing through the DCGAN architecture the image is fed into a Convolutional network which will act as a classifier to classify whether the generated images are good enough to get detected as a Malignant or Benign cell through an automated process. In this architecture, we have used 4 Convolutional layers, along with a dropout layer with 0.2 probabilities to minimize the over fitting in the model.

The basic architecture stands as then, the dataset is passed through the generator network architecture, it creates synthetic image as it's output. Then that image is passes through the discriminator network where it's been detected for being original or fake. Once the image is being approved then it's passed through our Convolutional network architecture. In that step it's being predicted whether the synthetic image which have been generated from our DCGAN architecture is malignant or benign.

Table 2. Variable and Meaning in the Proposed Algorithm

Variable	Meaning
Sy_{Smpl}	Synthetic Sample
Gn	Generate
$GUns$	Gaussian Noise
Lb, UB	Lower Bound, Upper Bound
$Diff$	Differentiate
N_{smpl}	New Sample
MLE	Maximum Likelihood Function
$Fdbk$	Feedback
$clts$	Calculates
$CNet$	CNN
$ConvoNN$	Convolutional Network
fv	Feature vector
$Mplng$	Max Pooling Layer
Dpl	Dropout Layer
St	Stride
$FlCl$	Fully Connected Layer
Opl	Output Layer

Source: Created by Author

RESULTS

In this section we discuss the results which have generated through our model.

At first, we show the images which are being generated from our DCGAN Architecture.

Table 3, clearly shows that our DCGAN model is creating images which is synthetic, but are good enough to pass the automated system of our CNN architecture. The training set contains around 30,000 blood cell image.

Table 4, shows the accuracy of our CNN architecture model.

Figure 5. Discriminator Architecture

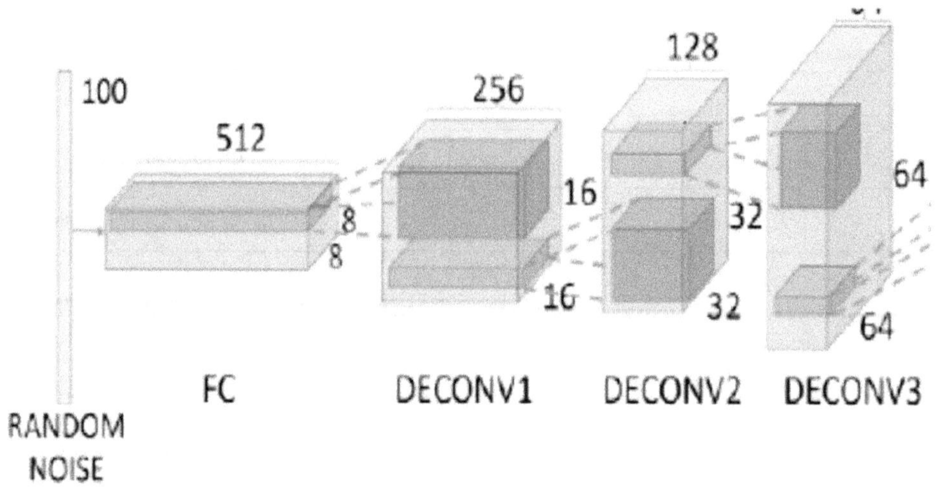

We have created our CNN architecture based on 32,000 training images which are different from the images which have been passed through the DCGAN architecture. The validation set consists of around 15,000 images along with 3,000 images in test set.

Figure 9, shows the predicted class of the correct class. The image which was predicted is a synthetic image which has been produced by DCGAN architecture. Our model has shown a probability of 0.71 it will belong to the malignant class of cancer cell.

This kind of results strengthens our algorithm's claim.

Figure 6. Generator Architecture

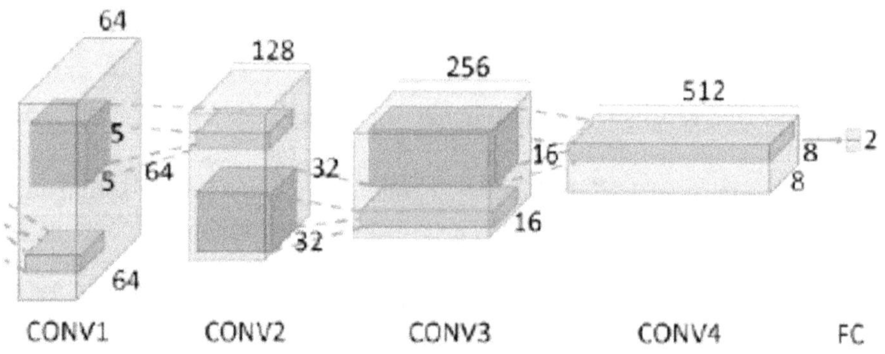

Table 3. Shows Image Formed after Epochs through our DCGAN Architecture

Parameter Value	Image Set
Image Size = 128 * 128 Noise Size = 10 Discriminator Learning Rate = 0.00034 Generator Learning Rate = 0.00036 After 50 Epochs.	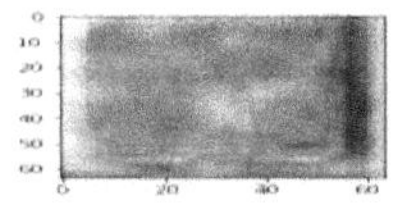
After 70 Epochs with constant parameter	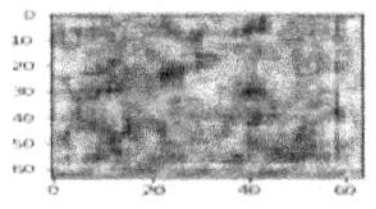
After 120 Epochs with constant parameter	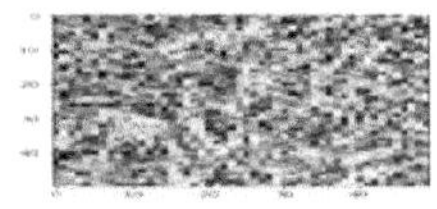
After 200 Epochs with constant parameter	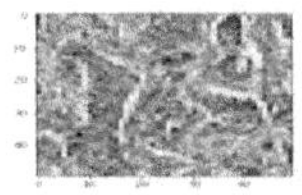
After 250 Epochs with constant parameter	
After 300 Epochs with constant parameter	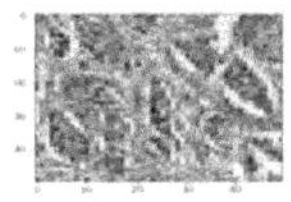

Source: Created by Author

CONCLUSION

In this paper, we have proposed a DCGAN based architecture which can generate synthetic blood cancer cell image using the two neural net architecture of it, namely the generator, and discriminator. Later, we have passed that newly generated image through a CNN architecture, which is trained to predict the output classes of the cancer cell images.

We have trained our DCGAN architecture on 30,000 blood cancer cell images, and our CNN architecture is being trained on another different set of 32,000 blood cell images. Both the classes malignant or benign were present in that dataset. The validation set consisted around 15,000 images along with 5,000 test set image to test the predictability of our model.

Table 5. Accuracy of CNN Architecture

Model	Accuracy
Validation Accuracy	94.56%
Training Accuracy	92.32%

Source: Created by Author

Figure 7. Accuracy curve for CNN Model
Source: Created by Author

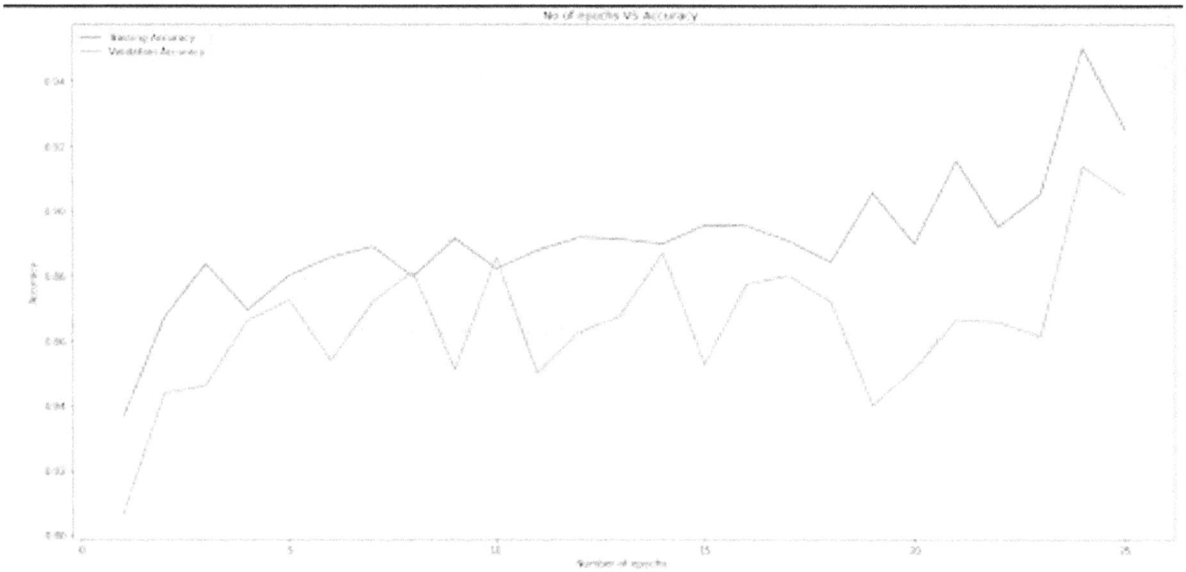

We have achieved a high accuracy of 92.32% on the validation set, and through which we get high probability of getting correct class in the output even through synthetic image.

Figure 8. Loss Curve of CNN Architecture
Source: Created by Author

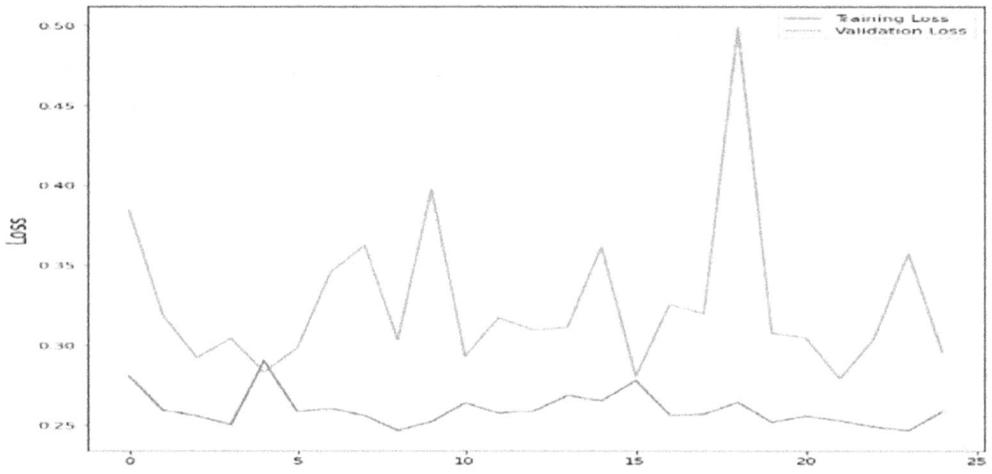

Figure 9. Predicted Class of Synthetic Image produced from DCGAN and predicted from CNN Architecture
Source: Created by Author

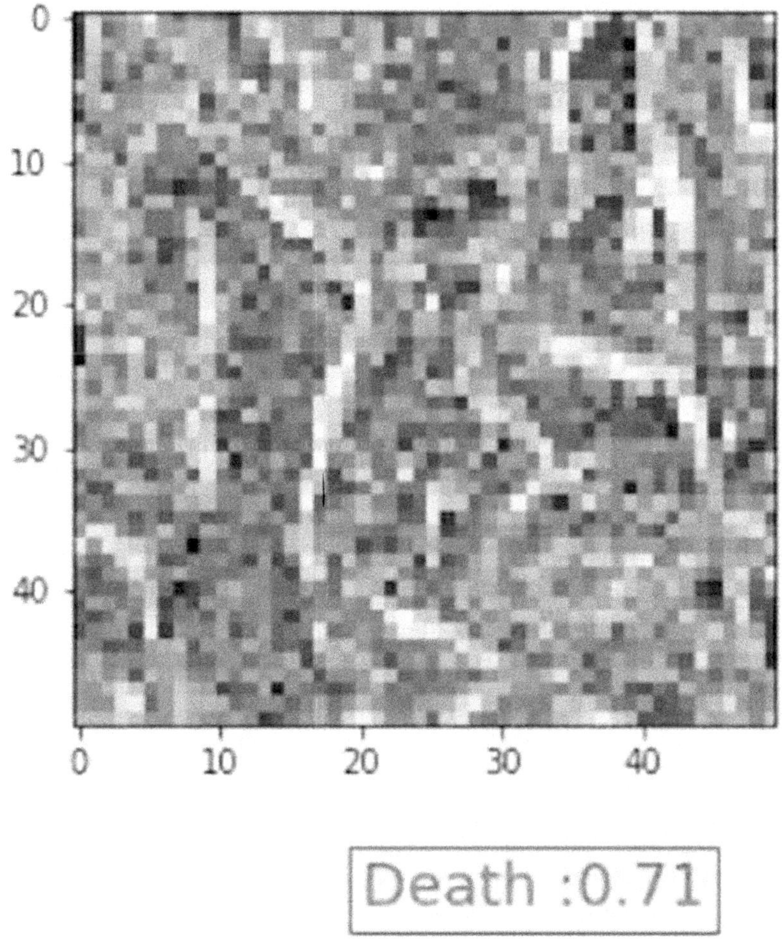

CREDIT AUTHOR STATEMENT

Ahan Chatterjee: Conceptualization, Methodology, Experiment, Original Draft, Editing
Swagatam Roy: Visualization, Original Draft, Experiment

REFERENCES

Bouvrie, J. (2006). *Notes on convolutional neural networks.* Academic Press.

Chatterjee, A., Roy, S., & Shrivastava, R. A Machine Learning Approach to Prevent Cancer. In *Handbook of Research on Disease Prediction Through Data Analytics and Machine Learning* (pp. 112–141). IGI Global.

Chaudhari, P., Agrawal, H., & Kotecha, K. (2019). Data augmentation using MG-GAN for improved cancer classification on gene expression data. *Soft Computing*, 1–11.

Chen, T., & Chefd'Hotel, C. (2014, September). Deep learning based automatic immune cell detection for immunohistochemistry images. In *International workshop on machine learning in medical imaging* (pp. 17-24). Springer. 10.1007/978-3-319-10581-9_3

Han, C., Hayashi, H., Rundo, L., Araki, R., Shimoda, W., Muramatsu, S., . . . Nakayama, H. (2018, April). GAN-based synthetic brain MR image generation. In *2018 IEEE 15th International Symposium on Biomedical Imaging (ISBI 2018)* (pp. 734-738). IEEE. 10.1109/ISBI.2018.8363678

Hu, H., Guan, Q., Chen, S., Ji, Z., & Yao, L. (2017). Detection and recognition for life state of cell cancer using two-stage cascade CNNs. *IEEE/ACM Transactions on Computational Biology and Bioinformatics*. PMID:29990223

Indolia, S., Goswami, A. K., Mishra, S. P., & Asopa, P. (2018). Conceptual understanding of convolutional neural network-a deep learning approach. *Procedia Computer Science*, *132*, 679–688. doi:10.1016/j.procs.2018.05.069

Kitrungrotsakul, T., Iwamoto, Y., Han, X. H., Takemoto, S., Yokota, H., Ipponjima, S., . . . Chen, Y. W. (2019, May). A cascade of CNN and LSTM network with 3D anchors for mitotic cell detection in 4D microscopic image. In *ICASSP 2019-2019 IEEE International Conference on Acoustics, Speech and Signal Processing (ICASSP)* (pp. 1239-1243). IEEE.

Lee, D. H., Seo, T. S., & Ko, Y. T. (2001). Spiral CT of the gastric carcinoma: Staging and enhancement pattern. *Clinical Imaging*, *25*(1), 32–37. doi:10.1016/S0899-7071(01)00245-5 PMID:11435037

Mani, N. B. S., Suri, S., Gupta, S., & Wig, J. D. (2001). Two-phase dynamic contrast-enhanced computed tomography with water-filling method for staging of gastric carcinoma. *Clinical Imaging*, *25*(1), 38–43. doi:10.1016/S0899-7071(99)00134-5 PMID:11435038

Pan, H., Xu, Z., & Huang, J. (2015, October). An effective approach for robust lung cancer cell detection. In *International Workshop on Patch-based Techniques in Medical Imaging* (pp. 87-94). Springer. 10.1007/978-3-319-28194-0_11

Perez, L., & Wang, J. (2017). *The effectiveness of data augmentation in image classification using deep learning*. arXiv preprint arXiv:1712.04621

Rashid, H., Tanveer, M. A., & Khan, H. A. (2019, July). Skin Lesion Classification Using GAN based Data Augmentation. In *2019 41st Annual International Conference of the IEEE Engineering in Medicine and Biology Society (EMBC)* (pp. 916-919). IEEE. 10.1109/EMBC.2019.8857905

Rubin, M., Stein, O., Turko, N. A., Nygate, Y., Roitshtain, D., Karako, L., Barnea, I., Giryes, R., & Shaked, N. T. (2019). TOP-GAN: Stain-free cancer cell classification using deep learning with a small training set. *Medical Image Analysis*, *57*, 176–185. doi:10.1016/j.media.2019.06.014 PMID:31325721

Rubin, M., Stein, O., Turko, N. A., Nygate, Y., Roitshtain, D., Karako, L., Barnea, I., Giryes, R., & Shaked, N. T. (2019). TOP-GAN: Stain-free cancer cell classification using deep learning with a small training set. *Medical Image Analysis*, *57*, 176–185. doi:10.1016/j.media.2019.06.014 PMID:31325721

Salehinejad, H., Valaee, S., Dowdell, T., Colak, E., & Barfett, J. (2018, April). Generalization of deep neural networks for chest pathology classification in x-rays using generative adversarial networks. In *2018 IEEE International Conference on Acoustics, Speech and Signal Processing (ICASSP)* (pp. 990-994). IEEE. 10.1109/ICASSP.2018.8461430

Thuy, M. B. H., & Hoang, V. T. (2019, December). Fusing of deep learning, transfer learning and gan for breast cancer histopathological image classification. In *International Conference on Computer Science, Applied Mathematics and Applications* (pp. 255-266). Springer.

Xie, Y., Xing, F., Kong, X., Su, H., & Yang, L. (2015, October). Beyond classification: Structured regression for robust cell detection using convolutional neural network. In *International conference on medical image computing and computer-assisted intervention* (pp. 358-365). Springer. 10.1007/978-3-319-24574-4_43

Yang, J., Zhao, J. X., Cao, Q., Hao, L., Zhou, D., Gan, Z., Ji, L.-N., & Mao, Z. W. (2017). Simultaneously inducing and tracking cancer cell metabolism repression by mitochondria-immobilized rhenium (I) complex. *ACS Applied Materials & Interfaces*, *9*(16), 13900–13912. doi:10.1021/acsami.7b01764 PMID:28368110

Zhang, J., Hu, H., Chen, S., Huang, Y., & Guan, Q. (2016, December). Cancer cells detection in phase-contrast microscopy images based on Faster R-CNN. In *2016 9th International Symposium on Computational Intelligence and Design (ISCID)* (Vol. 1, pp. 363-367). IEEE. 10.1109/ISCID.2016.1090

Chapter 8
The Rise of "Big Data" in the Field of Cloud Analytics

Dariusz Jacek Jakobczak
https://orcid.org/0000-0002-0297-6598
Koszalin University of Technology, Poland

Ahan Chatterjee
https://orcid.org/0000-0001-5217-4457
The Neotia University, India

ABSTRACT

The huge amount of data burst which occurred with the arrival of economic access to the internet led to the rise of market of cloud computing which stores this data. And obtaining results from these data led to the growth of the "big data" industry which analyses this humongous amount of data and retrieve conclusion using various algorithms. Hadoop as a big data platform certainly uses map-reduce framework to give an analysis report of big data. The term "big data" can be defined as modern technique to store, capture, and manage data which are in the scale of petabytes or larger sized dataset with high-velocity and various structures. To address this massive growth of data or big data requires a huge computing space to ensure fruitful results through processing of data, and cloud computing is that technology that can perform huge-scale and computation which are very complex in nature. Cloud analytics does enable organizations to perform better business intelligence, data warehouse operation, and online analytical processing (OLAP).

INTRODUCTION

The term "Big Data" used for explaining some detailed information of massive volume. Those data can be in both structured and unstructured form. This data can't be handled using traditional database methodologies and software technologies. The endless growth in volume is due to data which are being captured by organizations, social media sharing, and use of smart products i.e. IoT enabled devices has

DOI: 10.4018/978-1-7998-4706-9.ch008

Figure 1. Growth of Big Data Market Value
Source: Graph created by author, data collected from statista

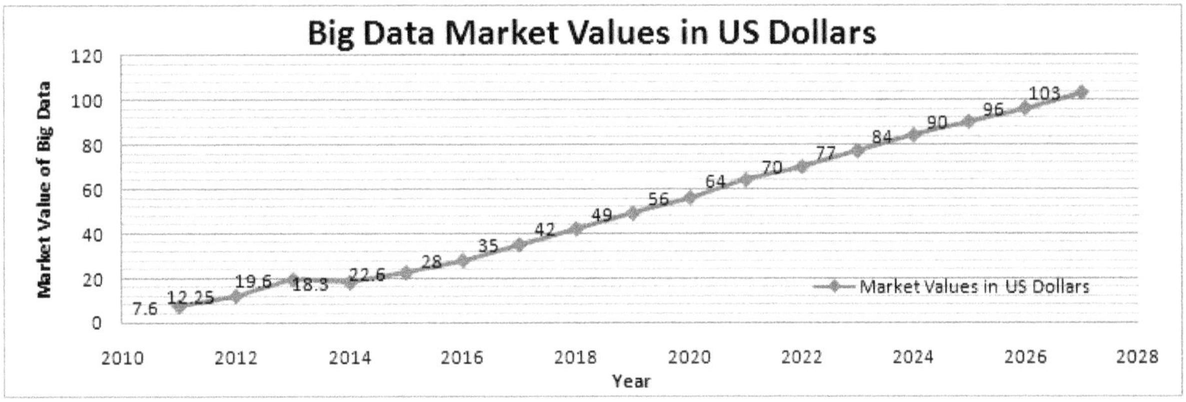

resulted in a massive data outbreak. The following graphical analysis will give a more clear view of the Big Data market scenario. (Sujitha & Praveen, 2015)

Here the graphical growth shows how the industry paced over the few years and how it will go for the next 8-10 years.

The following definition of Big Data is proposed on the basis of the 4 V's namely, Volume, Variety, Velocity and Value. *Big Data is a set of techniques and technology that require new forms of integration to uncover large hidden values from large datasets that are diverse, complex, and of massive scale.*

One of the most accepted definition of Cloud Computing is given by NIST it states as –

Figure 2. Four V's of Big Data
Source: Created by author

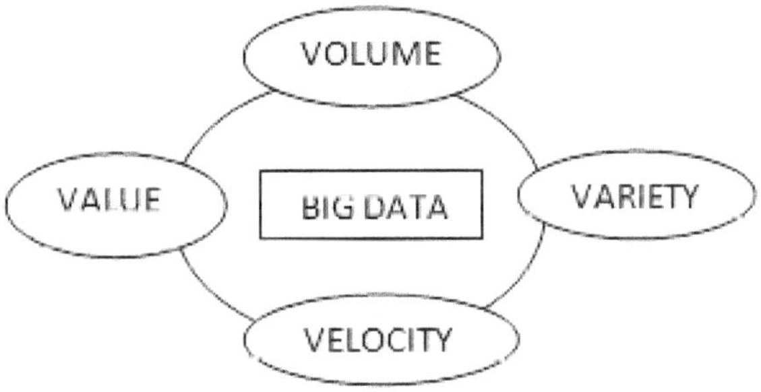

Cloud Computing is a model for enabling ubiquitous, convenient, on-demand network access to a shared pool of configurable computing resource (e.g. networks, servers, storage, applications and services) that can be rapidly provisioned and released with minimal management effort or service provider interaction.

In general various organizations adopt 3 (three) types of cloud deployment models namely Private Cloud, Public Cloud, Hybrid Cloud. (Deshmukh & Sumeet, 2015)

Cloud computing is a low cost, economical model that's being used for big data analytics. The convergence of Big Data and Cloud technology make big data analytics more successful outcomes. Various organizations moved towards dedicated servers for getting better analytical result. Hadoop can certainly deliver more accurate and critical results beyond some of the web scale companies such as Yahoo, Spotify. (Khan et al., 2013)

The Cloud computing delivers various service models to its user as service-based services in pay you go form. It delivers infrastructure, platform, and software as services. These services are namely, Infrastructure as a Service (IaaS), Platform as a Service (Paas), and Software as a Service (Saas). [4]

Traditional computing methods are getting out of the plate as the rate of growth of data is too high and with the passing of each day the data is becoming complex in nature, which triggered the big data application. Huge amount of intermediate data is being produced by these defacto softwares namely Google's map reduce framework and apache Hadoop. (Mayilvaganan & Sabitha, 2013)

With the rapid increase of Internet of Thing (IoT) and future internet the concept of smart city evolved in a pretty rapid pace. There is a generation of huge amount of data this data needs to be properly managed and analyzed using integrated Information Communication Technology (ICT) approach. With the arrival of ICT it brings some significant changes smart cities governance. Processing and integration of cross-disciplinary data is the key for knowledge and intelligence for sustainability, resilience of the governance of the city. (Ramamoorthy & Rajalakshmi, 2013)

Big Data Analysis

The definition of the big data comes from that 4 V's that will be elaborately discussed in this section.

1. **Volume:** The term "volume" here is referred to the humungous amount of data that is collected through various sources. The benefit of collecting a massive amount of data is we can find out various hidden patterns or information from these through data analysis. The said approach is called "mobile data challenge" by Nokia. This collection of data leads to various results as predictability of human behavior and human mobility by visualization of data.
2. **Variety:** Variety refers to the types of different data's which are being collected through various sources such as internet, social media, and sensors and other various devices. All these data can be classified into some of the mainstream types such as images, videos, text, and email. All of these collected data are unstructured that is they are in random space.
3. **Velocity:** It refers to the speed of data that is being transferred.
4. **Value:** The most important factor of the 4 V's it refers to the discoveries of hidden data or information from a huge data set collection.

Big Data can be classified into various blocks or parts depending on nature or type. The classification is based on the following 5 (Five) aspects.

1. Data Sources
2. Content Format
3. Data Stores
4. Data Staging
5. Data Processing (Yadav & Sohal, 2017)

Figure 3. Classification of Big Data
Source: Hashem, Anuar, Mokhtar (2014)

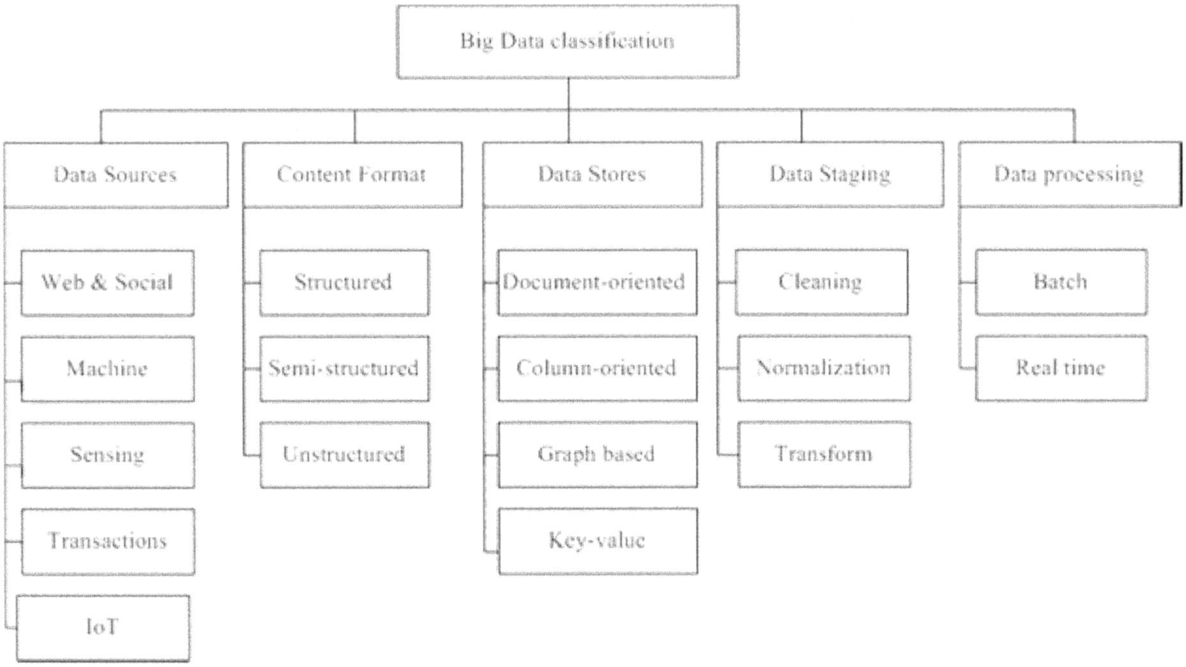

CLOUD COMPUTING

Cloud Computing is one of the fastest growing sectors in the IT industry in today's world. As every organization is seeking for connected storage place and enough good which can be analyzed to obtain results. Cloud service providers have started to provide integrated framework to support parallel data processing. (Sayeed et al., 2015)

Deployment Model

All the applications which are being used needs to be deployed in the cloud with the variable requirements, now all the deployment model have their own characteristics which are discussed.

Private Cloud

One organization operates and maintains the entire cloud infrastructure. People outside the organization don't have any access to the resources in the cloud.

Public Cloud

This type of cloud infrastructure is made available to the consumer on a subscription basis by the cloud service provider. This enables the consumers to invest a minimal amount in the cloud storage to deploy there system into cloud.

Hybrid Cloud

Hybrid cloud can be a merged form of public cloud and private cloud platform where the cloud provider can support some data of the organization and also provide some service in cloud with exchange of some subscription fees. Here consist different types of clouds of different types, and it allows data/ or application to be transfer from one to other cloud. (Vuppala et al., 2017)

Service Model

Cloud Computing provides various models to its consumer in exchange of subscription fees. This provides infrastructure, platform, and software as services. The following models are the models which are being use.

Software as a Service (Saas)

The Saas model provides user not to install any software, physically in their computer system all the required software is there on the cloud. A Saas provides access to both resource and application to its user.

Platform as a Service (Paas)

The Paas Platform provides the consumers as a service based platform. It provides access to the platform that the consumer need to develop to run their own applications.

Infrastructure as a Service (Iaas)

The Iaas Platform provides the consumer a computational infrastructure platform to its user. The consumer has total access over storage in cloud and resources which are available in the cloud both hardware and software on a proprietary basis. (Wang, 2017)

Big Data Analytics Adoption in Cloud

Big Data and Cloud Computing are co-joined. Big Data allows people to compute through distributed queries across multiple dataset and return results accurately. On the other hand cloud computing provides Hadoop, a class of distributed data-processing platforms. The data sources are being stored in a distributed fault-tolerance database and proceed through a programming model for large dataset with a parallel distributed algorithm.

Big Data utilizes the distributed storage technology service of Cloud Computing rather than the storages installed as hard drives in the computers. In the following table some big data cloud providers are being listed. (Prerna & Agarwal, 2017)

With the increase of the size of the data set it also gets more complex in nature; the complexity factor is being solved cloud computing infrastructure which generally addresses this kind of problem.

Map Reduce has become one of the most important big data processing in cloud environment platform. It provides parallel processing in cluster.

Figure 4. Cloud Computing Usage in Data Analyitcs
Source: Hashem, Anuar, Mokhtar (2014)

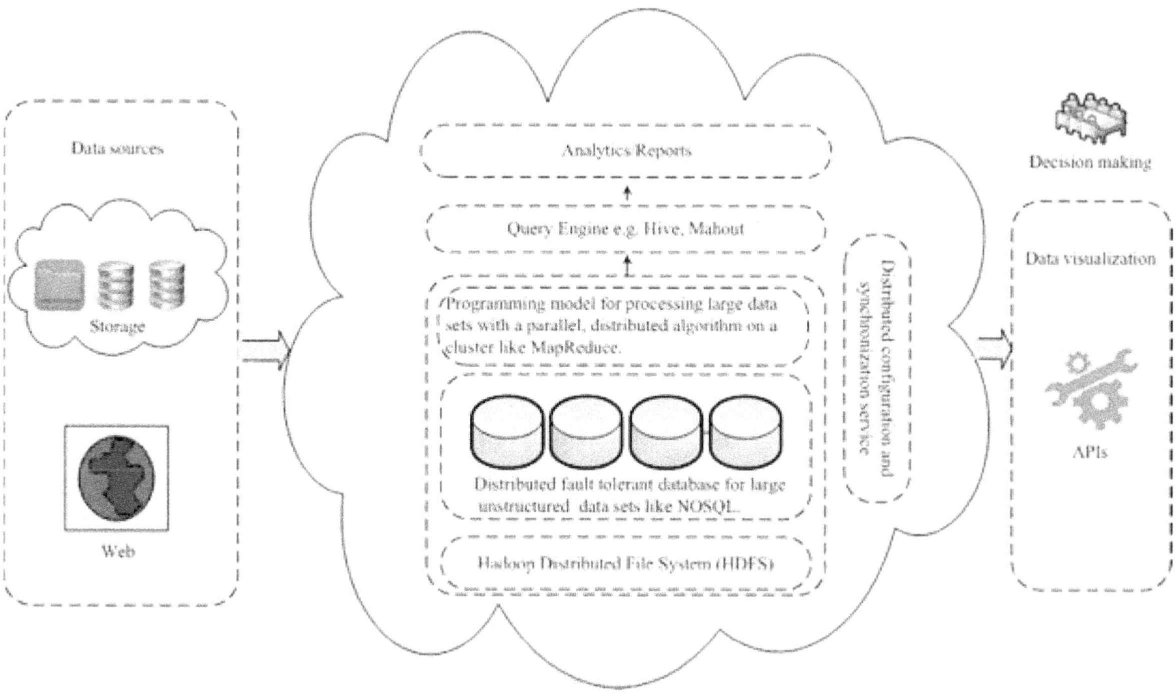

Cloud Analytics and Cloud Analytics Architecture

According to Gartner, an IT research organization "Analytics has been emerged as catch-all term for variety of different business intelligence (BI) - and application related initiatives". The word 'Analytics' means predicting the future outcome or score by means of analysis of previous or old data. And "Cloud Analytics" is referred as software platform which will be delivered on Software as a Service basis (SaaS).

Cloud Analytics Architecture

The utilization of cloud analytics is to get cloud service for a single or all component of an analytic solution this enables the organization to use the social media data, internet data, and all the third party data to get insights of customer preference and demands. The traditional data base model is no longer in use

Table 1. Comparison of various big data cloud provider

	Google Cloud	Microsoft Cloud	Amazon Cloud
Big Data Storage	Google Cloud Service	Azure	S3
Big Data Analytics	BigQuery	Hadoop on Azure	Elastic Map Reduce (Hadoop)
Map Reduce	App Engine	Hadoop on Azure	Elastic Map Reduce

here in cloud analytics. New database model is needed and used namely NoSQL which can efficiently store and analyze data from cloud.

There are mainly 4 (four) layers of cloud analytics architecture namely,

1. Infrastructure Layer
2. Data Storage and Management Layer
3. Analytics Layer
4. Visualization Layer

Infrastructure Layer

The infrastructure layer is the first layer of the cloud analytics. This is also known as the foundation layer. In this layer data is being stored and managed. Data present here are in both structured and unstructured form. Here the data is being collected from various sources like social data, sensor data, streaming data etc.

Data Storage and Management Layer

The second layer of the cloud analytics architecture is the data storage and management layer. Data Management layer is a repository layer here humongous amount of raw structured, raw unstructured, and raw semi structured data is being stored in flat file format. This particular layer is being managed using Hadoop oriented service. Here each data is being attached with a unique ID and this tag is called as Meta data tag. This layer is also known as "data lakes".

Analytics Layer

The third and most important layer of the cloud analytics architecture is Analytics Layer. This layer mainly holds the business intelligence applications. There are two tools which power this layer viz. pre-analytic tool and analytics tool. The pre-analytic tool chooses and organizes data collected from database, whereas the analytics tool is used to retrieve various patterns and retrieve useful insights from the database. Various Data mining algorithms mainly Supervised Learning where there is a continuous results available are used, algorithms used namely Regression, and Classification modeling is used. Moreover various analytics software is also used viz. Mat Lab, Map Reduce, SAS, and R etc.

Visualization Layer

The last layer of the cloud analytics architecture is the visualization layer. This layer provides user interface and it provides all the visualization and analytics results. Visualization Layer helps the Subject Matter Experts (SMS) to get through all the results without any help of IT personnel. (Kumar & Gobi, 2017)

Big Data Adoption

Various business organizations are moving toward cloud based system as they provide better insights and help them to boost their sales performance in the competitive and ruthless market where the "Survival of the Fittest Statement" stands true each and every time.

Figure 5. Cloud analytics architecture
Source: Kalpana, Chaitanya (2018)

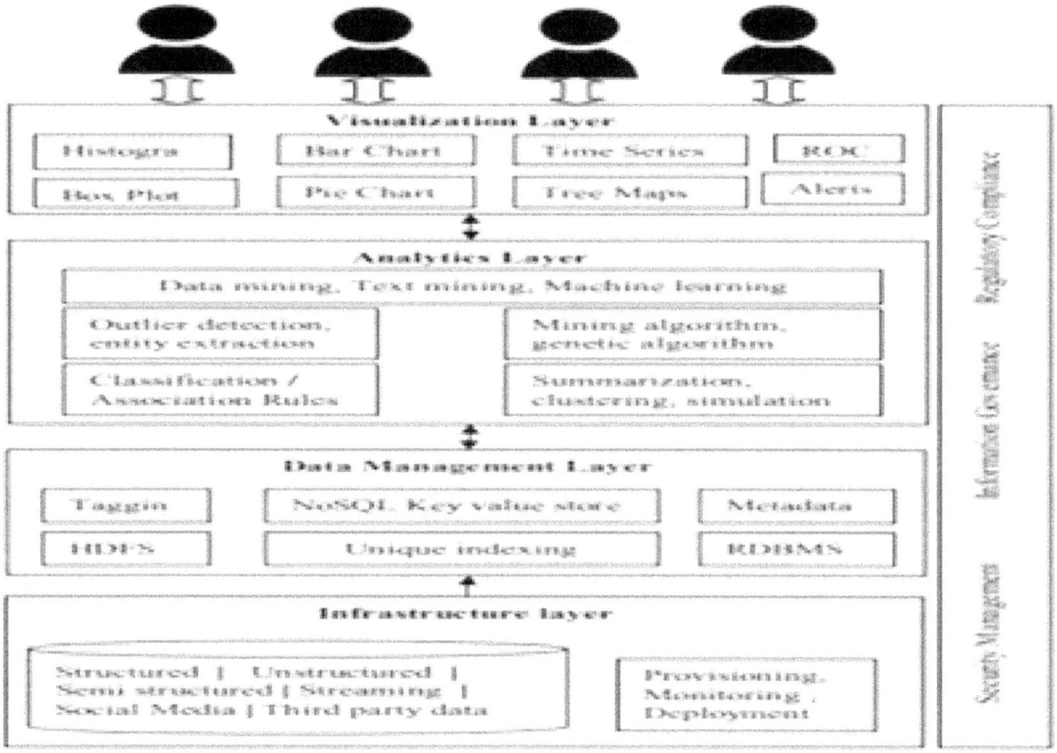

Business Drivers

Nowadays companies are focusing on business analytics trends to get their sales upright. Enterprise uses their private cloud infrastructure to improve risk and it can maintain control over it by analyzing load, cost, and security. The advantage of big data is being cost saving, revenue growth in an organization. Analytics service used by the companies easily boosts their scalability. Microsoft's solution enables user to analyze the Hadoop data from within the excel, adding new function abilities.

Cloud model being interleaved gives us the following benefits as discussed below,

Figure 6. Interleaving big data and cloud
Source: Created by the author

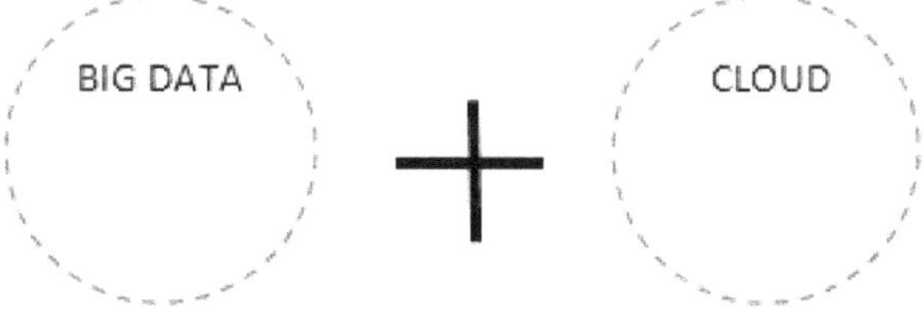

- **Speed:** Private clouds allow its users to do effective big data analytics and use internal resources from public cloud facilities. This let the users to speed up their infrastructure.
- **Extract Values:** Various organization gives emphasis on the Analytics as a Service (AaaS) model which is being supported by all the three models.
- **Cost Reduction:** When it comes storing huge amount of data cloud storage comes as an economical structure.
- **Improved Decision Making:** Memory analytics when gets combined with most modern technologies it gives more accurate analyzed results which improves decision making.
- **Data Valuation:** The unstructured or structured data which is collected in raw form is being analyzed and turned into useful data which will give insights of various hidden patterns.
- **Cloud Security:** Cloud service providers are now giving securities at different layers which enable a better security option for the organizations.
- **Innovation:** The convergence of big data and cloud gives innovative results. (Jaber & Zolkipli, 2013)

(i) Hadoop:

There are several software products that would facilitate the big data analytics and one such software is known as Hadoop, in general word it is the solution for most of the Big Data Problems.

This is the most available Java based programming framework which supports the processing of large amount of data in a distributed format. With the help of Hadoop we can analyze an over cluster of servers and applications.

The Hadoop model can be scaled up from a one server to many servers, where each and every single server acts as a local storage. When we work with such large scale of data often we face the problem of data redundancy, in hadoop data redundancy is minimized by *Hadoop Distributed File System (HDFS)* at the application layer itself. (Brindha & Jeyanthi, 2015)

Hadoop Distributed File System (HDFS)

The backbone of Hadoop, HDFS is a distributed file system which is designed to be procedure hardware. It's a low cost designed hardware and it's also fault-tolerant. It's very effective system when it comes to huge data set its file size is generally in gigabytes and terabytes. The HDFS system is more designed to be work on batch processing model and not in interactive use by user's model. Low Latency user model is more effective in this case.

HDFS works on Master-Slave method. It has only one master node which manages the file system namespace and regulates access to files, the master node is called NameNode. There are numbers of DataNodes, it usually in the count of one node per cluster. In internal system the file is split up into blocks and given the NameNode. The NameNode works on opening, closing, reading file systems.

Calculation of HDFS Nodes Storage

When we design the new Hadoop Cluster we need some storage place. To estimate the storage amount we need to calculate on the basis of some parameters. As miscalculation of storage place can lead to future problems where shortage of storage leads to non storage.

Figure 7. Hadoop Structure
Source: Yadav, Sohal (2017)

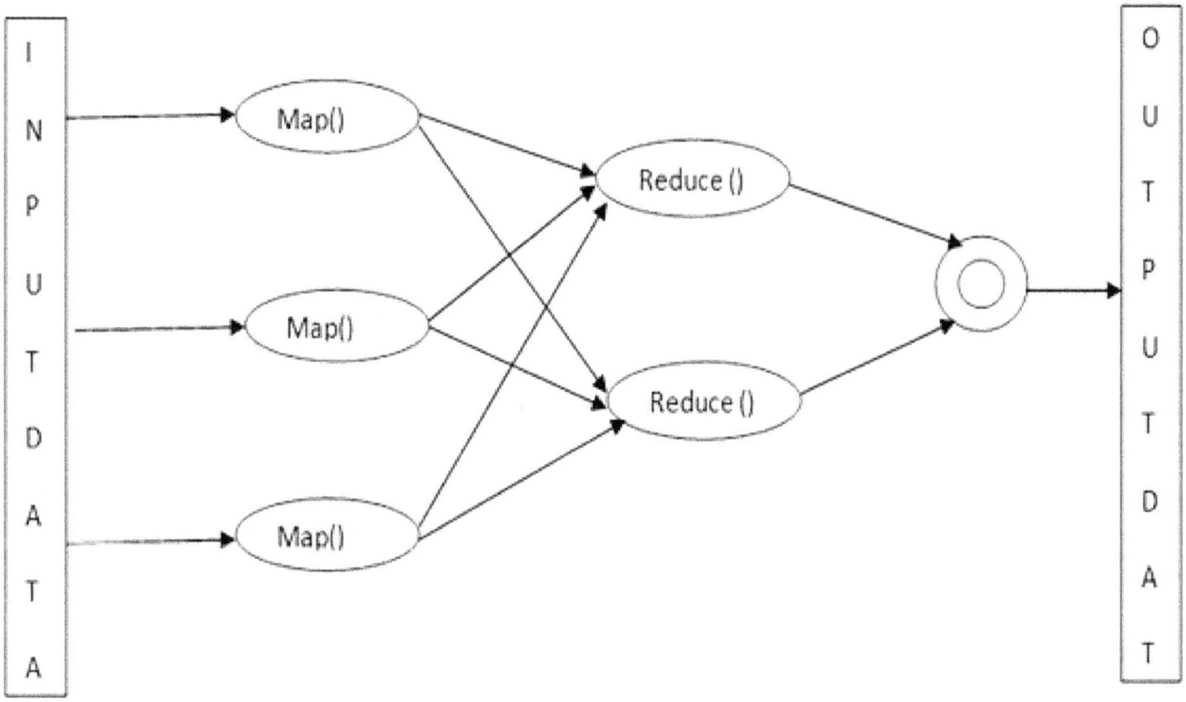

Figure 8. HDFS Architecture
Source: Created by Author

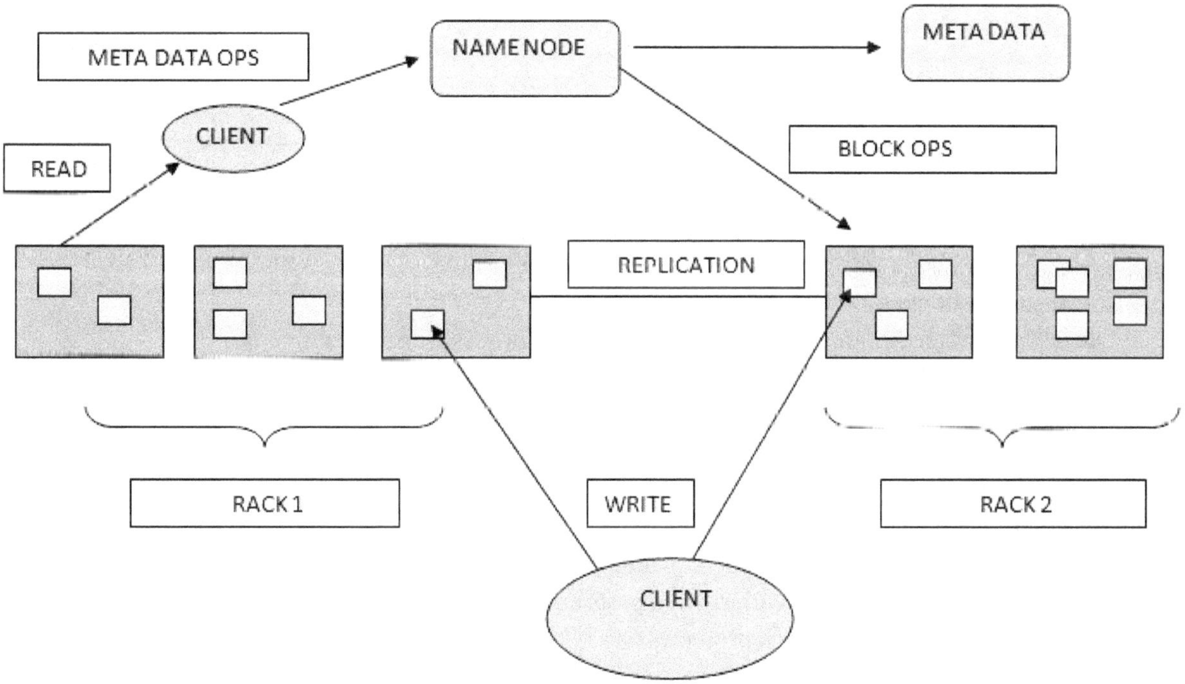

The storage is represented by 'H'

```
Formula:
Where:
C = Compression Ratio.
The type of compression used determines this factor.
C = 1 is taken when there is no compression.
R = Replication Factor.
In a production factor it's taken as 3.
S = Initial size of data that need to be moved to cluster.
This could be a combination of historical data and incremental data.
i = Incremental Data Factor.
It's usually 1/3 or 1/4; its hadoop's intermediate working storage space.
120% = It's 1.2 times the above total size
```

For Example, if the cluster size is of 2400 TB, but it is recommended to use up to only 2000 TB.

Calculation of Number of Data Nodes

The HDFS model follows the Mater-Slave Model. And in that model Data Nodes are the slave node. The Data Nodes actually stores data in the nodes. The Data Nodes provides replication and balancing. It also keeps an account of whether the slave nodes are alive/dead. The HDFS file is broken in small parts and these parts gets into Data Node.

```
The data node is represented by 'D'
Formula:
Where:
d = Disk Space available per node.
RAM, IOPS Bandwidth, CPU Configuration of nodes is taken into consideration.
H= HDFS Nodes Storage
Parameters that should be taken into consideration:
```

- There should be no failure in node.
- Other performance characteristics are not relevant (processor, memory)
- Adding new hardware is instentenous.

Map Reduce Framework

The map-reduce framework gets executed through parallel data processing. This framework makes small parts of the whole data i.e. dividing input data into small blocks and processes them all at a time into different machines. The Map-reduce works in two parts namely, Mapping and Reducing. The Map function reads the whole input data and then it produces pairs (key, value) it is similar to one of the data structure in python named as dictionary where inputs are there in the form of key and value. After producing pairs

Figure 9. Map reduce framework structure
Source: Yadav, Sohal (2017)

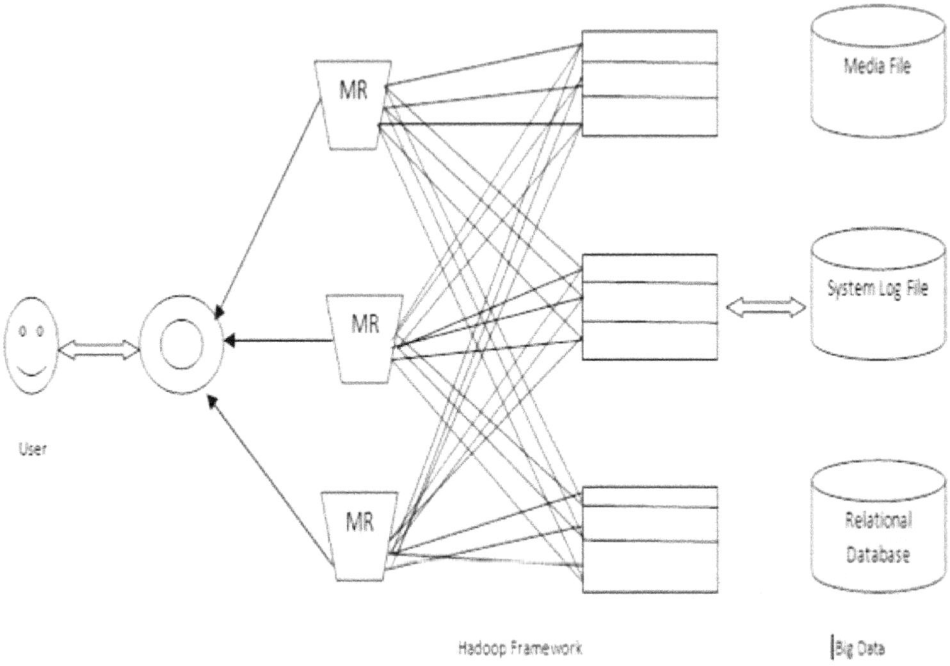

the map-reduce library takes up those pairs and group all the intermediate values associated with the same key and pass it to reduction function. Then the reduction function takes the respective key value and clusters them into smaller set of values. (Chakraborty et al., 2018)

INTRODUCTION OF EDGE COMPUTING, REAL TIME COMPUTING AND ANALYTICS

The main three paradigms of the industry of edge computing that are prevalent in the industry are namely Fog Computing, Mobile Edge Computing, and Mobile Cloud Computing. The concept of *fog computing* was introduced by Cisco; it was made by Cisco to make a bridge between the technology gaps of IoMT devices and Cloud. It comes with a distributed system and also its geographically enabled it's thoroughly connected through fog nodes such as access points, switches, and routers.

Similarly the concept of *Mobile Edge Computing* was introduced in the technology market by Nokia, the tele giant. It aimed to collect real-time data with the help of substations; they can act as intelligent service hubs. It can collect customer geo-locations.

Similarly the concept of *Mobile Cloud Computing* was introduced to eliminate the problem of shortage of capacity of mobile storage. As the storage capacity of mobiles getting shorter with every passing day due to the economic access to the internet. (Yadav & Sohal, 2017)

All these 3 parts of Edge Computing focuses on the fact of low latency of the data transfer between the IoMT devices to the core servers where it gets processed.

Table 2. Characteristics of fog computing, mobile edge computing, mobile cloud computing

	Fog Computing	**Mobile Edge Computing**	**Mobile Cloud Computing**
Owner and maintained by	Mobile and Cloud Provider	Mobile Network Provider	Private Organization
User	Open to all user	Only mobile users	Specific Users
Network	Both short and long range networks	Mobile networks	Short range networks
Geographical Distribution	Any location	Co-located with the base station	Static point location
Flexibility	Yes	No	No
Computing Environment	Both Indoor and Outdoor	Both Indoor and Outdoor	Indoor Only
Latency	Low	Low	Ranges from Low to High
Edge Analytics	No	No	Available

Source: Created by author

In the next 2we have uplifted a comparative study between these fog computing, mobile edge computing, and mobile cloud computing.

With the deployment of this model various challenges again comes, such as the scale of complexity of various networking protocols that are in use currently in the world of technology. The level of complexity is usually modeled up with the help of some basic parameters namely, high-performance of computation needs, data analytics which can be used in the deployment of edge computing mechanism which will suddenly yield better accurate result while calculating the data on its edge. As we wonder which applications need the use of data analytics to predict the closest result we have collected data and plotted a graph in which various models are plotted.(Sayeed et al., 2015)

To achieve a scalable and resourceful conclusion draw we need to use the two big technologies at a time that implementing cloud computing and edge computing. According to some research survey it's stated that in the context of edge computing the things that cover most ground are research in resource management (75%), security algorithms (68%), and algorithmic aspect (44%), and network technologies (50%).

Further conclusion that has been drawn from that paper was that there are three main model in edge computing from which the IoT users can pick them accordingly. Those models are namely, pure cloud service (In between cloud and end user edge service will act in the middle), edge services (Extension of cloud model), and coordinated fog-to-cloud-service (A mixed model of cloud and fog services)

FOG COMPUTING

The entire edge computing model is pretty big and a part of that edge computing is fog computing. These applications are pretty much associated with the IoT devices. Fog Computing is a decentralized computing model which acts as middle layer in between the IoT devices and the Cloud servers. While on the other hand mobile edge computing concentrates on the zone to solve delay-sensitive issues and contextual applications.

Figure 10. Applications plotted on the basis of complexity of edge analytics
Source: Cisco Global Index, 2015-2020

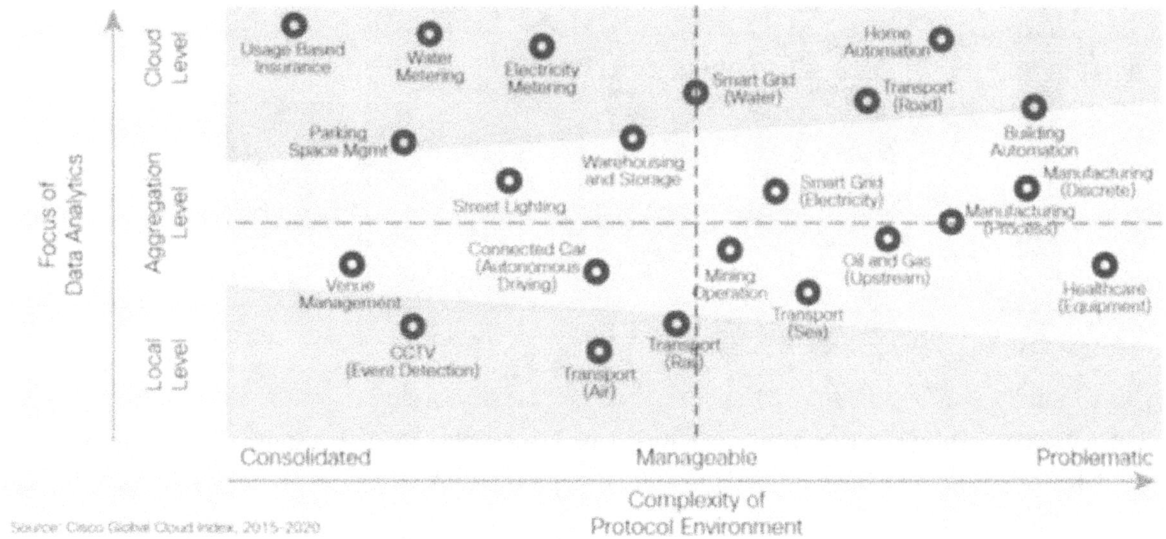

Transition From Cloud Computing to Fog Computing

The initial introduction of fog computing to the industry was done by the telecommunication industry. Consumers having IoT devices or will be having the premium network communication of lighting fast 5G technology will generate a humongous amount of data that need to be processed at that time on the edge of the network. On the basis of nodes available the fog computing gets converted to edge computing and cloud computing also.

The deployment model of fog computing consists of the client devices which is used by common people across the globe. Those devices are directly connected to the Edge devices which are deployed. Then an intermediary state is there which connects the edge devices with the cloud service.

To get the total use of edge computing we need to create a central zone, a *data centre in the middle of the box*. This data centers are known as *Cloudlets*. (Vuppala et al., 2017)

Cloudlets: Several technologies in the edge computing have lead to the non centralized deployment model. In which cloudlets are such a decentralized architecture which is mostly responsible for computing and enhancing the capabilities of IoT devices by reducing the communication delay. This small scale of clouds is also called as micro clouds and it's often installed in the public spaces to easily collect the data. The cloudlets acts as a layer of middle in between IoT enabled devices and cloud platforms.

The major goal of Fog computing is to provide seamless data movement in and through the IoT enabled devices. And it's expected to create a continuum between cloud and edge computing methodologies. And to provide this fluid data movement is much required.

While implanting this models various obstacles are also faced some of them are discussed below.

- **No Edge Service:** In the temporary time there is no such edge service provider who can provide on demand. Amazon taking just first steps by GreenGrass "services".

Figure 11. Deployment of Edge Computing Model

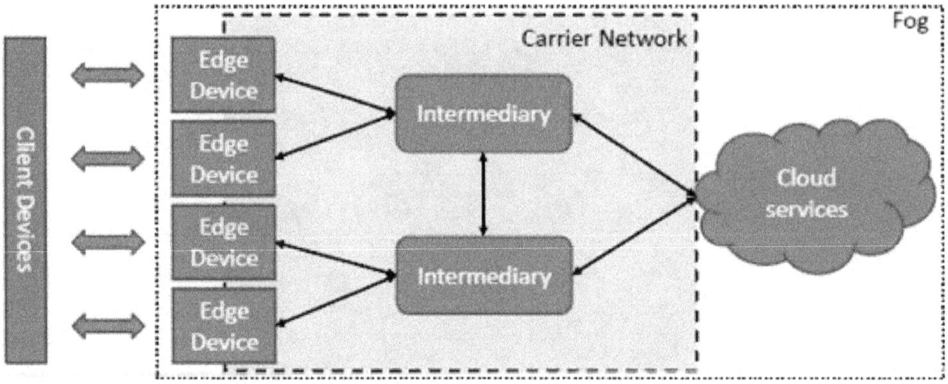

- **Lack of Hardware:** There is no such hardware availability. Custom solution is used in the form of Raspbery Pi, and BeagleBoards to state this problem. Application architecture needs to deal with this.
- **Management Error:** In Fog Computing the data and devices gets closed in which multiple data nodes are formed to use. Now no such manageable infrastructure present which handle those scaling, updating etc. tasks.
- **No- Network Transparency:** Distributed system need to hide distribution that is application can run only in a single machine and no such infrastructure is there.
- **Physical Security:** The traditional data centers are in a secured complex with some security personnel are there, but the fog devices need to be at more places thus physical security is definitely is a challenge. (Sayeed et al., 2015)

FOG AND EDGE COMPUTING SIMULATING TOOLS

To implement the fog and edge computing system one should take care of two levels of things. One is Software level and other is hardware level. In the software level parameters which concern are resource distribution, load balancing, and migration. On the other hand the concerned factors in hardware level are network capacity and capability, complexity, and scalability. To experiment with such high level of arrangement are costly thus simulation tools are used to create a virtual environment which is both

Figure 12. Roles in Fog Computing
Source: Created by the author

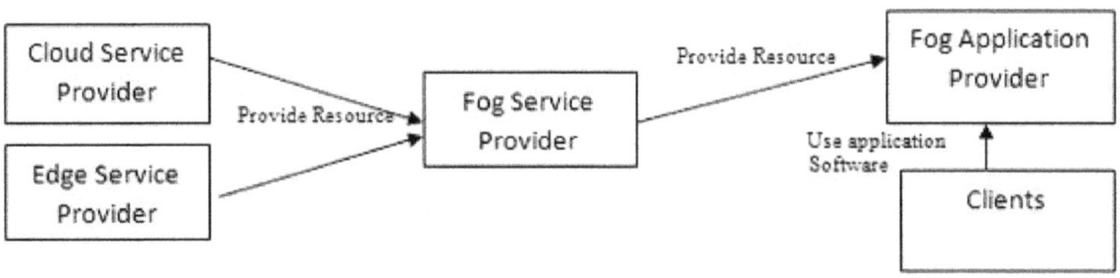

good for testing and implementing the design and it's also cost effective. We can get those results in a low cost testing model.

Simulation is used for a long time in the traditional technological market to easily retrieve the results. To simulate fog and edge computing models popular simulators are namely, *NS-2, TOSSIM, EmStar, OMNeT++, J-Sim,* and *Avrora.* These simulators are designed to test and develop network protocols in their budding step itself.

FogNetSim++ is a well known fog model simulator. It gives the option to simulate a large fog network. It enables the usage of fog nod scheduling and to tackle algorithms as well as handling mechanism. In this simulator it enables users to check the traffic management system that is scalability and usage of CPU and memory usage while simulating.

Another simulator which is used to deploy the fog environment and use it is *iFogSim* in this simulator the performance measure of latency, energy consumption and network complexity usage. This simulator measures performance results and simulate edge devices, cloud data-centers, sensors, network links, and stream processing application.

IOT Devices

IoT devices are some of the most available devices nowadays. These servers can produce a large amount of data at a point of time form those remote device. However the real time analysis can't be done if there is a delay in data processing which is solved by edge computing which a challenge in cloud was computing. A table is shown where the computing environment of IoT devices is shown. (Wang, 2017)

Figure 13. Taxonomy of IoT based edge device
Source: Hassan, Gillani, Ahmad, Imran, Yaqoob, 2019

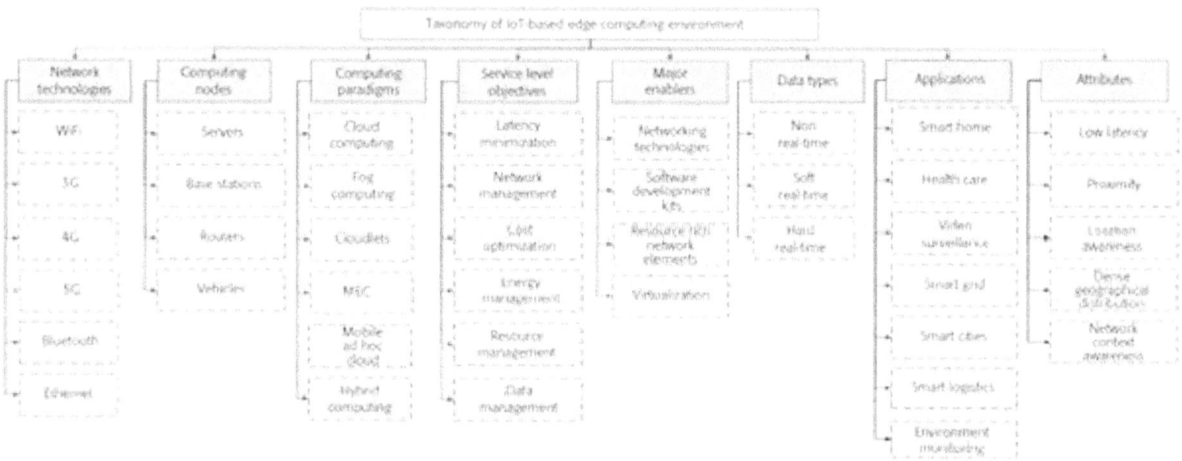

Service Levels

To implement this model various service levels are introduced some of them are discussed below.

- **Minimization of Latency:** High latency is the major problem to transfer and real time analysis in the IoT devices thus the latency should be minimized to fulfill the quality of service (QoS) requirement.
- **Network Management:** A well maintained and managed network service is required for seamless data analysis and transfer. Inefficient congestion control should be avoided as it degrades the networking condition.
- **Cost Optimization:** Implement such a high level infrastructure is also very cost effective it takes a huge costing to implement the system. Most of the expenses are in installing the nodes. Installing optimal number of nodes can reduce or lower the capital investment.
- **Energy Management:** It was found in a study that one trillion IoT device nodes need the sensing platform, thus energy is a big factor and we cant's afford to waste thus management is very much important.
- **Resource Management:** Optimal usage of resources is much required.
- **Data Management:** The large of IoT device present create a huge number of data thus managing and those data are important to draw conclusions from them.

Edge Analytics for Big Data

Advancing technology in the field of IoT led to the Smart Cities, it's one of the most modern innovative ideas in making and it's also turning into reality, where the entire city is wrapped with sensors and every movement is being sensored with the saving of natural to manmade resource could be possible. Consequently this huge amount of data has made the way for Big Data where the dataset is huge. The edge device has an automatic platform to support the deep learning procedure to analyze those data. It's used by localization of epileptogenicity using multi nodal rs-fMRI and EEG (electroencephalography) this system enables real time quick analysis of the data. (Yadav & Sohal, 2017)

Mobile Edge Computing (MEC)

With the huge wave of IoT devices and telecommunication hitting the 5G level it's much needed that the current data transfer gone to the edge, edge computing emerges and mobile edge computing is another peak in that revolution. The concept of the mobile edge computing is bursting. It pushes the computing so that the data can be transferred with low latency. It gives major hope in decrement of energy consumption which is an important factor nowadays. In the past decade the rise of cloud computing was at its peak.

But the modern time saw a major change; the paradigm is shifting from the traditional cloud computing towards edge computing. It's estimated that ten billion edge computing enabled device will be in use near future. The complexity of chip and the growth of IC in a chip is growing exponentially, following the Moore's Law. The main back draw of the cloud computing technology is the propagation delay, and proximity delay.

CLOUD SECURITY AND CRYPTOGRAPHY: A PROTECTIVE LAYER TO CLOUD ANALYTICS

Cloud Computing provides many facilities and advantages, but as everything coin has two sides it also has its own challenges and problems. Security issues are there, protection of users data is one of the most difficult challenges to be face and to be won. Various security issues are there as it merges a lot of technologies such as database, OS (Operating System), virtualization, resource scheduling, memory management, and concurrency problem.

- **Privacy of Data:** One of the biggest challenge, as everyone's personal data is very important. And consumers nowadays feel free to upload and save their data in a cloud platform. But they are not aware of where there data is going and how they can be bypassed by unauthorized persons to exploit the data. Thus throwing is as a big challenge.
- **Confidentiality of Data:** Another challenge of the modern cloud community, data confidentiality means that only authorized person can have the access to the owner's data. Encryption comes very handy here to ensure security.
- **Integrity of Data:** Data Integrity means that the data which has been stored in the cloud can be stored safely from being modified from any unauthorized access. Digital Signature and RAID technique are often adopted to solve this challenge.
- **Data Remanenece:** In order to maintain the cycle, it's recommended that the data from cloud should be deleted after a particular time cycle. The reformatting or deletion doesn't ensure complete deletion, those data can be accessed later by somehow means. Thus creating a challenge for the cyber security personals.
- **Transmission of Data:** Transmission of data means that the data when transferred from user to cloud it goes via a path, a client path, then the data is returned to cloud from the client server. While transmitting this data it is advised that the data should be encrypted so that it can be transferred safe. But as encryption-decryption process takes a lot of time this step is often neglected thus while transmitting the data, it remains unsafe and unauthorized access becomes easy. (Wang, 2017)

CRYPTOGRAPHY TECHNIQUE

Cryptography is a process of securing information through codes, so that no unauthorized person can have the access of those codes, and those data remains safe. In "cryptography", "crypt" means "hidden" and "graphy" means "writing". It simply secures the data. (Prerna & Agarwal, 2017)

In the field of technology of computer science, cryptography stands for secure information and data processing derived from various mathematical computation and set of rules called algorithms. To decrypt the data or that message so that no unauthenticated person can get those data. The components of cryptography are discussed below. (Jaber & Zolkipli, 2013)

- **Plaintext:** The raw or original form of data, that will be secured
- **Ciphertext:** After encryption done on plaintext, the unreadable form which is created caller ciphertext.

- **Encryption Algorithm:** The mathematical process in which plain text gets converted into ciphertext.
- **Decryption Algorithm:** The mathematical process in which the ciphertext gets converted into plaintext.
- **Encryption Key:** The value or key which is used by sender with the algorithm to convert plaintext into ciphertext.
- **Decryption Key:** The value or key which is used by receiver with algorithm to convert ciphertext into plaintext. (Chakraborty et al., 2018)

Figure 14. Cryptography
Source: Created by Author

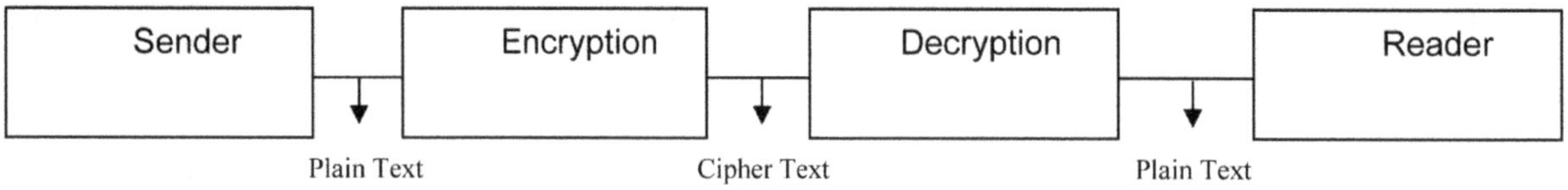

CRYPTOGRAPHIC ALGORITHMS

There are various algorithms which are being used to secure the data of the users. Cloud data security can be broadly classified into 3 (three) parts namely, Privacy Preservation, Storage Security, and Data Security. (Kumar & Gobi, 2017)

In Cryptography the plain text is encrypted and that text can't be decrypted by no computer or any person, only a correct cypher can do it.

The algorithms can be divided into 2 parts as given in next chart.

SYMMETRIC ALGORITHM

The symmetric algorithm is also known as symmetric key encryption or Secret Key Encryption. In this type of algorithm the encryption and decryption of data is done by one single key, and that key is kept secret thus it name as secret key algorithm. Power consumption in this algorithm is also less than the others. (Harmon, 2018)

ASSYMETRIC ALGORITHM

The asymmetric algorithm is also known as public key cryptography. In this type 2 different types of keys are being used, one is used for the encryption part and the other key is used for decryption. Key distribution problem is avoided by using this algorithm. Here the power consumption is high and thus slower than symmetric algorithm as huge data is being exchanged.

A comparative study is being presented by the next chart of various algorithms which are being used.

Figure 15. Classification of Cloud Storage Security System
Source: Created by Author

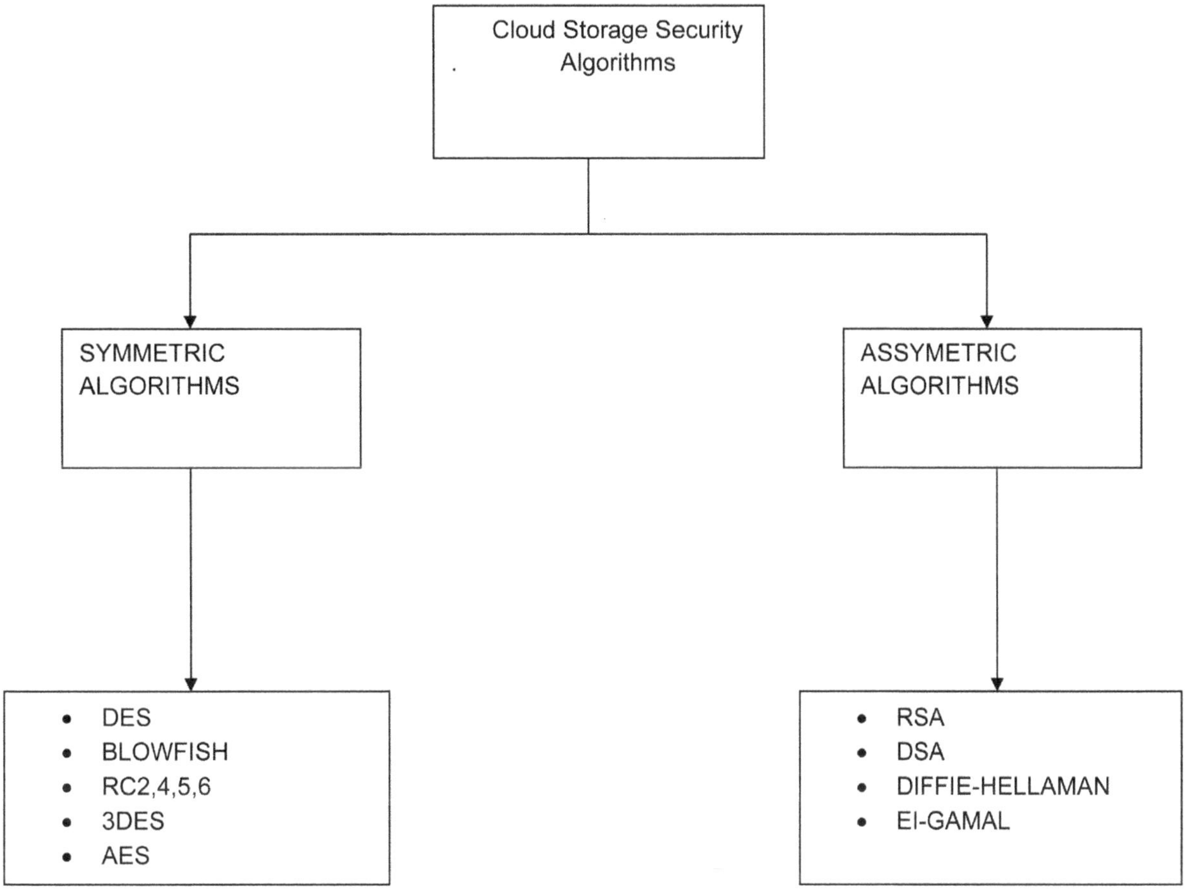

CONCLUSION

Various organizations are now investing in analysis and storing their data in cloud. Analysis making more informed choice decisions which is reflected as benefit in their revenue system. Cloud and Big Data combination can be used for network optimization and we can identify any system breach which leads to better security over data.

The introduction of various new concepts of IoT may lead to various growths in industry and can provide more jobs to data scientist and analysts.

Some of the applications of the Cloud Analytics are

- **Social Media:** The most important aspect of cloud analytics, from the data and analysis report of social media we can easily get an insight of a person's likes and dislikes.
- **Tracking Products:** Various Companies like Amazon uses data analytics to track their product and find out singular preference list for its entire user base.
- **Tracking Preference:** All the information's are stored in cloud and using analytics tool we can determine user preferences.

Table 3. Comparative Study of different algorithms

CHARECTERISTICS	AES	RSA	BLOWFISH	DES
Key Size (Bits)	128, 192, 256	1024	32-448	64
Key Used	Encryption and Decryption key is same.	Encryption is Public Key and Decryption is Private Key.	Encryption and Decryption key is same.	Encryption and Decryption key is same.
Scalability	Scalable	Non-Scalable	Scalable	Scalable
Vector Size (Initial)	128	1024	64	64
Security Provided	For Provider and User	Only for User	For Provider and User	For Provider and User
Capacity of Data Encryption	Large Amount of Data.	Small Amount of Data.	Less than AES	Less than AES
Authentication	Best Authentication.	Authentication is Robust.	Same as AES.	Less than AES
Memory Usage	Low RAM usage	Highest	Memory required less than 5 KB.	More than AES.
Run Time	Fastest	Slowest	Moderate	Same as AES

Source: Created by Author

REFERENCES

Brindha, K., & Jeyanthi, N. (2015, February). Securing cloud data using visual cryptography. In *International Confernce on Innovation Information in Computing Technologies* (pp. 1–5). IEEE. doi:10.1109/ICIICT.2015.7396073

Chakraborty, M., Jana, B., & Mandal, T. (2018, July). A Secure Cloud Computing Authentication Using Cryptography. In *2018 International Conference on Emerging Trends and Innovations In Engineering And Technological Research (ICETIETR)* (pp. 1-4). IEEE. 10.1109/ICETIETR.2018.8529100

Deshmukh, S., & Sumeet, S. (2015, December). Big Data Analytics Using Public Cloud Infrastructure: Use Cases and Cost Economics. In *2015 International Conference on Computational Intelligence and Communication Networks (CICN)* (pp. 782-784). IEEE. 10.1109/CICN.2015.159

Harmon, E. (2018). *Strategies used by cloud security managers to implement secure access methods.* Academic Press.

Jaber, A. N., & Zolkipli, M. F. B. (2013, November). Use of cryptography in cloud computing. In *2013 IEEE International conference on control system, computing and Engineering* (pp. 179-184). IEEE. 10.1109/ICCSCE.2013.6719955

Khan, Z., Anjum, A., & Kiani, S. L. (2013, December). Cloud based big data analytics for smart future cities. In *2013 IEEE/ACM 6th International Conference on Utility and Cloud Computing* (pp. 381-386). IEEE. 10.1109/UCC.2013.77

Kumar, G. K., & Gobi, D. M. (2017). Role of Cryptography & its Related Techniques in Cloud Computing Security. *International Journal for Research in Applied Science and Engineering Technology, 5.*

Manekar, A. K., & Pradeepini, G. (2015, December). Cloud based big data analytics a review. In *2015 International Conference on Computational Intelligence and Communication Networks (CICN)* (pp. 785-788). IEEE. 10.1109/CICN.2015.160

Mayilvaganan, M., & Sabitha, M. (2013, December). A cloud-based architecture for Big-Data analytics in smart grid: A proposal. In *2013 IEEE International Conference on Computational Intelligence and Computing Research* (pp. 1-4). IEEE. 10.1109/ICCIC.2013.6724168

Prerna, P., & Agarwal, P. (2017). Cryptography Based Security for Cloud Computing System. *International Journal of Advanced Research in Computer Science*, *8*(5).

Ramamoorthy, S., & Rajalakshmi, S. (2013, July). Optimized data analysis in cloud using BigData analytics techniques. In *2013 Fourth International Conference on Computing, Communications and Networking Technologies (ICCCNT)* (pp. 1-5). IEEE. 10.1109/ICCCNT.2013.6726631

Sayeed, Z., Liao, Q., Faucher, D., Grinshpun, E., & Sharma, S. (2015, June). Cloud analytics for wireless metric prediction-framework and performance. In *2015 IEEE 8th International Conference on Cloud Computing* (pp. 995-998). IEEE. 10.1109/CLOUD.2015.135

Sujitha, K., & Praveen, K. (2015, January). Analysing cloud simulation results using big data analytics model. In *2015 International Conference on Computer Communication and Informatics (ICCCI)* (pp. 1-6). IEEE.

Vuppala, S. K., Dinesh, M. S., Viswanathan, S., Ramachandran, G., Bussa, N., & Geetha, M. (2017, November). Cloud based big data platform for image analytics. In *2017 IEEE International Conference on Cloud Computing in Emerging Markets (CCEM)* (pp. 11-18). IEEE. 10.1109/CCEM.2017.11

Wang, R. (2017). Research on data security technology based on cloud storage. *Procedia Engineering*, *174*, 1340–1355. doi:10.1016/j.proeng.2017.01.286

Yadav, S. L., & Sohal, A. (2017). Review Paper on Big Data Analytics in Cloud Computing. *International Journal of Computer Trends and Technology*, *49*, 156–160.

Section 3

Case Studies From Business and Industry

Chapter 9
Analyzing EPQ Inventory Model With Comparison of Exponentially Increasing Demand and Verhult's Demand

Kuppulakshmi V.
Queen Mary's College, India

Sugapriya C.
Queen Mary's College, India

Jeganathan Kathirvel
Ramanujan Institute for Advanced Study in Mathematics, University of Madras, Chennai, India

Nagarajan Deivanayagampillai
https://orcid.org/0000-0003-1411-532X
Hindustan Institute of Technology and Science, India

ABSTRACT

This research investigates the comparison of inventory management planning in Verhult's demand and exponentially increasing demand. The working process is different in both the cases coupling the parameters and points out the constraints for the optimal total cost in both the cases. This analysis shows that rate of deterioration and percentage of reworkable items is considered as decision variable in both (1) exponentially increasing demand and (2) Verhult's demand. While comparing, the total cost in Verhult's demand pattern is more profitable production process. A substantial numerical example is considered to investigate the effect of change in the total cost in both the demand function. A sensitivity analysis is developed to study the effect of changes in total cost.

DOI: 10.4018/978-1-7998-4706-9.ch009

INTRODUCTION

In many researches EPQ model for deteriorating items are discussed to increase profit and reduce the required amount of total cost of the production so that the turnover ratio of the production increases. The main topics of this research study are percentage of defective items, reworking, inspection cost and pricing for difference parameters. So, only related researches on these topics are reviewed in this section. To achieve operation strategies goals, the industry must be able to effectively utilize resources and minimize total cost of the production plan. In the construction industry effective project planning and supervision is carried out and a deliberate implementation and enforcement of quality assurance put in place

Lushuli et al (2002) considers the problem in the presence of uncertain random and fuzziness. P. S. You (2003) extended the news paper boy problem with price dependent and allowing the customers to cancel their orders. This paper also used to determine how to price the stock. Zinjian Guo et al. (2005) describes the grey verhult's demand type for the forecasting and modified this to improved verhult's demand on time series. GM (1,1) model predict more accurate in the saturation stage.

Silas N. Onyango et al. (2010) introduced the price adjustment of the market when supply affecting the assert demand. Walvasian price adjustment assumption is made for excess demand. During the process they derived the Black- schools merton partial differential equation.

Deqiang zhou et al. (2012) analysed the grey verhulst model for oil forecasting, it resulted oil production increase according to the S- Type curved in neural network compared the result with previous pares. They concluded the optimal results are obtained in this paper. Neeraj kumar et al. (2012) presented the two ware inventory model with weibull deterioration of discount cash flow to minimize the total value cost. In this paper they assumed the total cost is affected by inflation rate and back ordering ratio. P. K. De et al. (2013) (conference) derived the concept of fuzzy based on lambda pessimistic and optimistic values to obtain the optimal total cost the expected total cost value is minimized for uncertain demand types. Dipak kumar Jana et al. (2013) solved the fuzzy stochastic genetic algorithm in random and fuzzy environment. To keep the inventory level high the holding cost and procurement cost remains high. A new concept of effect of inflation and learning effect are considered in stochastic and fuzzy stochastic environment.

D.K. Jana et al. (2013) explained briefly about the disposal and recycling of the product in fuzzy and bi fuzzy method. Genetic algorithm is also used for environment pollution. They considered the recycling products are again used as a new product. Dharmendra Yadav et al. (2015) explained about the credit period and overdue period for retailer's payment. Fuzzy profit functions, the interest yearned and the interest paid are used as triangular fuzzy number to maximize the profit. Asoke kumar Bhunia et al. (2015) discussed the two – storage inventory in rented ware house and own ware house. They compared the preservation technology of rented and own ware house and proved that the rented ware house is best in different scenario.

Anand Chauhan et al. (2015) compared verhult's demand rate with ramp type demand rate and trapezoidal type demand and proved that verhult's type is more profitable. They explained the demand pattern from the birth to the maturity stage with a mathematical concept. J.K. Syed et al. (2016) discussed the fuzzy inventory model using triangular fuzzy number they simply calculated the total cost in crisp and fuzzy model. Signed distance method is used to defuzzify the given concept.

S. Dari et al. (2017) explained the EPQ model with stockiest storages and deterioration of items are delayed. This paper is an extension of "An EPQ model for delayed deteriorating items with quadratic demand". Fabiano L.Ribeiro (2017) examined the verhult's demand type for the population growth of the molecules with logarithm and exponential functions. They concluded a universal pattern, common in all types of growth using verhult's demand.

Nita. H Shah et al. (2018) examined the article with price sensitive demand they introduced the preservation technology process to reduce the deterioration rate, so the total profit is increased. A brief analysis is given in the impact of decision variables between -20% to +20%. R.Udayakumar et al. (2018) compared the own warehouse and rented ware house with two levels of storage they compared the model I and model II under different scenario to get the uniqueness of optimal solution.

Valentin Pando et al. (2019) adopted the deterministic inventory model with holding cost in non – linear function of time and stock level. They conclude that paper with increase in profitability ratio and minimizing the inventory cost. G.S Mahapatra et al. (2019) considered the advertisement dependent demand with holding cost as triangular fuzzy number. The effect of inflation in stock allows the permissible delay in payments. They considered the three weibull distribution pattern to obtain the optimum solution.Yusuf Ibrahim Gwanda et al. (2019) flourished the production inventory model with verhult's demand to maximize profit. The amelioration rate is time dependent and three cases of production, growing and matured stage are briefly explained in this paper.

V. Kuppulakshmi et al. (2020) explained the paper with penalty maintenance cost and rework. This paper focused on the three different demand function and finally, optimum ordered quantity and total cost is derived when number of cycles increased. The maximization of profit is obtained by using fuzzy environment. In this paper comparison of two demand cases are discussed, finally it is proved that Verhul's demand is more profitable under the different production process with common parameters.

Assumptions

1. Shortages are allowed but they are fully satisfied by the reworked products.
2. Production of perfect products must be more than the demand.
3. Damaged goods are continuously collected by the labor team during the manufacturing period.
4. No machine breakdown occurs during the process.
5. During the shipment period $\left[0, T_2\right]$. products are sent to the customers at regular interval of period.
6. During the shipment period inventory level remains stable so, penalty maintenance cost is allowed in this period.
7. Rate of rework is constant.
8. Case I: exponential demand is considered.
   Case II: verhult's demand is considered.

Notations

I_{p_1} Operational inventory level in manufacturing period

I_{p_2} Operational inventory level in non-production period

I_{P_3} Operational inventory level in repairing time

I_{r_1} Retrievable inventory in production time

I_{r_2} Retrievable inventory in non-manufacturing duration

I_{r3} Retrievable inventory in rework time

$I_{\tau P_1}$ Total operational inventory in manufacturing time

$I_{\tau P_2}$ Net operational inventory in non-manufacturing time

$I_{\tau P_3}$ Net operational inventory in repairing time

$I_{\tau r_1}$ Net retrievable inventory in manufacturing time

$I_{\tau r_2}$ Net retrievable inventory in non-manufacturing time

$I_{\tau r_3}$ Net retrievable inventory in repairing time

I_{V_1} Net retrievable inventory in non-manufacturing time

I_{max} Maximum inventory level of retrievable products in manufacturing setup

I_{maxRR} Maximum inventory level of retrievable products in repairing procedure

P_r Rework process rate (per unit)

A_p Manufacturing setup (cost/setup)

A_r Repairing production setup (cost/setup)

H_p Operational products holding price (per unit)

H_r Retrievable item holding price (per unit)

ω Deteriorating rate

P Production rate

K Production cost per unit item

c Shortage cost per unit time

μ Time period at which the deterioration of the product starts

g Percentage of good quality products

ϕ Percentage of reworkable items

SHC Shortage cost in rework period

LBC Labor cost during production period

SPC Shipment cost

I_c Inspection cost

I_r Inspection rate

θ_T Total deteriorating unit

T_1 Total time in production period

T_2 Total time in shipment period

T_3 Total time in rework period

μ Time period at which the deterioration of the product starts

n Number of production setup in one cycle

D_F Verhult's demand

$D_{F(0)}$ Initial population of product

E_K Carrying capacity of the production environment

$\varphi \ \frac{F_r}{E_K}$ (constant of proportion)

S Shortage cost per cycle/\$

VERHULT'S DEMAND

During the production period e demand rate is directly proportional to the amount of manufacturing rate and it cannobe exceed the demand of the previous production rate of the machine.

Figure 1. Verhult's demand during production period

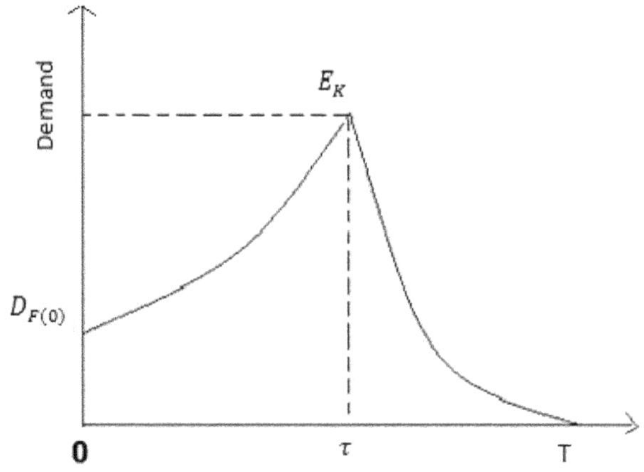

$$\frac{dD_F}{dt} = \frac{P}{E_K} D_F \left\{ E_K - D_F \right\} \qquad where \frac{P}{E_K} = \varphi = 0.01.$$

$$D_F = \frac{Ce^{\varphi\tau} E_K}{1 + Ce^{\varphi\tau}}.$$

When t=0

$$D_F(0) = \frac{CE_K}{1+C}.$$

From the above equation we get the constant

$$C = \frac{D_F(0)}{E_K - D_F(0)}.$$

Figure 2. Inventory control system using increasing exponential demand

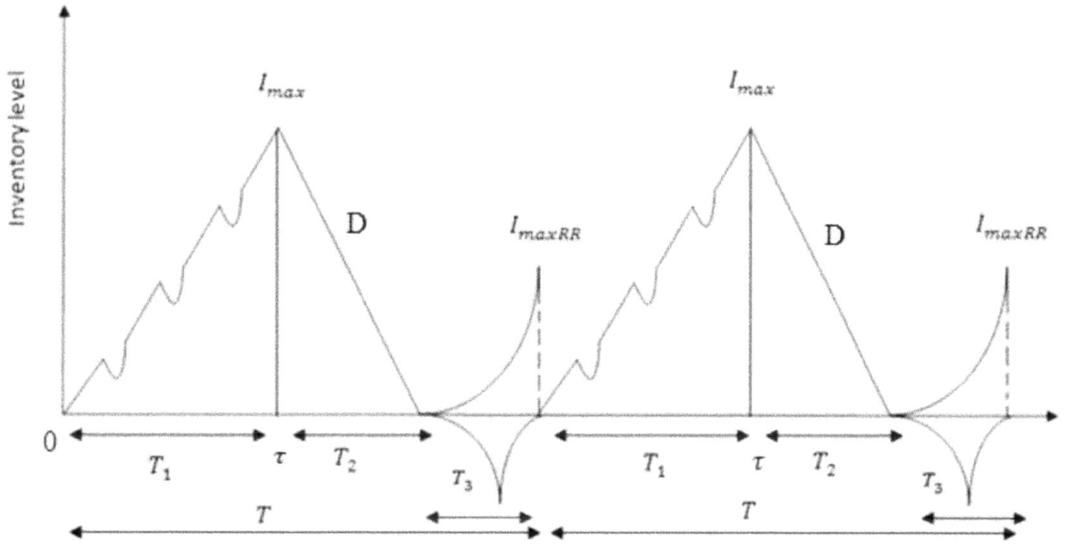

Figure 3. Inventory control system using Verhult's demand (D_F)

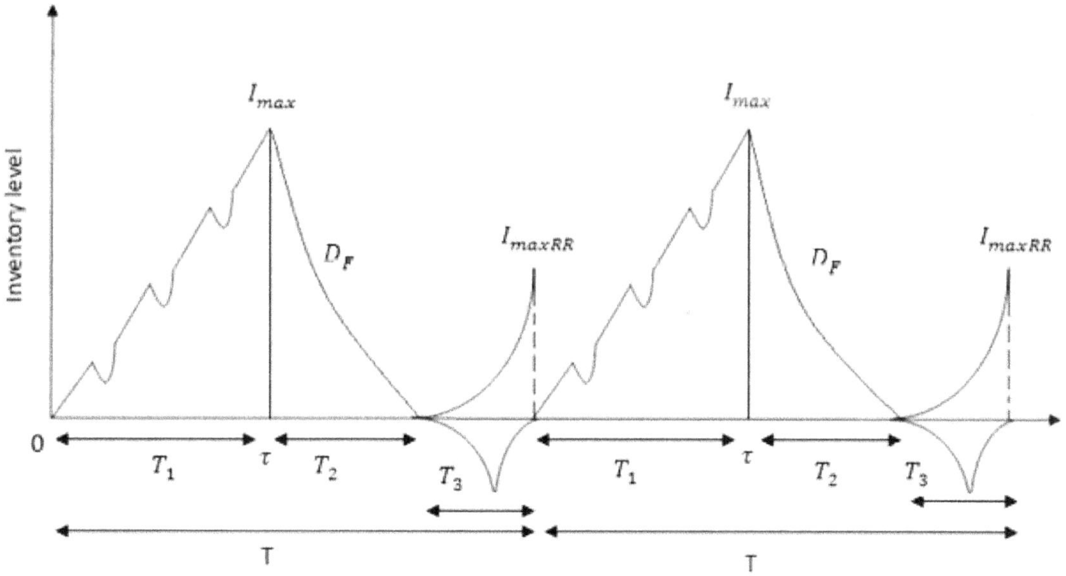

Problem Formulation

At the time $\tau = 0$. the production starts and the inventory is zero at this time. After some time damaged products are produced, and these damaged products are continuously reviewed by labors team by shipment and the damaged products are collected separately during the production time for rework. During the period $\left[0, T_1\right]$. production and supply start simultaneously and the inventory level reaches maximum I_{max}. at the time $\tau = T_1$. The production ends at this time and ohand inventory reduces by deterioration as well as by supply to customers. The demand of products takes place during the period $\left[0, T_2\right]$., and all good quality items are delivered (time uts). During $\left[0, T_3\right]$. period the rework production starts, and it satisfies the demand as well as shortages. During this rework period inventory level reaches I_{maxRR}.

Manufacturing Inventory level

Exponentially increasing demand is assumed for this manufacturing model. Were $\propto > 0 \, and \, 0 < \beta < .$ 1.

The stock level of produci products grows gradually from the period, and the damaged procts are collected separately. And so, the stock level in manufacturing time is given as

$$\frac{dI_{P_1}\left(\tau_1\right)}{d\tau_1} + \omega I_{P_1}\left(\tau_1\right) = \left(1 - \phi\right)P - \propto e^{\beta\tau_1} \qquad 0 \leq \tau_1 \leq T_1. \tag{1}$$

with the condition that $I_{P_1}\left(0\right) = 0$ .The stock level is reduced in the non production together with shipment in $\left[0, T_2\right]$. is given as,

$$\frac{dI_{P_2}\left(\tau_2\right)}{d\tau_2} + \omega I_{P_2}\left(\tau_2\right) = - \propto e^{\beta\tau_2} \qquad 0 \leq \tau_2 \leq T_2 \tag{2}$$

with the condition that $I_{P_2}\left(0\right) = 0$.

The stock level at the timef rework $\left[0, T_3\right]$. is given as

$$\frac{dI_{P_3}\left(\tau_3\right)}{d\tau_3} + \omega I_{P_3}\left(\tau_3\right) = p_r - \propto e^{\beta\tau_3} \qquad 0 \leq \tau_3 \leq T_3 \tag{3}$$

with the condition that $I_{P_3}\left(0\right) = 0$.

The Manufacturing model with the rework able items is derived by solving equation (1) using initial condition and the operational inventory level is obtained in the production period τ_1. during the interl $[0,T_1]$. as

$$\frac{dI_{P_1}(\tau_1)}{d\tau_1}+\omega I_{P_1}(\tau_1)=(1-\phi)P-\propto e^{\beta\tau_1} \tag{4}$$

$$I_{P_1}(\tau_1)=\left(\frac{(1-\phi)p}{\omega}\right)\left(1-e^{-(\omega)\tau_1}\right)-\left(\frac{\alpha}{\omega+\beta}\right)\left(e^{\beta\tau_1}-e^{-(\omega)\tau_1}\right) \tag{5}$$

By integrating the manufacturing stock level in the manufacturing time with time τ_1. gives the total inventory in a production period.

$$I_{\tau P_1}(\tau_1)=\int_0^{T_1}\left(\frac{(1-\phi)p}{\omega}\right)\left(1-e^{-(\omega)\tau_1}\right)d\tau_1-\int_0^{T_1}\left(\frac{\alpha}{\omega+\beta}\right)\left(e^{\beta\tau_1}-e^{-\omega\tau_1}\right)d\tau_1 \tag{6}$$

Considering the equation (IA.1), the third and the higher powers are neglected for small values of $\frac{(\omega)^3T_1^3}{3!}\ll 1$.

$$I_{\tau P_1}(\tau_1)=\frac{((1-\phi)p-\propto)}{2}T_1^2 \tag{7}$$

The operational inventory level is obtained in non production period τ_2 . during the interval $[0,T_2]$ as

$$\frac{dI_{P_2}(\tau_2)}{d\tau_2}+\omega I_{P_2}(\tau_2)=-\alpha e^{\beta\tau_2} \tag{8}$$

By integrating non production invtory level in the shipment period with time τ_2 gives the net stock level in a non manufacturing time

$$I_{P_2}(\tau_2)=\int_0^{T_2}\left(\frac{\alpha}{\omega+\beta}\right)\left(e^{(\beta+\omega)(T_2-\tau_2)}-1\right)d\tau_2 \tag{9}$$

$$I_{\tau p_2}\left(\tau_2\right)=\frac{\propto}{2}T_2^2 \tag{10}$$

when $I_{p_1}=I_{p_2}$., $\tau_1=T_1$. an $\tau_2=0$. then the following is obtained

$$\left(\frac{(1-\phi)p}{\omega}\right)\left(1-e^{-(\omega)\tau_1}\right)-\left(\frac{\alpha}{\omega+\beta}\right)\left(e^{\beta\tau_1}-e^{-(\omega)\tau_1}\right)=\left(\frac{\alpha}{\omega+\beta}\right)\left(e^{(\beta+\omega)T_2}-1\right) \tag{11}$$

then

$$T_2=\left(\frac{(1-\phi)p}{\alpha}\right)\left[T_1-\frac{\omega T_1^2}{2}\right]-\left[T_1-\frac{(\omega-\beta)T_1^2}{2}\right] \tag{12}$$

ck level in repairing te is given below

$$\frac{dI_{p_3}\left(\tau_3\right)}{d\tau_3}+\omega I_{p_3}\left(\tau_3\right)=p_r-\alpha e^{\beta\tau_3} \tag{13}$$

$$I_{\tau p_3}\left(\tau_3\right)=\frac{p_r-\alpha}{2}T_3^2 \tag{14}$$

At the time of production, some items are produced with defects and they are stored. Defective items are inspected by the team and those items are collected for rework process; these items can be calculated as

$$\frac{dI_{r_1}\left(\tau_{r_1}\right)}{d\tau_{r_1}}+\omega I_{r_1}\left(\tau_{r_1}\right)=\phi P,\quad 0\le\tau_{r_1}\le T_1 \text{ with } \quad I_{r_1}(0)=0 \tag{15}$$

Solving the above it is derived as

$$I_{\tau r_1}\left(\tau_{r_1}\right)=\frac{\phi P}{2}T_1^2 \tag{16}$$

At the Initial stage, the recoverable items are denoted by I_{max}

$$I_{max} = \frac{(\phi P)}{\omega}\left(1 - e^{-\omega T_1}\right) \tag{17}$$

Considering the equation (IA.2), the third and the higher powers are neglected for small values of $\frac{\omega^3 T_1^3}{2} \ll 1$

$$I_{max} = \phi P\left(T_1 - \frac{\omega T_1^2}{2}\right) \tag{18}$$

During the demand period the recoverable items are

$$\frac{dI_{r_2}\left(\tau_{r_2}\right)}{d\tau_{r_2}} + \omega I_{r_2}\left(\tau_{r_2}\right) = 0 \quad \text{with} \quad \tau_{r_2} \leq T_2 \tag{19}$$

initially,

$$I_{\tau r_2}\left(0\right) = I_{max} \tag{20}$$

The total stock level of retrievable products in the non manufacturing time is

$$I_{\tau r_2}\left(\tau_{r_2}\right) = \int_{\tau_{r_2}=0}^{(n-1)T_1+nT_2} I_{max}\, e^{-\omega\tau_{r_2}} d\tau_{r_2} \tag{21}$$

$$I_{V1} = \sum_{s=1}^{n} \int_{\tau_{r_2}=0}^{(s-1)T_1+sT_2} I_{max} e^{-\omega\tau_{r_2}}\, d\tau_{r_2} \tag{22}$$

$$I_{V1} = \sum_{s=1}^{n} I_{max}\left[\left((s-1)T_1 + sT_2\right) - \frac{\omega\left((s-1)T_1 + sT_2\right)^2}{2}\right] \tag{23}$$

At the completion of one manufacturing cycle, the stock level attains maximum

$$I_{maxRR=}\sum_{s=1}^{n} I_{max}\begin{bmatrix} 1-\omega\left(\left(s-1\right)T_1+sT_2\right)+ \\ \dfrac{\omega^2\left(\left(s-1\right)T_1+sT_2\right)^2}{2} \end{bmatrix} \tag{24}$$

The stock level of retrievable products in rework

$$\frac{dI_{r_3}\left(\tau_{r_3}\right)}{d\tau_{r_3}}+\omega I_{r_3}\left(\tau_{r_3}\right)=-P_r, \quad \text{with} \tau_{r_3}\leq T_3 \qquad I_{r_3}\left(0\right)=0 \tag{25}$$

$$I_{\tau r_3}\left(\tau_{r_3}\right)=\frac{P_r}{\omega}\left(e^{\omega\left(T_3-\tau_{r_3}\right)}-1\right) \tag{26}$$

The total recoverable items are

$$I_{\tau r_3}=\frac{P_{rT_3^2}}{2} \tag{27}$$

Equation (27) can be written as

$$I_{maxRR=}\frac{P_r}{\omega}\left(e^{\omega T_3}-1\right) \tag{28}$$

$$T_3=\frac{I_{maxRR}}{P_r} \tag{29}$$

The total operational and the recoverable inventory are given as

$$ToTI=nI_{\tau p_1}+nI_{\tau p_2}+I_{\tau p_3} \tag{30}$$

$$TRI=nI_{\tau r_1}+I_{V1}+I_{\tau r_3} \tag{31}$$

$$\theta_T=\left[n\phi PT_1+P_rT_3\right]-\left[n\left(\alpha+\beta I_{\tau p_1}\right)T_1+n\left(\alpha+\beta I_{\tau p_2}\right)T_2+\left(\alpha+\beta I_{\tau p_3}\right)T_3\right] \tag{32}$$

The Total Cost of the Manufacturing Cycle

The inspection amount during the production period is given as

$$INC = \frac{I_c I_r}{T_1} \int_0^{T_1} P d\tau_1 \tag{33}$$

$$INC = I_c I_r P \tag{34}$$

Price discount is allowed for bulk orders during the time $[\mu, T_2]$ is given as

$$PD = \frac{kr}{T_2} \int_{\mu}^{T_2} \alpha e^{\beta\tau_2} \, d\tau_2 \tag{35}$$

$$PD = \frac{kr}{T_2} \left[\frac{\alpha(e^{\beta T_2} - e^{\beta\mu})}{\beta} \right] \mu = 0.01 \tag{36}$$

The shortage cost is calculated during the rework period is given as

$$SHC = c \int_0^{T_3} - \alpha e^{\beta\tau_3} d\tau_3 \tag{37}$$

$$SHC = -c \left[\frac{\alpha(e^{\beta T_3} - 1)}{\beta} \right] \tag{38}$$

The total cost is given as

$$TC = \frac{nA_p + A_r + H_p(TOTI) + H_r(TRI) + \omega\theta_T + INC + PD + SHC + m(SPC)}{n(T_1 + T_2) + T_3} \tag{39}$$

Case (ii)

In verhult's demand labors are inspecting the damaged products during production time and separated the damaged products to the rework setup. Shortage products are satisfied by the reworked items. When the inventory level reaches I_{max} the products are delivered by equal shipments. Damaged goods are distributed by m shipment

Manufacturing Inventory Level for Verhult's Demand

The inventory level of producing items grows gradually from the period [0, T_1], and the damaged products are collected separately. Thus the inventory level in production period is given as

$$\frac{dI_{p_1}(\tau_1)}{d\tau_1} + \omega I_{p_1}(\tau_1) = (1-\phi)P - D_F, \quad 0 \le \tau_1 \le T_1 \tag{40}$$

with the condition that $I_{p_1}(0) = 0$

The inventory level is reduced in the non production together with shipment in $[0, T_2]$ is given as,

$$\frac{dI_{p_2}(\tau_2)}{d\tau_2} + \omega I_{p_2}(\tau_2) = -D_F \qquad 0 \le \tau_2 \le T_2 \tag{41}$$

with the condition that $I_{p_2}(0) = 0$ the inventory level in the rework period $[0, T_3]$ is given as

$$\frac{dI_{p_3}(\tau_3)}{d\tau_3} + \omega I_{p_3}(\tau_3) = p_r - D_F \qquad 0 \le \tau_3 \le T_3 \tag{42}$$

with the condition that $I_{p_3}(0) = 0$

Case (i)

By solving Equation (40) using initial condition, the operational inventory level is obtained in the production period τ_1 during the interval $[0, T_1]$ as

$$\frac{dI_{p_1}(\tau_1)}{d\tau_1} + \omega I_{p_1}(\tau_1) = (1-\phi)P - D_F \tag{43}$$

$$I_{p_1}(\tau_1) = \left(\frac{(1-\phi)p - D_F}{\omega}\right)\left(1 - e^{-\omega\tau_1}\right) \tag{44}$$

By integrating the manufacturing inventory level in the production period with time τ_1 gives the total inventory in a production period

$$I_{\tau p_1}(\tau_1) = \int_0^{\tau_1} \left(\left(\frac{(1-\phi)p - D_F}{\omega} \right) \left(1 - e^{-\omega \tau_1}\right) \right) d\tau_1 \tag{45}$$

Considering the equation (IA.1) the third and the higher powers are neglected for small values of $\frac{(\omega)^3 T_1^3}{3!} \ll 1$

$$I_{\tau p_1}(\tau_1) = \frac{p(1-\phi) - D_F}{2} T_1^2 \tag{46}$$

The operational inventory level is obtained in non production period τ_2 during the interval $[0, T_2]$ as

$$\frac{dI_{p_2}(\tau_2)}{d\tau_2} + \omega I_{p_2}(\tau_2) = -D_F \tag{47}$$

By integrating non production inventory level in the shipment period with time τ_2 gives the total inventory in a non production period

$$I_{p_2}(\tau_2) = \int_0^{T_2} \left(\frac{D_F}{\omega} \right) \left(e^{-\omega \tau_2} - 1\right) d\tau_2 \tag{48}$$

$$I_{\tau p_2}(\tau_2) = \frac{D_F}{2} T_2^2 \tag{49}$$

when $I_{p_1} = I_{p_2}$, $\tau_1 = T_1$ an $\tau_2 = 0$ then the following is obtained

$$\left(\frac{(1-\phi)p - D_F}{\omega} \right) \left(1 - e^{-\omega \tau_1}\right) = \left(\frac{D_F}{\omega} \right) \left(e^{-\omega \tau_2} - 1\right) \tag{50}$$

then

$$T_2 = \left(\left[T_1 - \frac{\omega T_1^2}{2}\right]\left[1 - \frac{(1-\phi)P}{D_F}\right]\right) \tag{51}$$

The inventory level in rework period is given below

$$\frac{dI_{P_3}(\tau_3)}{d\tau_3} + \omega I_{P_3}(\tau_3) = p_r - D_F \tag{52}$$

$$I_{\tau P_3}(\tau_3) = \frac{p_r - D_F}{2} T_3^2 \tag{53}$$

At the time of production, some items are produced with defects and they are stored. Defective items are inspected by the team and collected for rework process; these items can be calculated as

$$\frac{dI_{r_1}(\tau_{r_1})}{d\tau_{r_1}} + \omega I_{r_1}(\tau_{r_1}) = \phi P, \qquad 0 \le \tau_{r_1} \le T_1, I_{r_1}(0) = 0 \tag{54}$$

Solving the above it is derived as

$$I_{\tau r_1}(\tau_{r_1}) = \frac{\phi P}{2} T_1^2 \tag{55}$$

At the Initial stage, the recoverable items are denoted by I_{max}

$$I_{max} = \frac{(\phi P)}{\omega}\left(1 - e^{-\omega T_1}\right) \tag{56}$$

Considering the equation (IA.2) the third and the higher powers are neglected for small values of $\frac{\omega^3 T_1^3}{2} \ll 1$

$$I_{max} = \phi P\left(T_1 - \frac{\omega T_1^2}{2}\right) \tag{57}$$

During the shipment period the recoverable items are

$$\frac{dI_{r_2}\left(\tau_{r_2}\right)}{d\tau_{r_2}} + \omega I_{r_2}\left(\tau_{r_2}\right) = 0, \qquad \tau_{r_2} \leq T_2, \text{with} I_{r_2}\left(0\right) = 0 \tag{58}$$

Initially,

$$I_{r_2}\left(0\right) = I_{max} \tag{59}$$

The total inventory level of recoverable items in the non production period is same as the equation (21) to (29) for case (ii)

The total operational and the recoverable inventory are given as

$$ToTI = nI_{\tau P_1} + nI_{\tau P_2} + I_{\tau P_3} \tag{60}$$

$$TRI = nI_{\tau r_1} + I_{V1} + I_{\tau r_3} \tag{61}$$

$$\theta_T = \left[n\left(1-\phi\right)PT_1 + P_rT_3\right] - \left[nD_FT_1 + nD_FT_2 + nD_FT_3\right] \tag{62}$$

Inspection Cost

The inspection amount during the production period is given as

$$INC = nI_c = \frac{I_c}{T} \tag{63}$$

$$INC = \frac{I_cD_F}{\left(1-\omega\right)}I_{max} \tag{64}$$

Shortage Cost

The shortage of the products are fully satisfied by the reworked items

$$SHC = -\frac{P_r \log\left(\frac{S_0}{S}\right)}{\left(1-\omega\right)} I_{maxRR} \quad (65)$$

Shipment Cost

The cost is used to loading the goods for the yachting and unloading the fish products from the boat after completing the yachting is given under shipment cost.

$$SPC = \frac{A_P D_F}{\left(1-\omega\right)} I_{max} \quad (66)$$

The total cost of the production set up is given as,

$$TC = \frac{nA_P + A_r + H_p\left(TOTI\right) + H_r\left(TRI\right) + \omega\theta_T + INC + +SHC + m\left(SPC\right)}{n\left(T_1 + T_2\right) + T_3} \quad (67)$$

Numerical Illustration

To illustrate this inventory model the following parameters are used, P = \$1500/unit, $P_r = \$900$, $D_F = \$700$, $F_M = 0.005$, $F_s = 0.008$, $S_o = 1.5$,S=0.5 $\omega = 0.08$, $A_P = \$9 / unit$, $A_r = \$1.5 / unit$, $H_P = \$4.5 / unit$, k = 5, $H_r = \$0.6 / unit$, $\phi = 0.06$, $\pi = 3.414$, r = 0.1, $\mu = 0.001$, $I_c = \$0.5 / unit$, $I_r = 2$, m = 5, SPC = 100, $\alpha = 100$, $\beta = 0.05$, $c = 0.15$

For case (i) one production process with exponentially increasing demand is considered with $T_1 = 0.7158$ then TC = \$ 2,18,450

For case (ii) one production process with verhult's demand is considered with $T_1 = 0.0831$ then TC = \$ 43,586

While comparing the exponential increasing demand and verhult's demand this analysis shows better results exist in verhult's demand production process.

Algorithm

Considering the total cost for the production period alone in equation (39) for case (i) and equation (67) for case (ii)

Step 1: Substitute n = 1 in the equations (39) and (67) differentiate with the condition $\frac{\partial TC\left(T_1\right)}{\partial T_1} = 0$

Step 2: Find the value T_1 for equations (39) and (67) Increase the values of n to get the corresponding T_1 values.

Step 3: Substitute T_1 in equations (39) and (67) to get TC for case (i) and case (ii).

The optimal value for T_1 is obtained by iterative method. MATLAB 2014 is used to find the experimental result of equations (39) and (67).

Step 4: By iterating the values of T_1, the optimal solution is found as a nearby value of T_1 we get

the Total cost is given in the below table

By iterating the values of T_1 for case (i) and case (ii) the optimal total cost value is found is given as,

Table 1. Calculation of Total cost in exponentially increasing demand

T_1	0.7158	0.7029	0.6901	0.6772	0.6643	0.6515	0.6386	0.6257	0.6129
TC	218450	207900	197770	187900	178370	169230	160340	151770	143570

Figure 4. Graphical representation of exponential increasing demand (table 1)

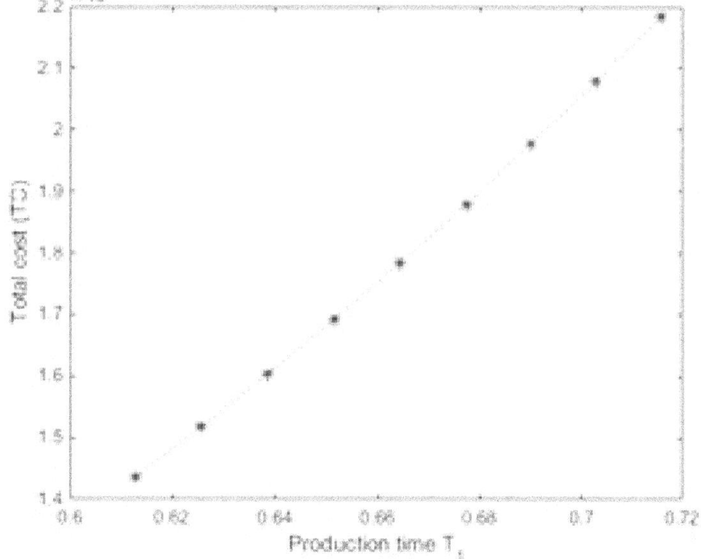

Table 2. Calculation of Total cost in Verhult's demand

T_1	0.0831	0.0816	0.0802	0.0787	0.0773	0.0758	0.0744	0.0729	0.0715
TC	43586	42030	40604	39103	37727	36280	34955	33563	32289
T	0.1753	0.1722	0.1692	0.1661	0.1631	0.1600	0.1570	0.1539	0.1509

Figure 5. Graphical representation of Verhult's demand (table 2)

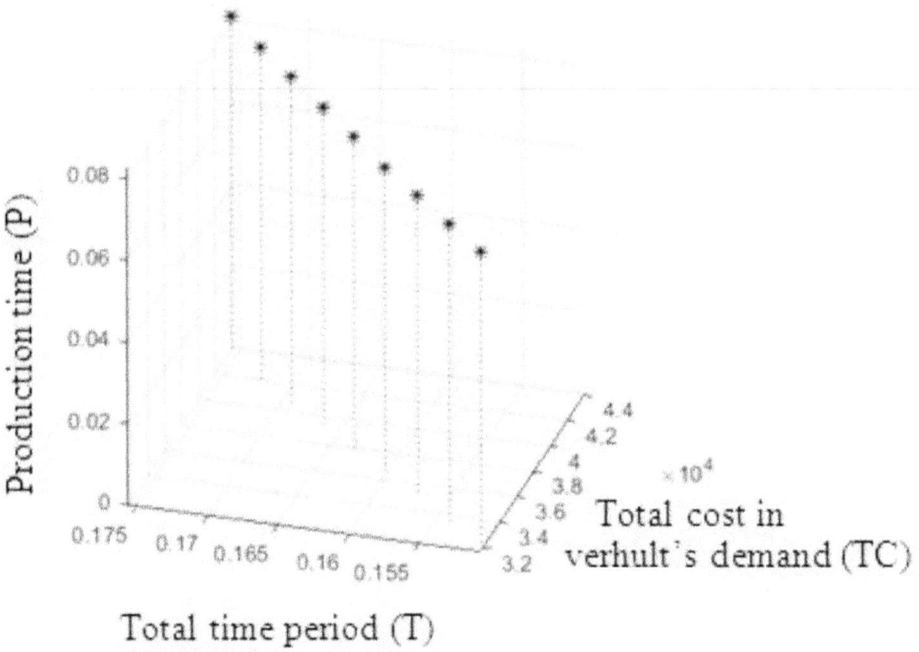

RESULT AND DISCUSSION OF SENSITIVE ANALYSIS

Sensitivity analysis for different parameters like ω, β, ϕ, P_r, P, I_c, SPC, A_p, H_r is analyzed with case (i) and for case (ii). Different parameter values can be substituted to find the total cost in equation (39) for case (i) and (67) for case (ii). In the sensitivity analysis the total cost is calculated for different parameters, one parameter value is varied to -30%, -20%, -10%, +10%, +20% and +30% to find the total cost. The importance of this analysis is one parameter is changed and the rest of the parameters are fixed to find the total cost is given in table 4 and table 5.

Table 3. Calculation of total cost in Exponential dependent demand

parameter	-30%	-20%	-10%	10%	20%	30%
ω	224400	221210	220420	216490	214560	212630
β	219090	218750	218660	218230	218020	217810
ϕ	229890	223740	222220	214720	211050	207420
P_r	218460	218450	218450	218450	218440	218440
P	77167	140430	160400	289080	373520	472970
I_c	214030	216390	216980	219920	221390	222860
SPC	216980	217760	217960	218940	219430	219920
A_p	218420	218430	218440	218460	218460	218470
H_r	217840	218200	218290	218750	218970	219200

Figure 6. Deterioration rate / Total cost

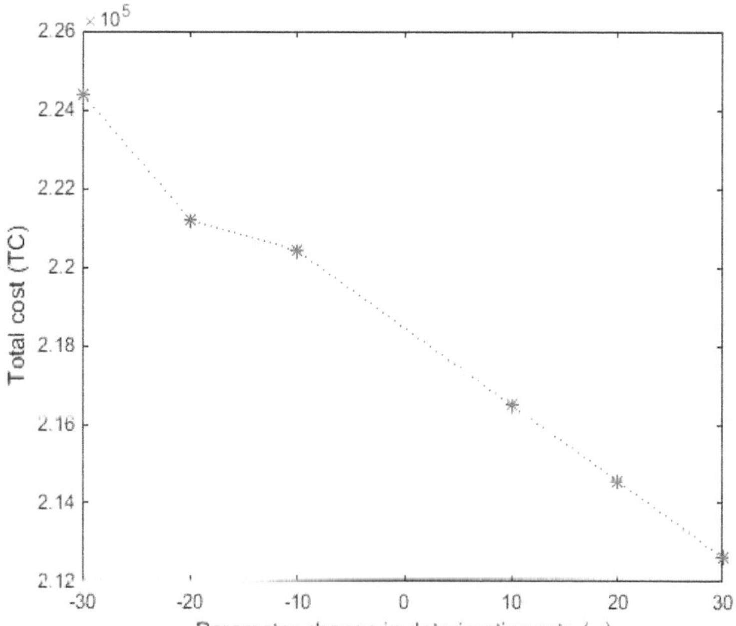

Figure 7. Exponential constant / Total cost

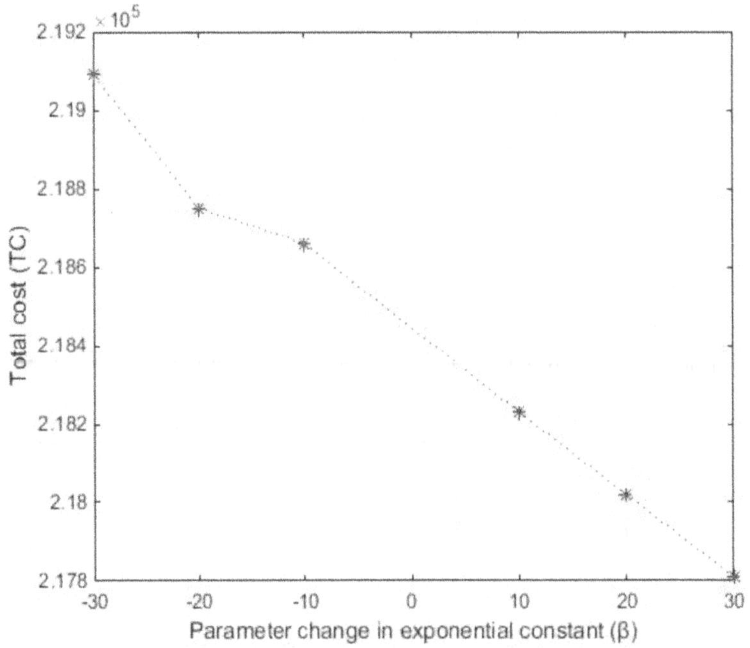

Figure 8. Percentage of reworkable item / Total cost

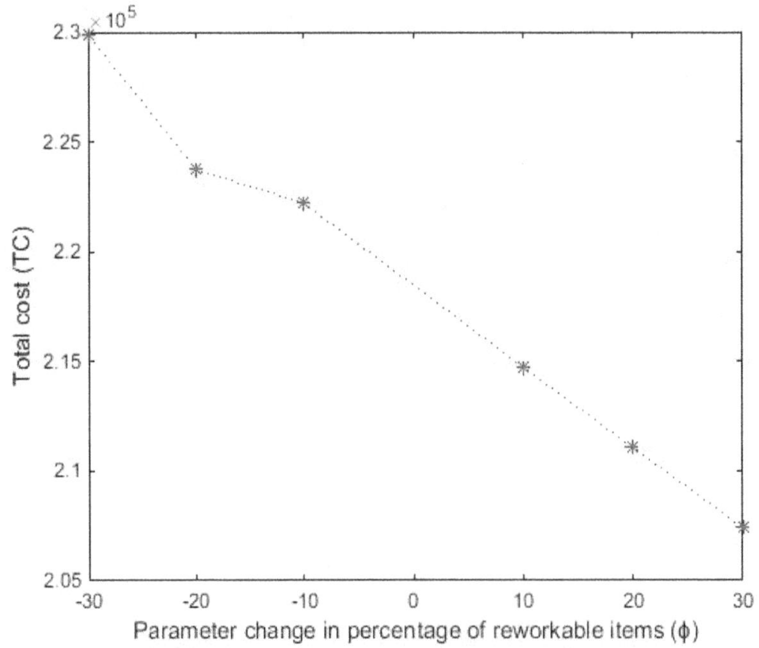

Figure 9. Rework process rate / Total cost

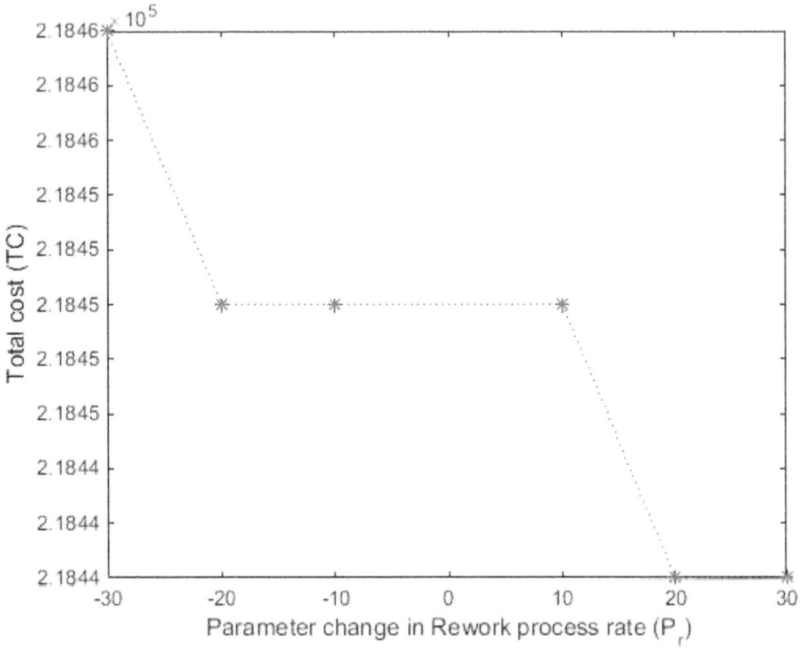

Figure 10. Production rate / Total cost

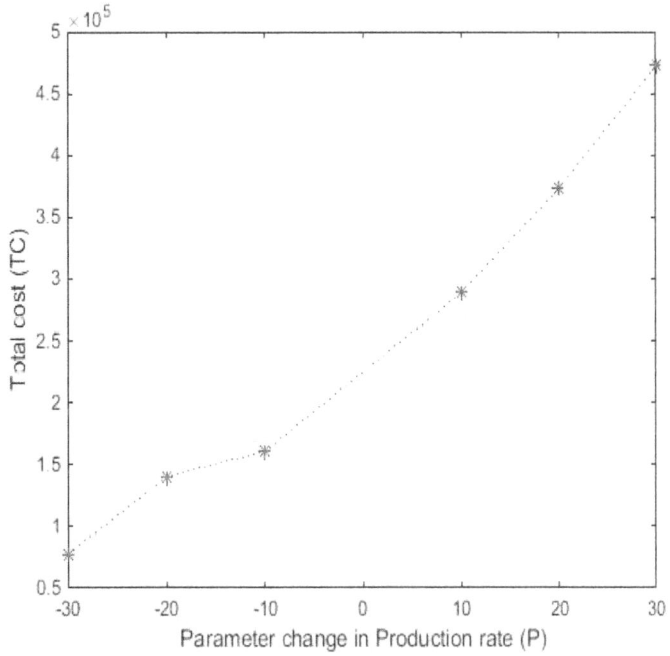

Figure 11. Inspection cost / Total cost

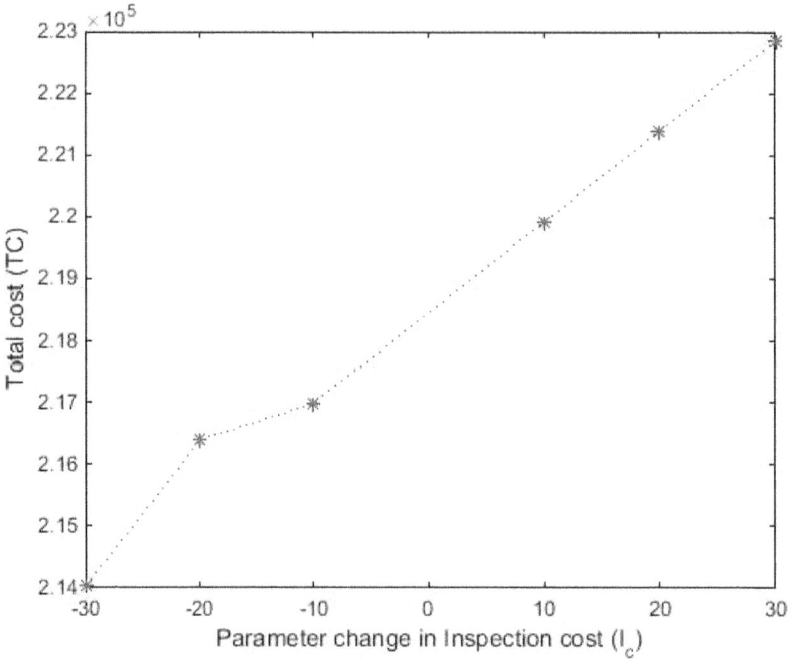

Figure 12. Shipment cost / Total cost

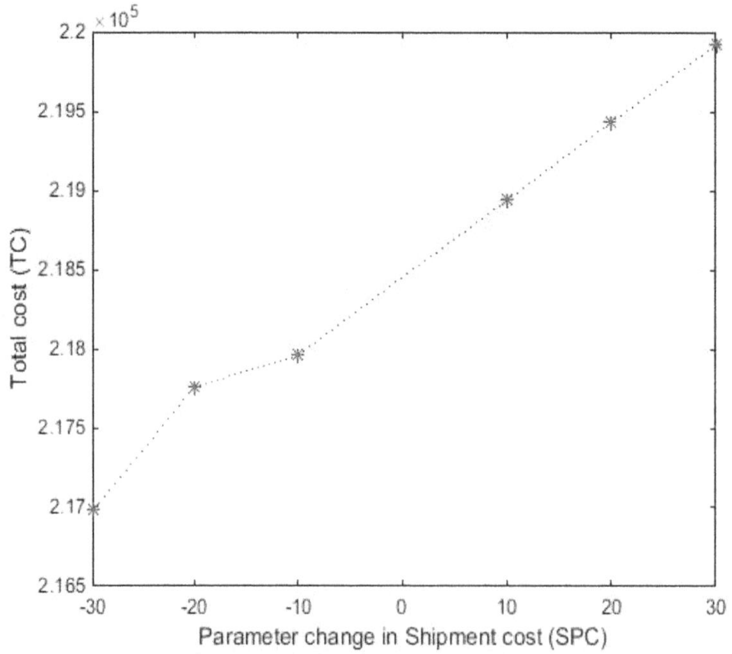

Figure 13. Manufacturing setup / Total cost

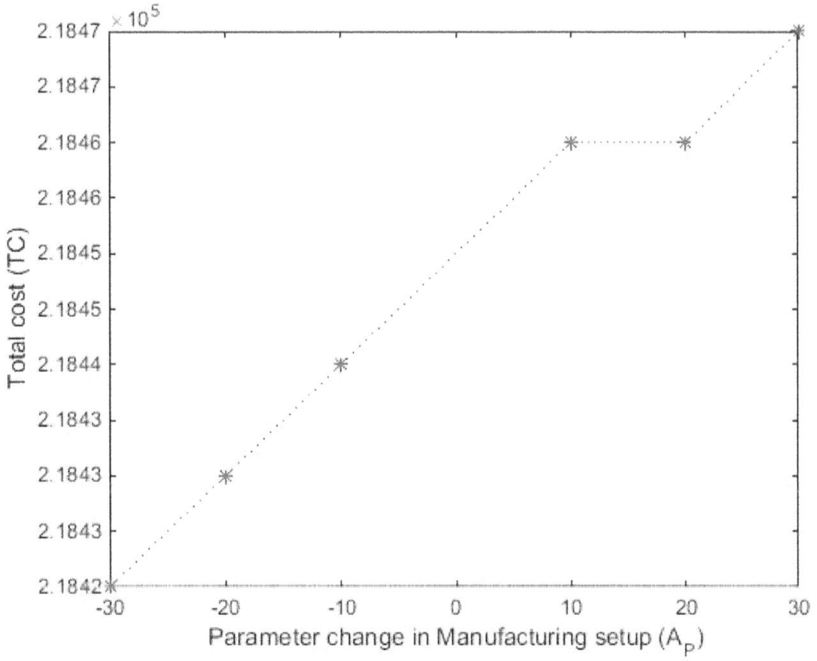

Figure 14. Rework holding cost / Total cost

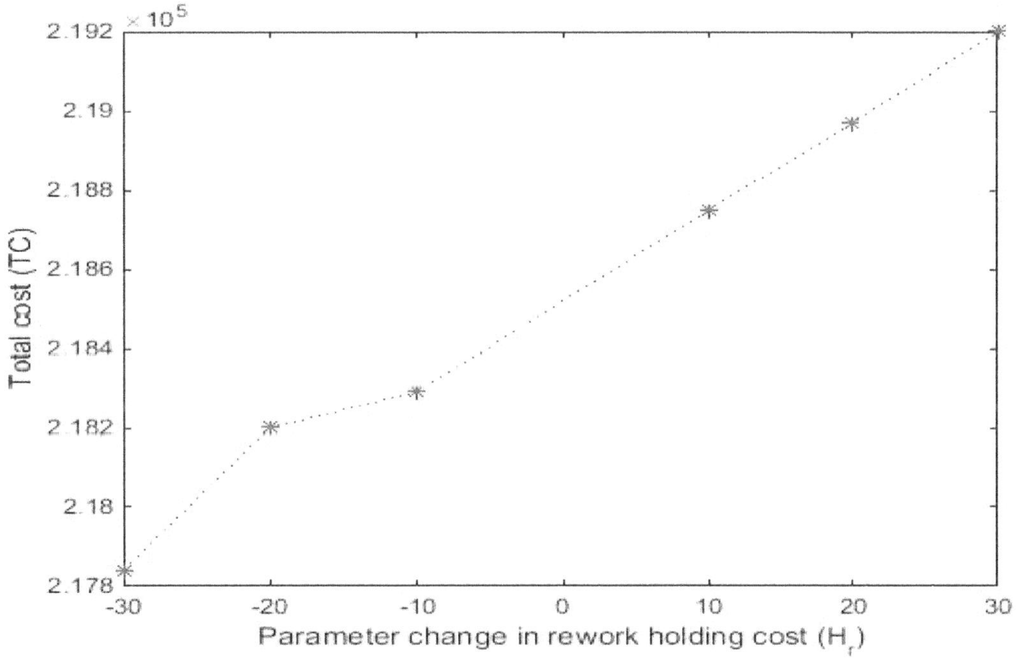

From table 4, result shows that the total cost is increased for all the parameters such as Production rate(P), Inspection cost $\left(I_c\right)$, Shipment cost (SPC), Production setup $\left(A_p\right)$. In exponential increasing demand, the total cost is decreases when the parameters exponential constant (β), deterioration rate (ω)., percentage of reworkable item(ϕ), rework process rate (P_r) are veried from -30% to +30%.

Table 4. Calculation of total cost in verhult's demand

parameter	-30%	-20%	-10%	10%	20%	30%
ω	42541	43092	43232	43947	44314	44687
ϕ	31095	37819	39478	47638	51635	55577
P_r	43733	43655	43635	43537	43488	43439
P	21371	32244	35310	52733	62750	73639
A_p	30791	37615	39321	47851	52116	56381
S_0	44018	43769	43714	43471	43365	43268
D_F	42868	43077	43119	43286	43351	43406
m	30424	37243	38948	47472	51734	55996

Figure 15. Deteriorating rate / Total cost

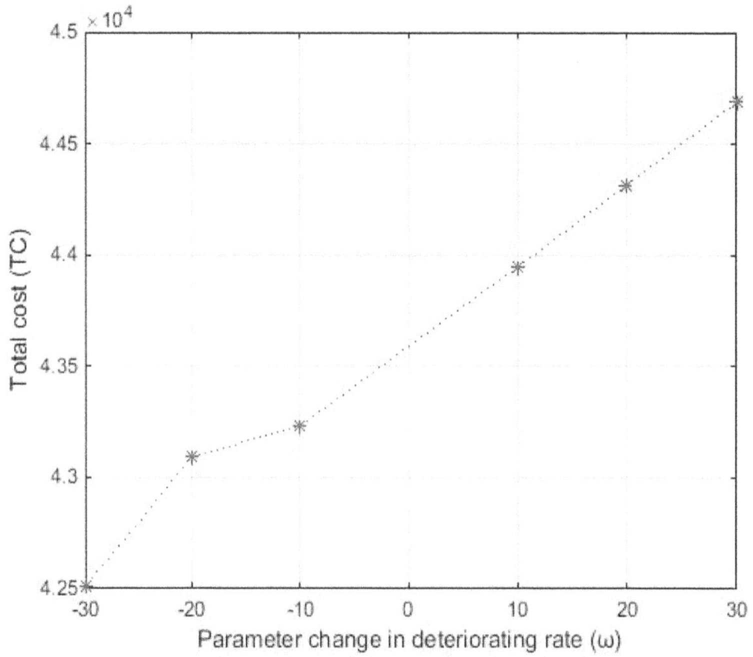

Figure 16. Percentage of reworkable item / Total cost

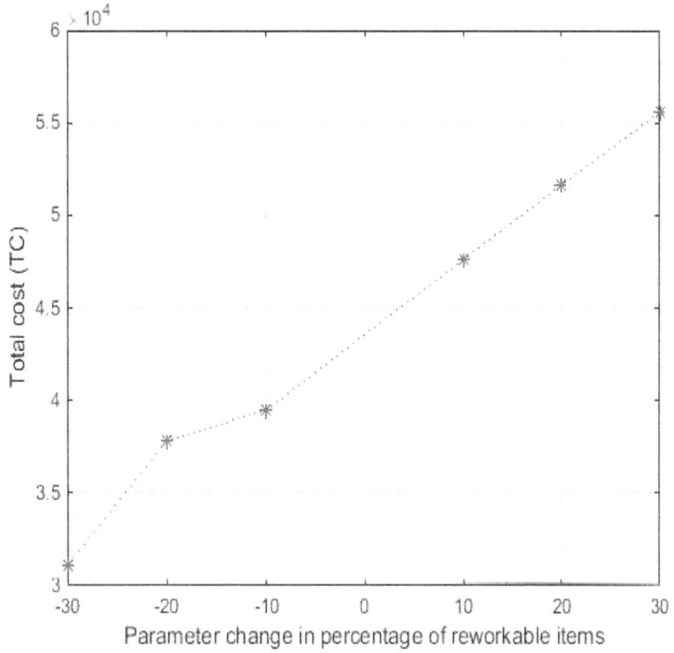

Figure 17. Rework process rate / Total cost

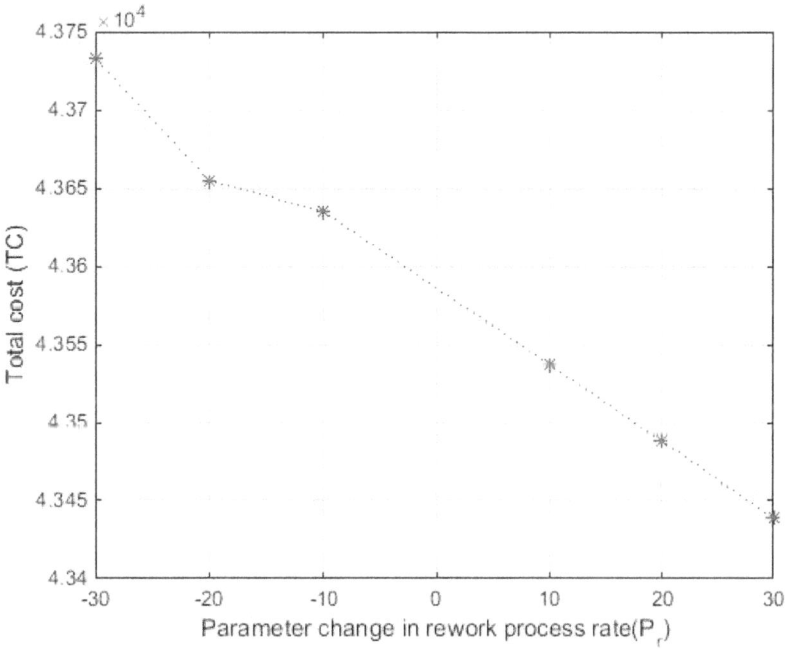

Figure 18. Production rate / Total cost

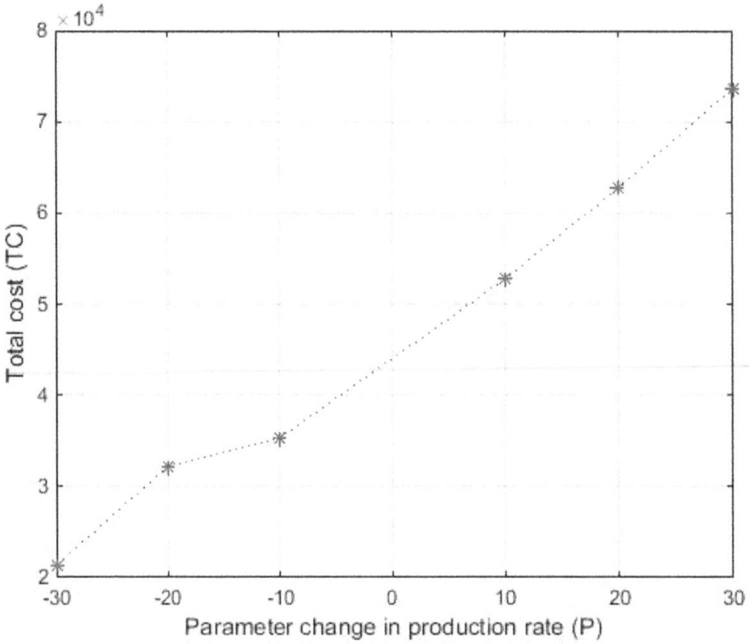

Figure 19. Manufacturing setup / Total cost

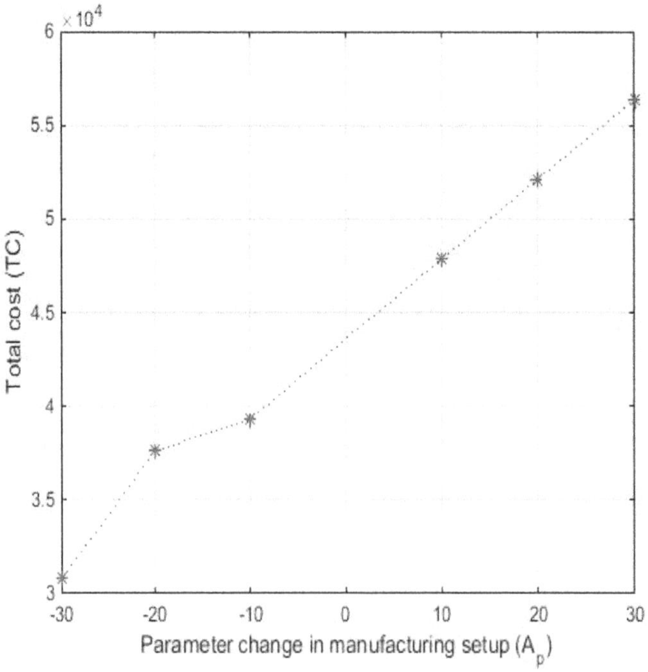

Figure 20. Soratge cost before rework / Total cost

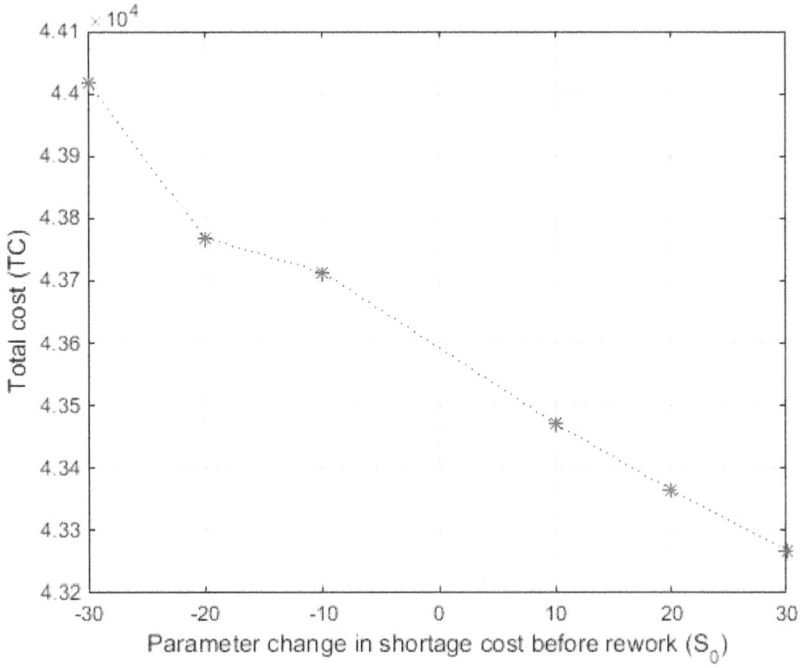

Figure 21. Verhult's demand / Total cost

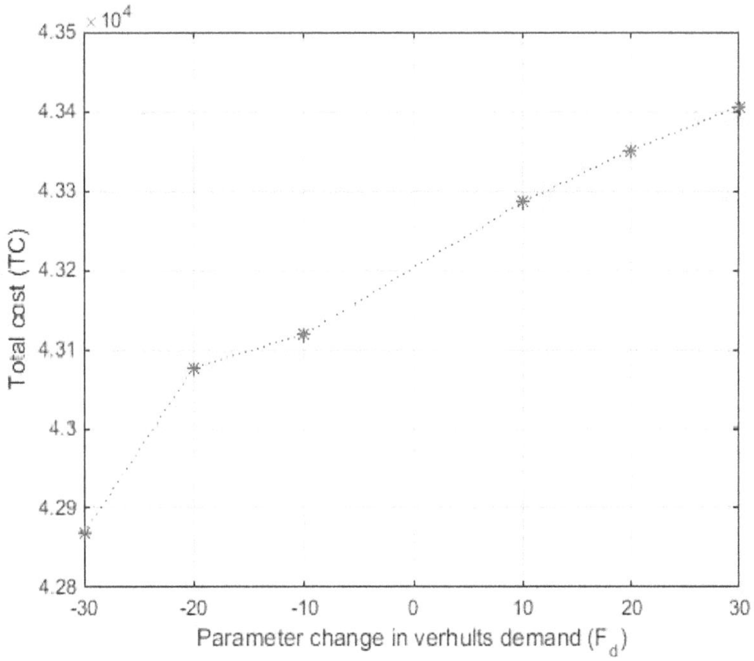

Figure 22. No of shipment / Total cost

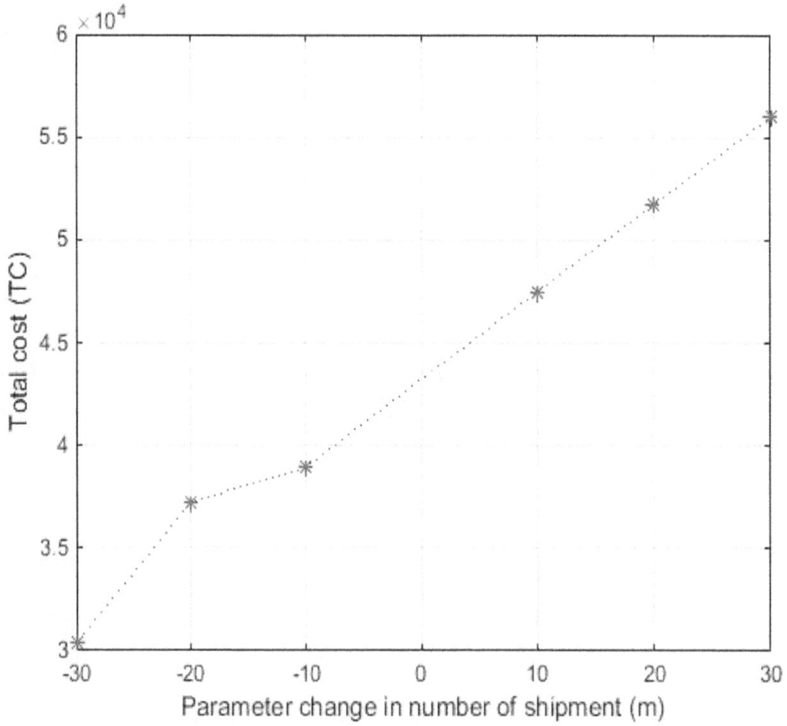

From table 5, result shows that the total cost is increased for all the parameters such as deterioration rate (ω), percentage of reworkable item (ϕ), Production rate(P), Production setup (A_P), number of shipment (m), In verhult's demand, the total cost is decreases when the parameters rework process rate (P_r) and Soratge cost before rework (S_0) are veried from -30% to +30%.

CONCLUSION

The result shows that the total cost is decreased in case(i) for the parameters deterioration rate (ω) and percentage of reworkable item (ϕ) increased. In case (ii) parameters such as deterioration rate (ω) and percentage of reworkable item (ϕ) increased so the total cost is incresed. Production setup (A_P) and Production rate(P) are linearly increased for increase in the total cost. Increasing the parameters in Production rate (P_r), the total cost decreases in both the cases. Even though the considering the parameters value is same in both the cases, this research shows that Verhult's demand pattern is quite effective. So, the total profit can be increased significantly. Finally sensitive analysis is revealed the variations in total cost for the different parameters with the numerical illustrations. Future result can also examine the production setup in different circumstances under fuzzy environment.

REFERENCES

Bhunia, A. K., Shaikh, A. A., Sharma, G., & Pareek, S. (2015). A two storage inventory model for deteriorating items with variable demand and partial backlogging. *J. Ind. Prod. Eng.*, *32*(4), 263–272. doi:10.1080/21681015.2015.1046508

Chauhan, A., & Singh, A. P. (2015). A note on the inventory models for deteriorating items with Verhulst's model type demand rate. *Int. J. Oper. Res.*, *22*(2), 243–261. doi:10.1504/IJOR.2015.067340

Dari, S., & Sani, B. (2017). An EPQ model for delayed deteriorating items with quadratic demand and shortages. Asian Journal of Mathematics and Computer Research, 22(2), 87–103.

De, P. K., & Rawat, A. (2013). Optimal order quantity of an EOQ model using expected value of a fuzzy function. *IEEE Int. Conf. Fuzzy Syst.* 10.1109/FUZZ-IEEE.2013.6622534

Guo, Z. (2005). a Verhulst Model on Time Series Error Corrected for Port Throughput Forecasting. *Journal of the Eastern Asia Society for Transportation Studies*, *6*, 881–891.

Gwanda, Y. I., Bari, A. N., & Singh, V. V. (2019). Optimal Production Model for Inventory Items with Verhulst's Demand and Time Dependent Amelioration Rate. *Palest. J. Math.*, *8*(2), 413–425.

Jana, D. K., Das, B., & Roy, T. K. (2013). *A partial backlogging inventory model for deteriorating item under fuzzy inflation and discounting over random planning horizon: A fuzzy genetic algorithm approach* (Vol. 2013). Adv. Oper. Res.

Jana, D. K., Maity, K., & Roy, T. K. (2013). A Bi-fuzzy Approach to a Production-Recycling-Disposal Inventory Problem with Environment Pollution Cost via Genetic Algorithm. *International Journal of Computers and Applications*, *61*(22), 975–8887.

Kumar, N., Singh, S. R., & Kumari, R. (2012). *An inventory model with time-dependent demand and limited storage facility under inflation* (Vol. 2012). Adv. Oper. Res.

Kuppulakshmi & Sugapriya. (2020). Effective Economic Production Quantity model for Penalty Maintenance cost with Rework Allowing Price Discount and Shortage. *Test Engineering and Management*, *83*, 16267-16286.

Li, L., Kabadi, S. N., & Nair, K. P. K. (2002). Fuzzy models for single-period inventory problem. *Fuzzy Sets and Systems*, *132*(3), 273–289. doi:10.1016/S0165-0114(02)00104-5

Mahapatra, G. S., Adak, S., & Kaladhar, K. (2019). A fuzzy inventory model with three parameter Weibull deterioration with reliant holding cost and demand incorporating reliability. *Journal of Intelligent & Fuzzy Systems*, *36*(6), 5731–5744. doi:10.3233/JIFS-181595

Pando, V., San-José, L. A., & Sicilia, J. (2019). Profitability ratio maximization in an inventory model with stock-dependent demand rate and non-linear holding cost. *Applied Mathematical Modelling*, *66*, 643–661. doi:10.1016/j.apm.2018.10.007

Ribeiro, F. L. (2017). An attempt to unify some population growth models from first principles. *Revista Brasileira de Ensino de Física*, *39*(1), 1–11. doi:10.1590/S0102-47442004000100001

Shah, N. H., & Naik, M. K. (2018). Inventory model for non – instantaneous deterioration and price-sensitive trended demand with learning effects. *Int. J. Inventory Research.*, *5*(1), 60–77. doi:10.1504/IJIR.2018.092356

Silas, N. (2010). Onyanga, Naftali Omollo- Ongati, Nyakinda Joseph Otula, "On the walraisian – Samuelson price adjustment model. *International Journal of Pure and Applied Mathematics*, *61*(2), 211–218.

Syed, J. K., & Aziz, L. A. (2016). Fuzzy Inventory Model without Shortages Using Triangular Fuzzy Numbers and Signed Distance Method. *International Journal of Scientific Research*, *5*(7), 1179–1182.

Udayakumar, R., & Geetha, K. V. (2018). An EOQ model for non-instantaneous deteriorating items with two levels of storage under trade credit policy. *J. Ind. Eng. Int.*, *14*(2), 343–365. doi:10.100740092-017-0228-4

Yadav, D., Singh, S. R., & Kumari, R. (2015). Retailer's optimal policy under inflation in fuzzy environment with trade credit. *International Journal of Systems Science*, *46*(4), 754–762. doi:10.1080/00207721.2013.801094

You, P. S. (2003). Dynamic pricing of inventory with cancellation demand. *The Journal of the Operational Research Society*, *54*(10), 1093–1101. doi:10.1057/palgrave.jors.2601619

Zhou, D. (2012). Grey Verhulst Model Based on BP Neural Network Optimization for Oil Production Forecasting. *Int. J. Energy Sci. IJES IJES*, *2*(3), 115–118.

APPENDIX

Consider the equation

$$I_{\tau p_1}\left(\tau_1\right) = \int_0^{T_1} \left(\frac{gp - \alpha}{\omega + \beta}\right)\left(1 - e^{-(\omega+\beta)\tau_1}\right) d\tau_1$$

$$e^{-(\omega+\beta)T_1} = 1 - \left(\omega + \beta\right)T_1 + \frac{\left(\left(\omega + \beta\right)T_1\right)^2}{2!} - \frac{\left(\left(\omega + \beta\right)T_1\right)^3}{3!} + \ldots$$

Considering the third and the higher powers are neglected for small values of $\frac{\left(\omega + \beta\right)^3 T_3^{\,3}}{3!} \ll 1$

We get,

$$I_{\tau p_1}\left(\tau_1\right) = \frac{gp - \alpha}{2} T_1^2$$

Similarly

$$\left(i\right) e^{-\omega T_1} = 1 - \left(\omega\right)T_1 + \frac{\left(\omega T_1\right)^2}{2!} - \frac{\left(\omega T_1\right)^3}{3!} + \ldots$$

$$\left(ii\right) e^{(\omega+\beta)T_1} = 1 + \left(\omega + \beta\right)T_1 + \frac{\left(\left(\omega + \beta\right)T_1\right)^2}{2!} + \frac{\left(\left(\omega + \beta\right)T_1\right)^3}{3!} + \ldots$$

Chapter 10
Statistics of an Appealing Class of Random Processes

Shaival Hemant Nagarsheth
https://orcid.org/0000-0001-9867-8167
Sardar Vallabhbhai National Institute of Technology, India

Shambhu Nath Sharma
Sardar Vallabhbhai National Institute of Technology, India

ABSTRACT

The white noise process, the Ornstein-Uhlenbeck process, and coloured noise process are salient noise processes to model the effect of random perturbations. In this chapter, the statistical properties, the master's equations for the Brownian noise process, coloured noise process, and the OU process are summarized. The results associated with the white noise process would be derived as the special cases of the Brownian and the OU noise processes. This chapter also formalizes stochastic differential rules for the Brownian motion and the OU process-driven vector stochastic differential systems in detail. Moreover, the master equations, especially for the coloured noise-driven stochastic differential system as well as the OU noise process-driven, are recast in the operator form involving the drift and modified diffusion operators involving an additional correction term to the standard diffusion operator. The results summarized in this chapter will be useful for modelling a random walk in stochastic systems.

INTRODUCTION

In 1905, Einstein pioneered the theory of the Brownian motion in which he derived the standard diffusion equation and proved that the probability density of the Brownian motion is normal density with zero mean and the variance $2Dt$, where the term $'D'$ has an interpretation as the diffusion coefficient (Einstein, 1956). The diffusion coefficient is also regarded as the stochastic perturbation term in the variance evolution of the stochastic differential system. The Brownian motion process and the Brownian motion-driven stochastic differential system have received considerable attention in the literature. The Brownian motion was initially considered for a study of microscopic particles in suspension. Subse-

DOI: 10.4018/978-1-7998-4706-9.ch010

quently, the Brownian motion process has found applications in stock pricing, control theory etc. to describe the effect of the random perturbation acting on the dynamical system. Ignoring the effect of random perturbations leads to inaccurate estimated state trajectory and may have its consequences in terms of poor decisions about control algorithms. Liptser and Shiryayev (1977) discussed stochastic calculus and non-linear filtering in the diffusion Markov setting. AN Kolmogorov, a legendary Russian mathematician, developed the backward and forward equations (Feller, 2000), which have left tremendous influence in theory of the dynamical system in the stochastic framework. Kiyoshi Itô, a famous Japanese mathematician, developed the stochastic differential rules for the Brownian motion process in a rigorous framework, especially from the mathematicians' viewpoint. Subsequently, the stochastic differential rule for the scalar function of the multi-dimensional state vector satisfying the Brownian motion-driven stochastic differential system was derived. These results were published in two seminal papers of Kiyoshi Itô in 1945, popularly known as the Itô calculus. The Itô calculus has found applications in mathematics as well as outside the mathematics world, especially in finance and technology. In Itô calculus, Kiyoshi Itô considered the term $dB_t = \dot{B}_t dt$, where B_t is the Brownian motion process. The results of the Itô calculus encouraged the mathematics community for further research on stochastic calculus. In 1975, Hida developed the theory of the white noise process, which sowed the seed of Hida calculus. The Hida calculus is alternatively known as the white noise calculus, an infinite-dimensional calculus. In the Hida setting, the term $\dot{B}_t$ is considered as a multiplication operator (Hida et al., 1993; Kuo and Russek, 1988; Kuo, 2009). On the other hand, RL Stratonovich developed the theory of coloured noise process, which involves the concept of the stochastic equation using the perturbation-theoretic expansion of a term involving the system non-linearity and the noise perturbation. He developed the theory of the OU process-driven stochastic differential system. The results of coloured noise calculus can also be derived using the functional calculus approach involving the concept of functional Taylor series, characteristic functional and cumulant generating functional. An excellent summary about the theory of the Brownian motion process and extending the theory for free particles, the harmonic oscillator can be found in (Ornstein and Uhlenbeck 1930; Wang and Uhlenbeck, 1945).

This chapter is intended to summarize succinctly standard results associated with the three noise processes, i.e. the Brownian noise process, coloured noise process, and the OU process. The statistical properties and the masters' equations for all the three noise processes are presented in this chapter. Moreover, stochastic differential rules for the Brownian motion and the OU process-driven 'vector' stochastic differential systems are formalized in detail. Notably, stochastic differential rules, the master's equation play a pivotal role to construct the theory of the noisy dynamical system. Furthermore, the filtering theory of the noisy dynamical system can be accomplished by a generalization of its corresponding master equation in which the observations accumulated up to the time t are conditioned. The evolution of conditional moment and the stochastic evolution of conditional moment are the standard formalisms to analyse stochastic differential systems. Some of the results of this chapter, especially stochastic differential rules and master's equations for the 'OU process-driven' stochastic differential system, are not sufficiently known in control and signal processing literature. It is believed that the results of this chapter will open the topic, i.e. random processes, to a broader audience by providing guidance for stochastic problems arising from 'diverse field', i.e. mathematical control theory, finance, technology, in which the OU process, coloured noise process, including the Brownian motion process are exploited to model the random perturbation.

BACKGROUND

Here, we present some standard structures of Stochastic Differential Equations (SDEs). The standard and famous structure of the stochastic differential equation is the Itô SDE, i.e.

$$dx_t = f(x_t, t)dt + G(x_t, t)dB_t. \tag{1}$$

Consider equation (1) in the Stratonovich sense, can be expressed alternatively in the Itô sense by introducing the notion of mean-square convergence, the Itô counterpart of the above SDE is

$$dx_i = (f_i(x_t, t)dt + \frac{1}{2}\sum_{j,\varphi} G_{j\varphi}(x_t, t)\frac{\partial G_{i\varphi}(x_t, t)}{\partial x_j})dt + \sum_{\varphi} G_{i\varphi}(x_t, t)dB_\varphi.$$

Equation (1) can be reformulated as

$$\dot{x}_t = f(x_t, t) + G(x_t, t)\dot{B}_t,$$

or

$$\dot{x}_t = f(x_t, t) + G(x_t, t)w_t. \tag{2}$$

Equation (2) is known as a Langevin SDE, a white noise-driven stochastic differential system, which is not defined from the mathematicians' viewpoint (Risken, 1984; p. 50). However, from the Practitioners' viewpoint, this structure has received attention in the literature. For this reason, the Itô SDE is a rigorous formalism to describe a stochastic differential system.

Alternatively, the stochastic differential equation can be reformulated in the *Hida sense*, the mathematicians' viewpoint, i.e.

$$dx_t = f(x_t, t)dt + G(x_t, t)\dot{B}_t dt,$$

where the term $\dot{B}_t$ is regarded as a multiplication operator δ_t^*, a rigorous mathematical object in the Hida sense. The Hida stochastic differential equation describes a white noise-driven stochastic differential system in the rigorous framework. In the Hida setting, the above stochastic differential equation can be recast as (Kuo, 2009)

$$dx_t = f(x_t, t)dt + \delta_t^*(G(x_t, t))dt.$$

The integral counterpart of the above equation can be recast as

$$x_t = x_{t_0} + \int_{t_0}^{t} f(x_s, s)ds + \int_{t_0}^{t} \delta_s^*(G(x_s, s))ds. \tag{3}$$

The last term of the right-hand side of equation (3) is also known as the 'Skorohod' integral, the white noise integral. A potential application of the Skorohod integral can be found in the 'Black-Scholes theory' for anticipatory markets in contrast to the Itô integral (Kuo, 2009; p. 69). A detail about the white noise calculus can also be found in Obata (1994). Equations (26)-(27) describe stochastic differential equations in the Markovian setting. On the other hand, the stochastic differential equation of the form

$$\dot{x}_t = f(x_t, t) + G(x_t, t)\xi_t,$$

where the input process ξ_t is coloured noise, describes the stochastic differential system in the non-Markovian setting.

Recently, the stochastic differential system in quantum framework involving the structure of Quantum Stochastic Differential Equations (QSDEs) has become the subject of investigation among the mathematics community. This has led to the non-commutative world of quantum theory. Hudson and Parthasarathy (1984) pioneered a famous paper formalizing the quantum Itô table and stochastic evolutions. Their results can be regarded as the quantum analogue of the Itô calculus. The theory of quantum stochastic calculus and filtering are grounded on the quantum Itô formula, master's equation and the Belavkin equation, see equation (4.15) of Bouten (2004, p. 78). For greater detail, an authoritative book authored by Parthasarathy (1992) as well as Belavkin (1980) can be consulted. It is interesting to note that the structure of the dynamical system decides a branch of control theory. Thus, the quantum stochastic differential system gives rise to the concept of quantum stochastic control (Belavkin, 1983). Here, this chapter discusses very briefly about the quantum stochastic differential system. The subject of quantum stochastic control is very young. It is believed that further investigations on the quantum analogue of stochastic differential equations will reveal exciting connections between different branches of sciences and the beautiful branch of mathematics.

STATISTICAL PROPERTIES, STOCHASTIC DIFFERENTIAL RULES, AND MASTER'S EQUATIONS

This section begins with the Brownian motion, and subsequently, the coloured noise process would be the subject of discussion. Some of the results for the OU process would be derived from the theory of coloured noise process.

Brownian Motion Process

The transition probability density $q_t(x, y)$ of the Brownian motion process is given by the normal density with the expectation x and the variance at (Feller, 2000) which satisfies the standard diffusion equation

$$\frac{\partial u(t,x)}{\partial t} = \frac{1}{2} a \frac{\partial^2 u(t,x)}{\partial x^2},$$

where $q_t(x,y) = \frac{1}{\sqrt{2\pi at}} e^{-\frac{(y-x)^2}{2at}}$.

The Brownian motion is a Markov process that can be demonstrated by introducing the notion of conditional probability density. The Brownian motion is a martingale; this can be proved using the definition of conditional expectation, $E(x_t \,|\, \Im_s) = x_s, t \geq s$, where the sigma-algebra $\Im_t = \bigcup_{s \leq t} \Im_s$. A rigorous definition of martingale can be found in Karatzas and Shreve (1991, p. 11). Furthermore, the Brownian motion has a continuous sample path; on the other hand, it does not have bounded variation. More notably, the Brownian motion process is differentiable nowhere. White noise can be regarded as an informal non-existent time derivative $\dot{B}_t$ of the Brownian motion process B_t. The Brownian motion process is not a stationary process, i.e. the covariance

$$\text{cov}\,(B_{t_1}, B_{t_2}) = R_{BB}(t_1, t_2) = \psi_w \min(t_1, t_2). \tag{4}$$

The above expression (4) suggests that the autocorrelation $R_{BB}(t_1, t_2)$ of the Brownian motion process depends on time instants rather than the time interval.

Here, we summarize the Itô stochastic differential rules for the Brownian motion process, (i) $dt\ dt = 0$ (ii) $dt\, dB_t = 0$ (iii) $dB_t\, dB_t = \psi_w dt$, where term 'ψ_w' denotes the intensity of the Brownian noise. A non-rigorous proof of the above differential rule is quite straightforward and can be found in Jazwinski [12]. The Itô differential rule for the scalar function of the multi-dimensional state vector satisfying the Itô stochastic differential equation, a multi-dimensional Itô differential rule, plays the pivotal role to develop the theory of stochastic differential systems, i.e. (i) to derive the expression of conditional moment (ii) to derive the Fokker-Planck equation (iii) to accomplish the stochastic evolution of energy function of the non-linear stochastic differential system. The stochastic evolution $d\varphi(x_t)$ of the scalar function of the n-dimensional state vector using the stochastic differential rule can be stated as

$$d\varphi(x_t) = tr(dx_t \frac{\partial \varphi(x_t)}{\partial x_t^T}) + \frac{1}{2} tr(dx_t dx_t^T \frac{\partial^2 \varphi(x_t)}{\partial x_t \partial x_t^T}),$$

where the scalar function $\varphi(x_t)$ is twice continuously differentiable. Note that the contributions to the term $d\varphi(x_t)$ come from the first and second-order derivatives of the state vector and higher derivative contributions vanish, which are attributed to the Brownian motion differential rule. After plugging the ith component of a stochastic differential equation, i.e. $dx_i = f_i(x_t, t)dt + \sum_{\phi} G_{i\phi}(x_t, t)dB_\phi$, in the above stochastic evolution, we have

$$d\varphi(x_t) = (f^T(x_t, t)\frac{\partial \varphi(x_t)}{\partial x_t} + tr(GG^T(x_t, t)\frac{\partial^2 \varphi(x_t)}{\partial x_t \partial x_t^T}))dt + \sum_{1\le i\le n, 1\le \phi \le r} \frac{\partial \varphi(x_t)}{\partial x_i} G_{i\phi}(x_t, t)dB_\phi, \quad (5)$$

where the size of the vector Brownian motion process is r, $dB_\phi dB_\gamma = \delta_{\phi\gamma} dt$, the terms f and G denote system nonlinearity and dispersion matrix respectively. Another version of the multi-dimensional Itô differential rule is summarized in Theorem (3.6) of Karatzas and Shreve (1991, p. 153). The integral counterpart of equation (5) can be written as

$$\varphi(x_t) = \varphi(x_{t_0}) + (\sum_i \int_{t_0}^{t} \frac{\partial \varphi(x_s)}{\partial x_i(s)} f_i(x_s, s)ds + \frac{1}{2}\sum_i \int_{t_0}^{t} (GG^T)_{ii}(x_s, s)\frac{\partial^2 \varphi(x_s)}{\partial x_i^2(s)} ds$$

$$+ \sum_{i<j} \int_{t_0}^{t} (GG^T)_{ij}(x_s, s)\frac{\partial^2 \varphi(x_s)}{\partial x_i(s)\partial x_j(s)} ds) + \sum_{1\le i\le n, 1\le \phi\le r} \int_{t_0}^{t} \frac{\partial \varphi(x_s)}{\partial x_i} G_{i\phi}(x_s, s)\, dB_\phi(s).$$

The Itô differential rule, multi-dimensional Itô differential rule, the Itô integration are encompassed in the Itô calculus. In the Itô calculus, the term 'dB_t' is regarded as a rigorous mathematical object.

Equation (5) can be exploited to arrive at the Fokker-Planck equation, the master's equation for the Itô stochastic differential system. The evolution $d\hat{\varphi}(x_t)$ of conditional moment is a consequence of equation (5) and the definition of conditional expectation

$$d\hat{\varphi}(x_t) = (\left\langle f^T(x_t, t)\frac{\partial \varphi(x_t)}{\partial x_t}\right\rangle + \frac{1}{2}\left\langle tr(GG^T(x_t, t)\frac{\partial^2 \varphi(x_t)}{\partial x_t \partial x_t^T})\right\rangle)dt,$$

where the notation '$\langle \; \rangle$' of this chapter denotes conditional expectation for the sake of notational convenience. Note that the expected value of the last term of the right-hand side of equation (5) vanishes, i.e. $\left\langle G_{i\phi}(x_t, t)dB_\phi \right\rangle = 0$. Consider $\varphi(x_t) = e^{S^T x_t}$, the evolution of the conditional characteristic function $\left\langle e^{s^T x_t} \right\rangle$ becomes

$$d\left\langle e^{s^T x_t}\right\rangle = (\left\langle f^T(x_t, t)\, S\, e^{S^T x_t}\right\rangle + \frac{1}{2}\left\langle tr(SS^T GG^T(x_t, t)e^{S^T x_t})\right\rangle)dt.$$

Making the use of the definition of the conditional characteristic function as well as the integration by part formula, we arrive at the Fokker-Planck equation, i.e.

$$\frac{\partial p}{\partial t} = -\sum_i \frac{\partial f_i(x, t)p}{\partial x_i} + \frac{1}{2}\sum_{i,j} \frac{\partial^2 (GG^T)_{ij}(x, t)p}{\partial x_i \partial x_j},$$

where $p = p(x, t | x_{t_0}, t_0)$. Other ways to arrive at the Fokker-Planck equation are writing the stochastic equation for the Itô stochastic differential system and the Kolmogorov forward equation (Feller, 2000). The filtering density evolution equation can be regarded as a generalization of the Fokker-Planck equation, see Liptser and Shiryayev (1977), Sharma (2009). The Fokker-Planck equation can be regarded as the prediction density evolution equation. Furthermore, particle filtering can be thought of a two-stage estimation procedure involving the prediction and filtering densities (Pitt and Shephard, 1999).

The Coloured Noise Process

The coloured noise process can be regarded as an arbitrary process, which generates a non-Markov process. Consider the process $x = \{x_t, \Im_t, t_0 \leq t \leq \infty\}$, which is sampled at two different time instants t and s, the process is zero mean, and stationary, then

$$\mathrm{cov}(x_t, x_s) = R_{xx}(\tau, \tau_{cor}),$$

where $\tau = t - s$ and τ_{cor} is the correlation time of the coloured noise process. The power spectral density of the coloured noise process is the function of the frequency term and correlation time, i.e. $S_{xx}(\omega, \tau_{cor}) = \Im R_{xx}(\tau, \tau_{cor})$. Furthermore, after introducing the notion of Gaussianity and Markovian character, the coloured noise process reduces to the OU process. Here, the Authors derive the master's equation for the coloured noise-driven stochastic differential system,

$$\dot{x}_t = f(x_t) + g(x_t)\xi_t, \tag{6}$$

where the process ξ_t is coloured noise, the term g denotes the process noise coefficient. Define the conditional probability density

$$p_t(x) = \langle \delta(x_t - x) \rangle,$$

where $p_t(x) = p(t, x | x_{t_0}, t_0)$. The time derivative of the conditional probability density

$$\dot{p}_t(x) = -\frac{\partial}{\partial x}\langle \delta(x_t - x)\dot{x}_t \rangle, \tag{7}$$

equation (6) in combination with equation (7) leads to

$$\dot{p}_t(x) = -\frac{\partial}{\partial x}\left\langle \delta(x_t - x)f(x_t) + g(x_t)\xi_t \right\rangle = -\frac{\partial f(x)p}{\partial x} - \frac{\partial g(x)}{\partial x}\left\langle \delta(x_t - x)\xi_t \right\rangle. \quad (8)$$

The correlation between the Dirac-delta function and the noise process, i.e. $\left\langle \delta(x_t - x)\xi_t \right\rangle$, can be obtained using the functional Taylor series, characteristic functional, and cumulant generating functional involving the functional derivative. Making the use of equation (22) of Hänggi (1995, p. 79), i.e.

$$\left\langle x_t g[x] \right\rangle = \sum_{n=0}^{\infty} \frac{1}{n!} \int_{t_0}^{t} \cdots \int_{t_0}^{t} C_{n+1}(t, t_1, t_2, \ldots, t_n) \left\langle \frac{\delta^n g[x]}{\delta x_{t_1} \delta x_{t_2} \ldots \delta x_{t_n}} \right\rangle dt_1 dt_2 \ldots dt_n. \quad (9)$$

Consider $x_t = \xi_t$, and $g[x] = \delta(x_t - x)$, equation (9) can be recast as

$$\left\langle \delta(x_t - x)\xi_t \right\rangle = \sum_{n=0}^{\infty} \frac{1}{n!} \int_{t_0}^{t} \cdots \int_{t_0}^{t} C_{n+1}(t, t_1, t_2, \ldots, t_n) \left\langle \frac{\delta^n \delta(x_t - x)}{\delta\xi_{t_1} \delta\xi_{t_2} \ldots \delta\xi_{t_n}} \right\rangle dt_1 dt_2 \ldots dt_n. \quad (10)$$

After plugging equation (10) into equation (8), we have

$$\dot{p}_t(x) = -\frac{\partial f(x)p}{\partial x} - \frac{\partial g(x)}{\partial x} \left(\sum_{n=0}^{\infty} \frac{1}{n!} \int_{t_0}^{t} \cdots \int_{t_0}^{t} C_{n+1}(t, t_1, t_2, \ldots, t_n) \left\langle \frac{\delta^n \delta(x_t - x)}{\delta\xi_{t_1} \delta\xi_{t_2} \ldots \delta\xi_{t_n}} \right\rangle dt_1 dt_2 \ldots dt_n\right),$$

furthermore,

$$\dot{p}_t(x) = -\frac{\partial f(x)p}{\partial x} - \left\langle \xi_t \right\rangle \frac{\partial g(x)p_t(x)}{\partial x} - \left(\frac{\partial g(x)}{\partial x}\right.$$

$$\left. \times \left(\sum_{n=1}^{\infty} \frac{1}{n!} \int_{t_0}^{t} \cdots \int_{t_0}^{t} C_{n+1}(t, t_1, t_2, \ldots, t_n) \left\langle \frac{\delta^n \delta(x_t - x)}{\delta\xi_{t_1} \delta\xi_{t_2} \ldots \delta\xi_{t_n}} \right\rangle dt_1 dt_2 \ldots dt_n\right)\right). \quad (11)$$

Note that equation (11) is the master's equation for the coloured noise-driven stochastic differential system. A detailed proof about the master's equation, equation (11), involving the concept of the functional derivative is given in Hänggi (1995). Equation (11) assumes the structure of infinite series and also termed as a stochastic equation in literature (Stratonovich, 1963). An alternative 'proof' of the stochastic equation can be demonstrated by introducing the notion of transition probability density and conditional characteristic function. Furthermore, the conditional characteristic function can be regarded as the conditional moment generating function, i.e.

$$\left\langle e^{iu(x_\tau - x)} \right\rangle = \sum_{0 \le n} \frac{(iu)}{n!} \left\langle (x_\tau - x)^n \right\rangle, \text{ where } \left\langle (x_\tau - x)^n \right\rangle = E((x_{t+\tau} - x_t)^n \,|\, x_t = x).$$

The OU Process

The OU process is a coloured noise process and found attention in the sense of analysing and accomplishing 'filtering theory' in which the observation noise is the OU process. A rigorous discussion about the OU process as the observation noise can be found in Bhatt and Karandikar (1999), see references therein, especially probability Journals. The OU process can be seen as the Brownian motion with friction term (Revuz and Yor, 1991). Here, the Authors state statistical properties, stochastic differential rules and the master's equation for the OU process-driven stochastic differential system. In this section, we begin with the bilinear stochastic differential system with constant coefficients in the Itô and Stratonovich senses and arrive at the statistical properties of the OU process.Consider the bilinear stochastic differential system of the form

$$dx_t = (\mu + \alpha x_t)dt + (\beta + \gamma x_t)dB_t.$$

Its solution (Terdik, 1989) can be stated as

$$x_t = (\mu - \beta\gamma)\int_{-\infty}^{t} e^{(\alpha - \frac{\gamma^2}{2})(t-s)+\gamma(B_t - B_s)} ds + \beta \int_{-\infty}^{t} e^{(\alpha - \frac{\gamma^2}{2})(t-s)+\gamma(B_t - B_s)} dB_s. \tag{12}$$

The above solution corresponds to the bilinear system in the Itô setting. Consider the bilinear stochastic differential system of this chapter in the Stratonovich setting that can be brought to the Itô formalism using the 'mean square convergence', i.e.

$dx_t = ((\mu + \frac{\beta\gamma}{2}) + (\alpha + \frac{\gamma^2}{2})x_t)dt + (\beta + \gamma x_t)dB_t$ and the solution

$$x_t = (\mu - \frac{\beta\gamma}{2})\int_{-\infty}^{t} e^{\alpha(t-s)+\gamma(B_t - B_s)} ds + \beta \int_{-\infty}^{t} e^{\alpha(t-s)+\gamma(B_t - B_s)} dB_s,$$

see equation (8) also. For $\gamma = 0$, the bilinear stochastic differential system reduces to $dx_t = (\mu + \alpha x_t)dt + \beta\, dB_t$, where the solution

$$x_t = \mu\int_{-\infty}^{t} e^{\alpha(t-s)+\gamma(B_t - B_s)} ds + \beta \int_{-\infty}^{t} e^{\alpha(t-s)+\gamma(B_t - B_s)} dB_s \tag{13}$$

becomes a special case of equation (12). As a consequence of equation (13), we have

$$Ex_t = \frac{-\mu}{\alpha} \text{ and } \mathrm{cov}(x_t,\, x_\tau) = \frac{\beta^2 e^{-|\alpha||t-\tau|}}{2|\alpha|}.$$

For the stochastic differential system

$$dx_t = (\mu - \alpha x_t)dt + \beta\, dB_t,\, \alpha > 0,\, Ex_t = \frac{\mu}{\alpha},\, \mathrm{cov}(x_t, x_\tau) = \frac{\beta^2 e^{-\alpha|t-\tau|}}{2\alpha}.$$

The solution x_t of the bilinear stochastic differential system of this chapter with $\mu = 0, \gamma = 0$ becomes the OU process, where

$$Ex_t = 0,\, \mathrm{cov}(x_t, x_\tau) = R_{xx}(t,\tau) = R_{xx}(s) = \frac{\beta^2 e^{-\alpha|s|}}{2\alpha},\; s = t - \tau. \tag{14}$$

Equation (14) suggests that the OU process is *zero* mean and *stationary* process. The power spectral density of the OU process can be obtained by taking the Fourier transform $\Im$ of the autocorrelation of the OU process, i.e.

$$S_{xx}(\omega) = = \Im \frac{\beta^2 e^{-\alpha|s|}}{2\alpha} = \frac{\beta^2}{\omega^2 + \alpha^2}.$$

The Gaussian character of the OU process can be demonstrated using the Kolmogorov backward equation. A rigorous discussion about the backward equation can be found in Karatzas and Shreve (1991, p. 281). The backward equation for the OU process (Feller, 2000) can be stated as

$$\frac{\partial u(t,x)}{\partial t} = -\alpha x \frac{\partial u(t,x)}{\partial x} + \frac{1}{2}\frac{\partial^2 u(t,x)}{\partial x^2}. \tag{15}$$

By introducing the notion of change of variables, equation (15) reduces to the standard diffusion equation

$$\frac{\partial w(\tau,\, y)}{\partial \tau} = \frac{1}{2}\frac{\partial^2 w(\tau,\, y)}{\partial y^2}, \text{where } y = xe^{-\alpha t},\; \tau = \frac{1 - e^{-\alpha t}}{2\alpha}. \tag{16}$$

Equation (16) suggests that the transition probability density of the OU process is given by the *normal* density with the expectation $y = xe^{-\alpha t}$, and the variance $\tau = \frac{1 - e^{-\alpha t}}{2\alpha}$. Thus, the OU process is a

Gaussian process. The OU process satisfies the Itô stochastic differential equation confirming the *Markovian* property. The OU process has *continuous* sample-path. Furthermore, the OU process is not a *martingale*; this can be demonstrated by using the definition of martingale (Karatzas and Shreve, 1991; p. 11) and the solution stated in equation (13) of this chapter.

The OU process x_t satisfies the stochastic differential equation $dx_t = -\alpha x_t dt + \beta dB_t$, thus the stochastic differential rules for the OU process can be formalized using the Itô differential rule, i.e.

(i) $dt\,dt = 0$, (ii) $dt\,dx_t = 0$, (iii) $dx_t\,dx_t = \beta^2 dt$.

The OU process x_t is differentiable nowhere, which is an immediate consequence of the above-stated third property. Here, we derive the stochastic differential rule for a non-linear function $\varphi(\varsigma_t, t)$ of the state ς_t satisfying the SDE,

$$\dot{\varsigma}_t = f(\varsigma_t, t) + g(\varsigma_t, t)x_t, \tag{17}$$

where x_t is the OU process. The function φ is twice continuously 'differentiable', then the stochastic evolution

$$d\varphi(\varsigma_t, t) = \frac{\partial \varphi(\varsigma_t, t)}{\partial \varsigma_t} d\varsigma_t + \frac{\partial \varphi(\varsigma_t, t)}{\partial t} dt + \frac{1}{2} \frac{\partial^2 \varphi(\varsigma_t, t)}{\partial \varsigma_t^2} (d\varsigma_t)^2.$$

The last term of the right-hand side of the above equation vanishes as a direct consequence of the property $dt\,dt = 0$, thus

$$\frac{\partial \varphi(\varsigma_t, t)}{\partial t} = \frac{\partial \varphi(\varsigma_t, t)}{\partial \varsigma_t} \dot{\varsigma}_t + \frac{\partial \varphi(\varsigma_t, t)}{\partial t}. \tag{18}$$

A rearrangement of the terms of equation (18) in conjunction with equation (17) leads to the following stochastic differential equation:

$$\frac{\partial \varphi(\varsigma_t, t)}{\partial t} = \frac{\partial \varphi(\varsigma_t, t)}{\partial \varsigma_t} f(\varsigma_t, t) + \frac{\partial \varphi(\varsigma_t, t)}{\partial t} + \frac{\partial \varphi(\varsigma_t, t)}{\partial \varsigma_t} g(\varsigma_t, t) x_t.$$

Here, we extend the stochastic differential rule for the OU process-driven *vector* stochastic differential system. Suppose the solution vector $\varsigma_t \in U$, where the phase space $U \subset R^n$ and the scale non-linear function $\varphi : U \to R$. The function φ is twice continuously differentiable, the stochastic evolution $d\varphi(\varsigma_t, t)$ of the scalar non-linear function of n-dimensional state vector can be cast as

$$d\varphi(\varsigma_t, t) = (d\varsigma_t)^T \frac{\partial \varphi(\varsigma_t, t)}{\partial \varsigma_t} + \frac{\partial \varphi(\varsigma_t, t)}{\partial t} dt + \frac{1}{2} tr(d\varsigma_t d\varsigma_t^T \frac{\partial^2 \varphi(\varsigma_t, t)}{\partial \varsigma_t \partial \varsigma_t^T})$$

$$= (\sum_i f_i(\varsigma_t, t) \frac{\partial \varphi(\varsigma_t, t)}{\partial \varsigma_i} + \frac{\partial \varphi(\varsigma_t, t)}{\partial t} + \sum_{i,\gamma} g_{i\phi}(\varsigma_t, t) \frac{\partial \varphi(\varsigma_t, t)}{\partial \varsigma_i} x_\phi)dt,$$

where the ith component of the state vector ς_t satisfies $\dot{\varsigma}_i = f_i(\varsigma_t, t) + \sum_\phi g_{i\phi}(\varsigma_t, t) x_\phi$ and

$$\frac{\partial \varphi(\varsigma_t, t)}{\partial t} = \sum_i f_i(\varsigma_t, t) \frac{\partial \varphi(\varsigma_t, t)}{\partial \varsigma_i} + \frac{\partial \varphi(\varsigma_t, t)}{\partial t} + \sum_{i,\phi} g_{i\phi}(\varsigma_t, t) \frac{\partial \varphi(\varsigma_t, t)}{\partial \varsigma_i} x_\phi . \tag{19}$$

The integral counterpart of equation (19) can be recast as

$$\varphi(\varsigma_t, t) = \varphi(\varsigma_{t_0}, t_o) + \sum_i \int_{t_o}^{t} f_i(\varsigma_s, s) \frac{\partial \varphi(\varsigma_s, s)}{\partial \varsigma_i(s)} ds + \int_{t_o}^{t} \frac{\partial \varphi(\varsigma_s, s)}{\partial s} ds$$

$$+ \sum_{i,\varphi} \int_{t_0}^{t} g_{i\phi}(\varsigma_s, s) \frac{\partial \varphi(\varsigma_s, s)}{\partial \varsigma_i} x_\phi (s) ds.$$

Master's Equations for the OU Process

The 'exact' master's equation for the OU process-driven stochastic differential system becomes a special case of equation (7), i.e.

$$\dot{p}_t(x) = -\frac{\partial f(x)p}{\partial x} - \frac{\partial g(x)}{\partial x} (\int_{t_0}^{t} C_2(t, s) \left\langle \frac{\delta\delta(x_t - x)}{\delta\xi_s} \right\rangle ds)$$

$$= -\frac{\partial f(x)p}{\partial x} + \frac{\partial g(x)}{\partial x} \frac{\partial}{\partial x} (\int_{t_0}^{t} C_2(t, s) \left\langle \delta(x_t - x) \frac{\delta x_t}{\delta \xi_s} \right\rangle ds). \tag{20}$$

Note that equations (11) and (20) exploit the structure of a stochastic differential system described by equation (6) in contrast to equation (17). This was chosen only for notational convenience. One can arrive at equation (20) by introducing zero mean, stationarity, and Gaussianity conditions into equation (11). The Gaussianity allows to replace the second-order cross cumulant with the second-order covariance, i.e.

$$C_2(t,\ s) = R_{\xi\xi}(t-s) = \frac{D}{\tau_{cor}} e^{-\frac{|t-s|}{\tau_{cor}}}, \tag{21}$$

where the higher-order cumulants vanish, and the terms D and τ_{cor} are associated with the OU process stochastic differential equation $d\xi_t = -\frac{1}{\tau_{cor}}\xi_t\, dt + \frac{\sqrt{2D}}{\tau_{cor}} dB_t$. The functional derivative $\frac{\delta x_t}{\delta \xi_s}$ of the state x_t with respect to the noise process ξ_s can be stated as (Fox, 1986; Hänggi and Jung, 1995)

$$\frac{\delta x_t}{\delta \xi_s} = g(x_t)\exp(\int_s^t (f'(x_\tau) - g'(x_\tau)\frac{f(x_\tau)}{g(x_\tau)})d\tau). \tag{22}$$

After plugging equations (21)-(22) into equation (20), we have

$$\dot{p}_t(x) = -\frac{\partial f(x)p}{\partial x} + \frac{D}{\tau_{cor}}\frac{\partial g(x)}{\partial x}(\frac{\partial g(x)}{\partial x}$$

$$\times(\int_{t_0}^t e^{-\frac{|t-s|}{\tau_{cor}}}\left\langle \delta(x_t - x)\exp(\int_s^t (f'(x_\tau) - g'(x_\tau)\frac{f(x_\tau)}{g(x_\tau)})d\tau)\right\rangle ds)). \tag{23}$$

The further simplification of equation (23) is not possible without introducing a condition on the input noise. This expression can be regarded as the '*exact*' master's equation for the OU process-driven stochastic differential system. Equation (23) can be recast in the 'operator' form, i.e.

$$dp_t(x) = (A(p) + D_t^g)dt,$$

and can be regarded as a parabolic linear 'non-homogeneous' equation of order two in partial differentiation. The 'non-homogeneous property' associated with the above equation is attributed to the generalized diffusion term D_t^g. The operator $A(.)$ describes the drift operator. The first and second terms of the right-hand side of equation (19) interpret the drift and diffusion terms, respectively. Here, the term 'generalized' implies an extension of the concept of the standard diffusion operator. Equation (19) of this chapter suggests that it is intractable to decouple the conditional probability density term from the diffusion operator, since the conditional probability density term is embedded within the 'noise-Dirac-delta' correlation term, see equation (23). The conditional probability density term can be decoupled from the correlation term by 'exploring' some additional assumptions and approximations. Furthermore, the Taylor series expansion of the functional derivative $\frac{\delta x_t}{\delta \xi_s}$ in powers of $(s-t)$, where $s \le t$, can be stated as

$$\frac{\delta x_t}{\delta \xi_s} = \frac{\delta x_t}{\delta \xi_t} + (s-t)(\frac{\partial}{\partial s}(\frac{\delta x_t}{\delta \xi_s}))\Big|_{s=t} + O((s-t)^2). \tag{24}$$

From equation (22), we have

$$\frac{\delta x_t}{\delta \xi_t} = g(x_t), \tag{25}$$

$$\frac{\partial}{\partial s}\frac{\delta x_t}{\delta \xi_s}\Big|_{s=t} = -g(x_t)(f'(x_t) - g'(x_t)\frac{f(x_t)}{g(x_t)}). \tag{26}$$

Equations (25)-(26) in combination with equation (20) lead to

$$\frac{\delta x_t}{\delta \xi_s} = g(x_t) + (t-s)\ g(x_t)(f'(x_t) - g'(x_t)\frac{f(x_t)}{g(x_t)})$$

$$= g(x_t)(1 + (t-s)\ (\frac{f'(x_t)g(x_t) - g'(x_t)f(x_t)}{g(x_t)}) + O((s-t))^2, \tag{27}$$

thus, equation (23) in combination with equation (27) becomes

$$\dot{p}_t(x) = -\frac{\partial f(x)p}{\partial x} + \frac{D}{\tau_{cor}}\frac{\partial g(x)}{\partial x}(\frac{\partial g(x)}{\partial x}$$

$$\times \int_{t_0}^{t} e^{-\frac{|t-s|}{\tau_{cor}}} \left\langle \delta(x_t - x)g(x_t)(1 + (t-s)g(x_t)(\frac{f(x_t)}{g(x_t)})' + O((s-t)^2)) \right\rangle ds),$$

after ignoring the higher-order terms, i.e. $O((s-t)^2)$, the above equation reduces to

$$\dot{p}_t(x) = -\frac{\partial f(x)p}{\partial x} + \frac{D}{\tau_{cor}}\frac{\partial g(x)}{\partial x}(\frac{\partial g(x)}{\partial x}$$

$$\times \int_{t_0}^{t} e^{-\frac{|t-s|}{\tau_{cor}}} \left\langle \delta(x_t - x)g(x_t)(1 + (t-s)g(x_t)(\frac{f(x_t)}{g(x_t)})') \right\rangle ds), \tag{28}$$

after a few steps of calculations and introducing the notion of long-time limit, equation (28) can be recast as

$$\dot{p}(x) = -\frac{\partial}{\partial x} fp + D\frac{\partial}{\partial x}(g\frac{\partial}{\partial x} g(1 + \tau_{cor} g(\frac{f}{g})')p). \tag{29}$$

Thus, equation (29) can be regarded as the *approximate* master equation for the OU process, since term 'approximate' is attributed to the first-order Taylor series expansion of the functional derivative $\frac{\delta x_t}{\delta \xi_s}$, which is exploited to arrive at equation (29). Note that equation (29) describes a parabolic linear homogeneous differential equation of order two in partial differentiation for the transition probability density in contrast to equation (23) describing a linear non-homogeneous equation. The equation can be further recast as the operator form

$$dp = A_a(p)dt = (A_d(p) + A_m(p))dt,$$

where the operator $A_a(.)$ is a linear homogeneous operator and can be regarded as the sum of the drift operator $A_d(.)$ and the modified diffusion operator $A_m(.)$. The modified diffusion operator $A_m(.)$ for the approximate master's equation, equation (29), is

$$A_m(.) = D\frac{\partial}{\partial x}(g\frac{\partial}{\partial x} g(1 + \tau_{cor} g(\frac{f}{g})'))(.),$$

here the additional correction term $\tau_{cor} g(\frac{f}{g})'$ gives rise to the term 'modified'. The OU process with zero correlation time can be regarded as the Gaussian white noise process. Thus, equation (29) with the zero correlation time τ_{cor} leads to the master's equation for the white noise-driven stochastic differential system, i.e. $\dot{p}(x) = -\frac{\partial}{\partial x} fp + D\frac{\partial}{\partial x}(g\frac{\partial}{\partial x} gp)$. The statistical properties of the white noise are direct and special cases of the OU process statistics.

CONCLUDING REMARKS

In this chapter, the authors have summarized the three noise processes, i.e. the Brownian motion, coloured noise, and the OU process, encompassing their statistics, stochastic differential rules and master's equations. This chapter derives the exact and approximate master's equations for the OU process-driven the stochastic differential system as well as recasts them in the 'operator' form. Another contribution of this chapter is to formalize the stochastic differential rule for the OU process-driven *vector* stochastic differential system in detail. The noise processes discussed in this chapter with their representation can directly be used to model the noisy parameter in a stochastic system.

This chapter reviews some standard structures of stochastic differential equations as well as discusses briefly stochastic calculus associated with them. The Authors have listed standard references on quantum stochastic calculus as well, which are not widely known, that will be of great value to specialists. This chapter will be of interest to mathematical physicists, control theorists, and economists broadly looking for investigations on the theory of noisy dynamical systems as well as control of stochastic differential equations. A good source of stochastic processes and their bibliographic details can be found in Hänggi and Jung (1995), Wax (1954) as well.

REFERENCES

Belavkin, V. (1980). Quantum filtering of Markov signals with white quantum noise. *Radiotechnika I Electronica, 25*, 1445-1453.

Belavkin, V. (1983). Theory of control of observable quantum systems. *Automation and Remote Control, 44*, 178–188.

Bhatt, A. B., & Karandikar, R. L. (1999). *On filtering with Ornstein-Uhlenbeck process as observation noise*. Retrieved from www.isid.ac.in/~statmath/eprints/2002/isid200232.ps

Bouten, L. (2004). *Filtering and control in quantum optics* (PhD Thesis Dissertation). Nottingham University, UK.

Einstein, A. (1956). *Investigations on the theory of the Brownian movement*. Dover Publications, Inc.

Feller, W. (2000). *An Introduction to Probability Theory and its Applications* (Vol. 2). John Wiley and Sons.

Fox, R. F. (1986). Uniform convergence to an effective Fokker-Planck equation for weakly colored noise. *Physical Review A, 34*(5), 4525–4527. doi:10.1103/PhysRevA.34.4525 PMID:9897829

Hänggi, P. (1995). The functional derivative and its use in the description of noisy dynamical systems. In L. Pesquera & M. Rodriguez (Eds.), Stochastic Processes Applied to Physics (pp. 69-95). World Scientific.

Hänggi, P., & Jung, P. (1995). Colored noise in dynamical systems. In I. Prigogine & S. A. Rice (Eds.), *Advances in Chemical Physics* (pp. 239–323). John Wiley and Sons.

Hida, T., Kuo, H.-H., Pontho, J., & Streit, L. (1993). *White Noise: An Infinite Dimensional Calculus*. Kluwer, Academic Publishers. doi:10.1007/978-94-017-3680-0

Hudson, R. L., & Parthasarathy, K. R. (1984). Quantum Itô formula and stochastic evolutions. *Communications in Mathematical Physics, 93*(3), 301–323. doi:10.1007/BF01258530

Jazwinski, A. H. (1970). *Stochastic Processes and Filtering Theory*. Academic Press.

Karatzas, I., & Shreve, S. E. (1991). Brownian Motion and Stochastic Calculus (graduate text in mathematics). New York: Springer.

Kuo, H.-H. (2009). White noise stochastic integration. Stochastic Analysis: Classical and Quantum, eProceedings, 57-71.

Kuo, H.-H., & Russek, A. (1988). White noise approach to stochastic integration. *Journal of Multivariate Analysis, 24*(2), 218–236. doi:10.1016/0047-259X(88)90037-1

Liptser, R. S., & Shiryayev, A. N. (1977). *Statistics of Random Processes* (Vol. 1). Springer. doi:10.1007/978-1-4757-1665-8

Obata, N. (1994). White Noise Calculus and Fock Space. *Lecture Notes in Mathematics*, ▪▪▪, 1577.

Ornstein, L. S., & Uhlenbeck, G. E. (1930). On the theory of the Brownian motion. *Physical Review*, *36*, 823–841.

Parthasarathy, K. R. (1992). *An Introduction to Quantum Stochastic Calculus*. Birkhäuser.

Pitt, M. K., & Shephard, N. (1999). Filtering via simulations: Auxiliary particle filtering. *Journal of the American Statistical Association*, *94*(446), 590–599. doi:10.1080/01621459.1999.10474153

Protter, P. E. (1991). *Stochastic Integration and Differential Equations*. Springer, Berlin.

Revuz, D., & Yor, M. (1991). *Continuous Martingales and Brownian Motion*. Springer-Verlag. doi:10.1007/978-3-662-21726-9

Risken, H. (1984). *The Fokker-Planck Equation: Method of Solutions and Applications*. Springer-Verlag. doi:10.1007/978-3-642-96807-5

Sharma, S. N. (2009). A Kushner approach for small random perturbations of a stochastic Duffing-van der Pol system. *Automatica*, *45*(4), 1097–1099. doi:10.1016/j.automatica.2008.12.010

Stratonovich, R. L. (1963). *Topics in the Theory of Random Noise* (Vol. 1). Gordan and Breach.

Terdik, G. Y. (1990). Stationary solutions for bilinear systems with constant coefficients. In E. Cinlar, K. L. Chug, & R. K. Getoor (Eds.), *Seminar on Stochastic Processes* (pp. 196–206). Birkhäuser. doi:10.1007/978-1-4612-3458-6_12

Wang, M. C., & Uhlenbeck, L. S. (1945). On the theory of the Brownian motion II. *Reviews of Modern Physics*, *17*(2-3), 323–342. doi:10.1103/RevModPhys.17.323

Wax, N. (Ed.). (1954). *Selected Papers on Noise and Stochastic Processes*. Dover Publications, Inc.

Chapter 11
The Universality of the Kalman Filter:
A Conditional Characteristic Function Perspective

Sandhya Rathore
Sarvajanik College of Engineering and Technology, India

Shambhu Nath Sharma
Sardar Vallabhbhai National Institute of Technology, India

Shaival Hemant Nagarsheth
https://orcid.org/0000-0001-9867-8167
Sardar Vallabhbhai National Institute of Technology, India

ABSTRACT

The universality of the Kalman filtering can be found in the control theory. The Kalman filter has found its applications in sophisticated autonomous systems and smart products, which are attributed to its realization in a single complex chip. In this chapter, considering the Kalman filter from the perspective of conditional characteristic function evolution and Itô calculus, three Kalman filtering theorems and their formal proof are developed. Most notably, this chapter reveals the following: (1) Kalman filtering equations are a consequence of the 'evolution of conditional characteristic function' for the linear stochastic differential system coupled with the linear discrete measurement system. (2) The Kalman filtering is a consequence of the 'stochastic evolution of conditional characteristic function' for the linear stochastic differential system coupled with the linear continuous measurement system. (3) The structure of the Kalman filter remains invariant under two popular stochastic interpretations, the Itô vs Stratonovich.

DOI: 10.4018/978-1-7998-4706-9.ch011

INTRODUCTION

This chapter is inspired from the following two papers: (i) A simple concept with new perspectives leads to surprising results (Holbrook and Bhatia, 2006) (ii) Anderson and Moore (1991) published an interesting paper, "Kalman filtering: whence, what and whither?" Here, they explained the Kalman filtering by extending it in non-linear filtering perspective with some discussions (Anderson and Moore, 1991: p. 52). This chapter aims to answer the question, whither of the Kalman filtering? in greater detail. This chapter sketches the answer to the question in greater detail with a conditional characteristic perspective. The conditional characteristic function evolution, a new perspective in the Kalman filtering setting, is not available yet in literature. Secondly, the Kalman filtering can be sketched from the stochastic evolution of the conditional characteristic function for the linear Stochastic Differential Equation (SDE) with linear observation equation as well.

Notation: For the action of the conditional expectation operator, we adopt three notations, i.e., $\hat{x}_t = E(x_t | Y_t) = \langle x_t \rangle$. A choice of the notation hinges on the structure of conditional expectation evolution equations.

BACKGROUND

Norbert Wiener, in 1921, proved that the Brownian motion has a continuous sample path and non-differentiable with probability one (Sharma and Gawalwad, 2016). Norbert Wiener developed the filtering equation using the convolution integral equation for linear observation equations. The Wiener filter is a signal estimation method. Suppose the Wiener filter is the LTI system, the Wiener filtering can be regarded as an infinite-dimensional LTI system; the transfer function would be irrational. Kalman (1960) sowed the seed of the filtering of dynamic systems in which he considered linear stochastic differential equation coupled with linear observation equation (Kalman, 1960). This attracted attention in greater circles via its Apollo mission application. Kalman filtering has found appealing applications in smartphones, as well. Thus, the Kalman filtering is ubiquitous and is in everybody's life. For Kalman filter as an LTI system, where the observation is the input and the filtered estimate is the output then it can be regarded as finite-dimensional, and the transfer function of the Kalman filter would be rational. The Kalman filtering and its discoverer's legacy can be found in Anderson and Moore (1991). Kalman meets Shannon (Gattami, 2014) in noisy Gaussian channels, the ubiquitous Riccati equation is a part of the coupled Kalman filtering equations for the continuous state-continuous measurement systems. Combining the celebrated Wiener-Höpf equation, Wiener filter, linear stochastic differential equation and linear measurement equation coupled with the orthogonal projection Lemma, Kalman gain and coupled filtering equations are obtained. Furthermore, the notion of non-linear Kalman filtering was introduced. The non-linear filtering has greater conceptual depth as well as involves greater mathematical rigor. Non-linear filtering theory hinges on the filtering density evolution equation and the stochastic evolution of the conditional expectation of the scalar function. A remarkable paper on non-linear filtering can be traced back to a pioneering work of 1967, H. J. Kushner. That is popularly known as the Kushner-Stratonovich equation (Lipster and Shiryaev, 1977). An alternative interpretation of the Kushner-Stratonovich is FKK filtering (Fujisaki et al., 1972).

THE KALMAN FILTER: A CONDITIONAL CHARACTERISTICS FUNCTION PERSPECTIVE

Here, Kalman filter is constructed from the conditional characteristic function perspective by utilizing the Itô differential rule for the Wiener process (Karatzas and Shreve, 1991), which is not available yet in literature. This chapter is inspired from the fact that the older, but the celebrated result with new perspectives, produces greater insights. The Itô calculus (Kunita, 2010) and conditional characteristic function (Lipster and Shiryaev, 1977) are the cornerstone formalisms of the construction of the coupled Kalman filtering equations. The Itô calculus, a 1945 discovery, sowed the seed of stochastic calculus. The Itô calculus was adjudged for the inaugural Gauss prize. The conditional characteristic function is a conditional moment generating function that forms the cornerstone formalism for the Central Limit Theorem of stochastic processes.

Consensus on the Itô-Stratonovich dilemma recommends that 'the Stratonovich SDE is expected for stochastic processes in typical, continuous and real physical systems' (Rathore and Sharma, 2018). On the other hand, the Itô SDE is the correct stochastic calculus for stochastic finance and theoretical biology. The non-linear filtering would be different for two perspectives, the Itô and the Stratonovich. Interestingly, Kalman filtering is invariant under the Itô and Stratonovich perspectives (Mannella and McClintock, 2012). Three Theorems are constructed here. The first is about the *evolution* of conditional characteristic function for the Kalman filtering for the continuous state-discrete measurement system. The second Theorem is about the *stochastic evolution* of the conditional characteristic function (Lipster and Shiryaev, 1977) for the Kalman filtering for the continuous state-continuous measurement systems. The third Theorem is about the stochastic evolution of the bilinear Kalman filter for the continuous state-continuous measurement system, where the Kalman filtering is a special case of the bilinear Kalman filtering. This chapter depends on the evolution of conditional characteristic function, and the evolution of conditional characteristic function holds for the continuous state. Thus, the continuous state obeying the Itô SDE is the subject of investigations in place of the stochastic difference equation.

Theorem 1. Consider the filtering model for the continuous state-discrete measurement system, i.e.,

$$dx_i(t) = \sum_{\alpha} A_{i\alpha}(t)x_{\alpha}(t)dt + \sum G_{i\varphi}(t)dW_{\varphi}(t),\ y_i(t_k) = \sum C_{i\varphi}(t_k)x_{\varphi}(t_k) + V_i(t_k),$$

where the vector Brownian motion process $W = \left\{ W_t \in R^m, 0 \le t \le \infty \right\}$. The terms x_t and A_t denote the state vector and system state matrix, respectively. The term $G(t)dW_t$ denotes the vector stochastic correction term with $\operatorname{var}(dW_t) = Idt$. The term y_{t_k} indicates a noisy observation vector. The observation noise V_{t_k} is $N(0, R_k)$, where R_k is a real-valued observation noise variance matrix. Then, the evolution of conditional characteristic function of the above filtering model between the observation instants, τ, $t_{k-1} \le \tau < t_k$ becomes

$$d\left\langle \exp(s^T x_\tau) \right\rangle = \left\langle \sum_p \sum_\alpha A_{p\alpha}(\tau)x_\alpha \frac{\partial \exp(s^T x_\tau)}{\partial x_p} + \frac{1}{2}\sum_{p,q}(GG^T)_{pq}(\tau)\frac{\partial^2 \exp(s^T x_\tau)}{\partial x_p \partial x_q} \right\rangle d\tau.$$

Alternatively, $d\left\langle \exp(s^T x_\tau)\right\rangle = (\sum_p \sum_\alpha A_{p\alpha}(\tau)\left\langle x_\alpha(\tau)\frac{\partial \exp(s^T x_\tau)}{\partial x_p}\right\rangle$

$$+\frac{1}{2}\sum_{p,q}(GG^T)_{pq}(\tau)\left\langle \frac{\partial^2 \exp(s^T x_\tau)}{\partial x_p \partial x_q}\right\rangle)d\tau, \tag{1}$$

where the characteristic function $\left\langle \exp(s^T x_\tau)\right\rangle = E(\exp(s^T x_\tau)\big|Y_{t_{k-1}})$. The conditional expectation operator E accounts for the accumulated observation $Y_{t_{k-1}} = \left\{y_{t_n}, 0 \le n \le k-1\right\}$, $t_{k-1} < \tau \le t_k$ and t_{k-1}, t_k are the observation instants. The Kalman filtering *between the observations* becomes a special case of the conditional characteristic function evolution. On the other hand, the Kalman filtering at the observation is not a consequence of the evolution of conditional characteristic function.

Proof. The proof of the conditional characteristic function evolution not accounting for noisy observables can be sketched by utilizing the property of the differential and the expectation operators, Itô calculus. For the action of the conditional expectation operator $E(.)$ on lengthier terms, we adopt another notation $\langle\ \rangle$ as well.

$$\left\langle d\exp(s^T x_t)\right\rangle = \left\langle d\phi(x_t, s)\right\rangle = \left\langle \phi(x_{t+dt}, s) - \phi(x_t, s)\right\rangle = \psi(t+dt, s) - \psi(t, s).$$

In the sense of characteristic function, the variable s becomes $j\omega$, the input argument of the Fourier Transform. Now,

$$\left\langle \exp(s^T x_t)\right\rangle = \psi(t,s) = \left\langle \phi(s, x_t)\right\rangle,\ d\left\langle \exp(s^T x_t)\right\rangle = \psi(t+dt, s) - \psi(t,s) = \left\langle d\exp(s^T x_t)\right\rangle. \tag{2}$$

The above relation reveals the fact that the differential and the conditional expectation operator can be interchanged. Since we wish to develop the conditional characteristic function evolution at the instant τ, where observations are not available, we adopt the notation τ in place of the time variable t. Thus,

$$d\exp(s^T x_\tau) = (\sum_p \sum_\alpha A_{p\alpha}(\tau)x_\alpha(\tau)\frac{\partial \exp(s^T x_\tau)}{\partial x_p} + \frac{1}{2}\sum_{p,q}(GG^T)_{pq}(\tau)\,\frac{\partial^2 \exp(s^T x_\tau)}{\partial x_p \partial x_q})d\tau$$

$$+\sum_{p,\varphi} G_{p\varphi}(\tau)\frac{\partial \exp(s^T x_\tau)}{\partial x_p}dW_\varphi. \tag{3}$$

The above equation is a consequence of the Itô differential rule (Karatzas and Shreve, 1991) for the scalar function $\exp(s^T x_t)$. From (2)-(3), we have

$$d\left\langle \exp(s^T x_t) \right\rangle = \left\langle \sum_p \sum_\alpha A_{p\alpha}(\tau) x_\alpha \frac{\partial \exp(s^T x_\tau)}{\partial x_p} + \frac{1}{2} \sum_{p,q} (GG^T)_{pq}(\tau) \frac{\partial^2 \exp(s^T x_\tau)}{\partial x_p \partial x_q} \right\rangle d\tau$$

$$+ \left\langle \sum_{p,\varphi} G_{p\varphi}(t) \frac{\partial \exp(s^T x_t)}{\partial x_p} dW_\varphi \right\rangle. \tag{4}$$

Since the conditional expectation operator is a linear operator and

$$\left\langle \sum_{p,\varphi} G_{p\varphi}(t) \frac{\partial \exp(s^T x_t)}{\partial x_p} dW_\varphi \right\rangle = 0.$$

Equation (4) boils down to

$$d\left\langle \exp(s^T x_\tau) \right\rangle = \left\langle \sum_p \sum_\alpha A_{p\alpha}(\tau) x_\alpha(\tau) \frac{\partial \exp(s^T x_\tau)}{\partial x_p} + \frac{1}{2} \sum_{p,q} (GG^T)_{pq}(\tau) \frac{\partial^2 \exp(s^T x_\tau)}{\partial x_p \partial x_q} \right\rangle d\tau$$

$= (\sum_p \sum_\alpha A_{p\alpha}(\tau) \left\langle x_\alpha(\tau) \frac{\partial \exp(s^T x_\tau)}{\partial x_p} \right\rangle + \frac{1}{2} \sum_{p,q} (GG^T)_{pq}(\tau) \left\langle \frac{\partial^2 \exp(s^T x_\tau)}{\partial x_p \partial x_q} \right\rangle) d\tau$. Thus, we arrive at (1) of the Theorem 1 of the chapter. *QED*

Now, consider $\exp(s^T x_t) = x_i(t)$. Then, the first term of the right-hand side will contribute, and the last term of the right-hand side will vanish. As a result of this, equation (1) becomes

$$d\left\langle x_i(\tau) \right\rangle = \sum_\alpha A_{i\alpha}(\tau) \left\langle x_\alpha(\tau) \right\rangle d\tau,$$

Alternatively

$$d\left\langle x(\tau) \right\rangle = A_\tau \left\langle x_\tau \right\rangle d\tau. \tag{5a}$$

Consider $\exp(s^T x_t) = x_i . x_j$, we have

$$d\left\langle x_i x_j \right\rangle = (\sum_\alpha A_{i\alpha}(\tau) \left\langle x_\alpha(\tau) x_j(\tau) \right\rangle + \sum_\alpha A_{j\alpha}(\tau) \left\langle x_\alpha(\tau) \right\rangle \left\langle x_j(\tau) \right\rangle + \frac{1}{2} \sum_{p,q} (GG^T)_{pq}(\tau) \left\langle \frac{\partial^2 x_i x_j}{\partial x_p \partial x_q} \right\rangle) d\tau$$

$$= (\sum_\alpha A_{i\alpha}(\tau) \left\langle x_\alpha(\tau) x_j(\tau) \right\rangle + \sum_\alpha A_{j\alpha}(\tau) \left\langle x_\alpha(\tau) x_i(\tau) \right\rangle + \frac{1}{2} \sum_{p,q} (GG^T)_{pq}(\tau) \left\langle \delta_{ip}\delta_{jq} + \delta_{iq}\delta_{jp} \right\rangle) d\tau$$

$$= (\sum_{\alpha} A_{i\alpha}(\tau)\langle x_{\alpha}(\tau)x_{j}(\tau)\rangle + \sum_{\alpha} A_{j\alpha}(\tau)\langle x_{\alpha}(\tau)x_{i}(\tau)\rangle + (GG^{T})_{ij}(\tau))d\tau. \tag{5b}$$

After combining (5b) with (5a), we get

$$dP_{ij} = d\langle x_i x_j\rangle - \langle x_i\rangle d\langle x_j\rangle - \langle x_j\rangle d\langle x_i\rangle - d\langle x_i\rangle d\langle x_j\rangle$$

$$= (\sum_{\alpha} A_{i\alpha}(\tau)\langle x_{\alpha}(\tau)x_{j}(\tau)\rangle + \sum_{\alpha} A_{j\alpha}(\tau)\langle x_{\alpha}(\tau)x_{i}(\tau)\rangle + (GG^{T})_{ij}(\tau) - \sum_{\alpha} A_{i\alpha}(\tau)\langle x_{j}(\tau)\rangle\langle x_{\alpha}(\tau)\rangle$$

$$-\sum_{\alpha} A_{j\alpha}(\tau)\langle x_{\alpha}(\tau)\rangle\langle x_{i}(\tau)\rangle)d\tau$$

$$= (\sum_{\alpha} A_{i\alpha}(\tau)P_{j\alpha}(\tau) + \sum_{\alpha} P_{i\alpha}(\tau)A_{j\alpha}(\tau) + (GG^{T})_{ij}(\tau))d\tau.$$

Alternatively,

$$dP_{\tau} = (A_{\tau}P_{\tau} + P_{\tau}A_{\tau}^{T} + (GG^{T})_{ij}(\tau))d\tau. \tag{6}$$

Equation (1) describes the evolution of characteristic function for the Kalman filtering *between the observations* and (5)-(6) are the coupled Kalman filtering equations *between the observations*. The Kalman filtering *at the observation* can be sketched using the conditional probability density, and conditional expectation *at the observations* instants t_{k-1} and t_k. A proof of the Kalman filtering equations *at the observation* can be found in (Jazwinski, 1970; Patel and Sharma, 2014). That is not a consequence of the evolution of conditional characteristic function. For the vector case,

$$\hat{x}_{t_k}^{t_k} = \hat{x}_{t_k}^{t_{k-1}} + P_{t_k}^{t_{k-1}}C_{t_k}^{T}(C_{t_k}P_{t_k}^{t_{k-1}}C_{t_k}^{T} + R_k)^{-1}(y_{t_k} - C_{t_k}\hat{x}_{t_k}^{t_{k-1}}), \tag{7}$$

$$P_{t_k}^{t_k} = P_{t_k}^{t_{k-1}} - P_{t_k}^{t_{k-1}}C_{t_k}^{T}(C_{t_k}P_{t_k}^{t_{k-1}}C_{t_k}^{T} + R_k)^{-1}C_{t_k}P_{t_k}^{t_{k-1}}. \tag{8}$$

Note that $\hat{x}_{t_k}^{t_k} = E(x_{t_k}|Y_{t_{k-1}}), \hat{x}_{t_k}^{t_k} = E(x_{t_k}|Y_{t_k})$. The Kalman filtering for the continuous state-discrete measurement system is given by the following set of four equations by adopting the vec function setting as well, i.e.

$$dvec(\langle x(\tau)\rangle) = vec(A_{\tau}\langle x_{\tau}\rangle)d\tau = (\langle x_{\tau}\rangle^{T} \otimes I)vec(A_{\tau})d\tau, \tag{9a}$$

$$dvec(P_\tau) = ((I \otimes A_\tau + A_\tau \otimes I)vec(P_\tau) + vec((GG^T)(\tau)))d\tau, \tag{9b}$$

$$vec(\hat{x}_{t_k}^{\ t_k}) = vec(\hat{x}_{t_k}^{\ t_{k-1}}) + ((y_{t_k} - C_{t_k}\hat{x}_{t_k}^{t_{k-2}})^T (C_{t_k}P_{t_k}^{t_{k-1}}C_{t_k}^T + R_k)^{-1}C_{t_k} \otimes I)\,\mathrm{vec}(P_{t_k}^{t_{k-1}}), \tag{9c}$$

$$vec(P_{t_k}^{\ t_k}) = vec(P_{t_k}^{\ t_{k-1}}) - (P_{t_k}^{t_{k-1}}C_{t_k}^T(C_{t_k}P_{t_k}^{t_{k-1}}C_{t_k}^T + R_k)^{-1}C_{t_k} \otimes I)vec(P_{t_k}^{t_{k-1}}). \tag{9d}$$

Equations (9a)-(9d) describe the Kalman filtering for the continuous state-discrete measurement system using the *vec* function interpretation. This chapter is intended to develop the Kalman filtering in the conditional characteristic function evolution setting. On the other hand, the conditional characteristic function evolution setting is not applicable to the discrete state-discrete measurement system. Thus, in this chapter, the Kalman filtering for the discrete state-discrete measurement system is not the subject of investigations. A good source on the Kalman filtering for the discrete state-discrete measurement can be found in Anderson and Moore (1991).

Remark 1. The stationary Kalman filtering for the continuous state-discrete measurement has algebraic variance equations and constant coefficients. Conditional variance equations for *the between the observation case* and at *the observation case* become

$$0 = (I \otimes A + A \otimes I)vec(P_{t_{k-1}}^{t_{k-1}}) + vec(GG^T),$$

$$vec(P_{t_k}^{\ t_k}) = vec(P_{t_{k-1}}^{\ t_{k-1}}) - (P_{t_{k-1}}^{t_{k-1}}C_{t_k}^T(C_{t_k}P_{t_{k-1}}^{t_{k-1}}C_{t_k}^T + R_k)^{-1}C_{t_k} \otimes I)vec(P_{t_k-1}^{t_{k-1}}),$$

respectively.

Theorem 2. Consider the *filtering model* for the continuous state-continuous measurement

$$dx_i(t) = \sum_\alpha A_{i\alpha}(t)x_\alpha(t)dt + \sum G_{i\varphi}(t)dW_\varphi(t)$$

$$dz_i(t) = \sum_\alpha C_{i\alpha}(t)x_\alpha(t)dt + d\eta_i,$$

where the vector Brownian motion process $W = \{W_t \in R^m, 0 < t < \infty\}$. Consider x_t and A_t denote the state vector and system state matrix, respectively. The term $G(t)dW_t$ denotes the vector stochastic correction term with $\mathrm{var}(dW_t) = Idt$. The term z_t indicates the noisy observation vector that is available continuously. The observation noise statistics is $N(0, \phi_\eta dt)$. Then, the stochastic evolution of conditional characteristic function of the above filtering model assumes the structure of an SDE, i.e.

$$d\left\langle \exp(s^T x_t)\right\rangle = \left\langle \sum_p \sum_\alpha A_{p\alpha}(t)x_\alpha \frac{\partial \exp(s^T x_t)}{\partial x_p} + \frac{1}{2}\sum_{p,q}(GG^T)_{pq}(t)\frac{\partial^2 \exp(s^T x_t)}{\partial x_p \partial x_q}\right\rangle dt$$

$$+(\left\langle \exp(s^T x_t)x_t^T C_t^T\right\rangle - \left\langle \exp(s^T x_t)\right\rangle\left\langle x_t^T C_t^T\right\rangle)\phi_\eta^{-1}(dz_t - C_t\left\langle x_t\right\rangle dt).$$

$$= (\sum_p \sum_\alpha A_{p\alpha}(t)\left\langle x_\alpha(\tau)\frac{\partial \exp(s^T x_t)}{\partial x_p}\right\rangle + \frac{1}{2}\sum_{p,q}(GG^T)_{pq}(t)\left\langle \frac{\partial^2 \exp(s^T x_t)}{\partial x_p \partial x_q}\right\rangle)dt$$

$$+\sum_{\alpha,\beta}(\left\langle \exp(s^T x_t)\sum_\gamma C_{\alpha\gamma}(t)x_\gamma(t)\right\rangle - \left\langle \exp(s^T x_t)\right\rangle\left\langle \sum_\gamma C_{\alpha\gamma}(t)x_\gamma(t)\right\rangle)$$

$$\times(\phi_\eta^{-1})_{\alpha\beta}(dz_\beta - \sum_\gamma C_{\beta\gamma}(t)\left\langle x_\gamma\right\rangle dt), \tag{10}$$

where the conditional characteristic function $\left\langle \exp(s^T x_t)\right\rangle = E(\exp(s^T x_t)|Y_t)$. The conditional expectation operator E accounts for the accumulated observation $Y_t = \{z_\tau, t_0 \le \tau \le t\}$. The Kalman filtering becomes a special case of stochastic evolution of the conditional characteristic function, see (10).

Proof. For the filtering model of Theorem 2, the stochastic evolution of the conditional characteristic function assumes the structure of a stochastic integro-differential equation. One can construct the stochastic evolution of conditional characteristic function using Itô differential rule and the multiplication theorem of densities. As a result of this, we get (10) of Theorem 2 of the chapter.

Notably, we can construct the stochastic evolution of the conditional characteristic function by an alternative backward method. First, the method involves the construction of the stochastic evolution of conditional characteristic function for the Itô SDE using Itô differential rule and the multiplication theorem of densities (Liptser and Shiryaev, 1977). As a result of this, the stochastic evolution of conditional characteristic function for the non-linear Itô SDE becomes

$$d\left\langle \exp(s^T x_t)\right\rangle = (\left\langle \sum_i f_i(t,x_t)\frac{\partial \exp(s^T x_t)}{\partial x_i}\right\rangle + \frac{1}{2}\left\langle \sum_{i,j}(GG^T)_{ij}(t,x_t)\frac{\partial^2 \exp(s^T x_t)}{\partial x_i \partial x_j}\right\rangle)dt$$

$$+(\left\langle \phi h^T\right\rangle - \left\langle \phi\right\rangle\left\langle h^T\right\rangle)\phi_\eta^{-1}(dz_t - \left\langle h\right\rangle dt), \tag{11}$$

and the stochastic evolution for the linear Itô SDE is

$$d\left\langle \exp(s^T x_t)\right\rangle = (\sum_p \sum_\alpha A_{p\alpha}(t)\left\langle x_\alpha(\tau)\frac{\partial \exp(s^T x_t)}{\partial x_p}\right\rangle + \frac{1}{2}\sum_{p,q}(GG^T)_{pq}(t)\left\langle \frac{\partial^2 \exp(s^T x_t)}{\partial x_p \partial x_q}\right\rangle)dt$$

$$+\sum_{\alpha,\beta}(\left\langle \exp(s^T x_t)\sum_{\gamma} C_{\alpha\gamma}(t)x_\gamma(t)\right\rangle - \left\langle \exp(s^T x_t)\right\rangle\left\langle \sum_{\gamma} C_{\alpha\gamma}(t)x_\gamma(t)\right\rangle)$$

$$\times(\phi_\eta^{-1})_{\alpha\beta}(dz_\beta - \sum_{\gamma} C_{\beta\gamma}(t)\left\langle x_\gamma\right\rangle dt).$$

QED

The above is a special case of (11). Equation (10) can be regarded as the stochastic evolution of conditional characteristic function for the Kalman filtering for the continuous state-continuous measurement system. The term

$$(\left\langle \exp(s^T x_t)x_t^T C_t^T\right\rangle - \left\langle \exp(s^T x_t)\right\rangle\left\langle x_t^T C_t^T\right\rangle)\phi_\eta^{-1}$$

of (10) can be regarded as the gain of the Kalman filter from the conditional characteristic function perspective.

Consider $\exp(s^T x_t) = x_i(t)$. Then, the first term of the right-hand side will contribute, and the second term of the right-hand side will vanish. The last term of the above will contribute. Thus, equation (10) reduces to

$$d\left\langle x_i\right\rangle = (\sum_{\alpha} A_{i\alpha}(t)\left\langle x_\alpha(t)\right\rangle + \sum_{\alpha,\beta}(\sum_{\gamma} C_{\alpha\gamma}(t)(\left\langle x_i(t)x_\gamma(t)\right\rangle - \left\langle x_i\right\rangle\left\langle x_\gamma\right\rangle)$$

$$\times(\phi_\eta^{-1})_{\alpha\beta}(dz_\beta - \sum_{\gamma} C_{\beta\gamma}(t)\left\langle x_\gamma\right\rangle dt)$$

$$= (\sum_{\alpha} A_{i\alpha}(t)\left\langle x_\alpha(t)\right\rangle dt + \sum_{\alpha,\beta}(\sum_{\gamma} C_{\alpha\gamma}(t)P_{i\gamma}) \times(\phi_\eta^{-1})_{\alpha\beta}(dz_\beta - \sum_{\gamma} C_{\beta\gamma}(t)\left\langle x_\gamma\right\rangle dt). \tag{12}$$

Consider $\exp(s^T x_t) = x_i.x_j$, equation (10) boils down to

$$d\left\langle x_i x_j\right\rangle - (\sum_{\alpha} A_{i\alpha}(t)\left\langle x_\alpha(t)x_j(t)\right\rangle + \sum_{\alpha} A_{j\alpha}(t)\left\langle x_\alpha(t)x_i(t)\right\rangle + \frac{1}{2}\sum_{p,q}(GG^T)_{pq}(t)\left\langle \frac{\partial^2 x_i x_j}{\partial x_p \partial x_q}\right\rangle)dt$$

$$+\sum_{\alpha,\beta}(\left\langle x_i x_j \sum_{\gamma} C_{\alpha\gamma}(t)x_\gamma(t)\right\rangle - \left\langle x_i x_j\right\rangle\left\langle \sum_{\gamma} C_{\alpha\gamma}(t)x_\gamma(t)\right\rangle) \times(\phi_\eta^{-1})_{\alpha\beta}(dz_\beta - \sum_{\gamma} C_{\beta\gamma}(t)\left\langle x_\gamma\right\rangle dt). \tag{13}$$

After embedding (A.2) in (13), the above boils down to

$$(\sum_{\alpha} A_{i\alpha}(t)\langle x_{\alpha}(t)x_j(t)\rangle + \sum_{\alpha} A_{j\alpha}(t)\langle x_{\alpha}(t)x_i(t)\rangle \quad +(GG^T)_{ij}(t))dt \quad +\sum_{\alpha,\beta}\sum_{\gamma}(P_{i\gamma}\langle x_j\rangle + P_{j\gamma}\langle x_i\rangle)C_{\alpha\gamma}(t)$$
$$\times(\phi_{\eta}^{-1})_{\alpha\beta}(dz_{\beta} - \sum_{\gamma} C_{\beta\gamma}(t)\langle x_{\gamma}\rangle dt),$$

$$\langle x_i(t)\rangle d\langle x_j(t)\rangle = (\sum_{\alpha} A_{j\alpha}(t)\langle x_i(t)\rangle\langle x_{\alpha}(t)\rangle \qquad +\sum_{\alpha,\beta}(\sum_{\gamma} C_{\alpha\gamma}(t)P_{j\gamma}\langle x_i(t)\rangle)$$
$$(\phi_{\eta}^{-1})_{\alpha\beta}(dz_{\beta} - \sum_{\gamma} C_{\beta\gamma}(t)\langle x_{\gamma}\rangle dt), \tag{14}$$

$$\langle x_j(t)\rangle d\langle x_i(t)\rangle = (\sum_{\alpha} A_{i\alpha}(t)\langle x_j(t)\rangle\langle x_{\alpha}(t)\rangle \qquad +\sum_{\alpha,\beta}(\sum_{\gamma} C_{\alpha\gamma}(t)P_{i\gamma})\langle x_j(t)\rangle$$
$$(\phi_{\eta}^{-1})_{\alpha\beta}(dz_{\beta} - \sum_{\gamma} C_{\beta\gamma}(t)\langle x_{\gamma}\rangle dt), \tag{15}$$

$$d\langle x_i(t)\rangle d\langle x_j(t)\rangle = (\sum_{\alpha} A_{i\alpha}(t)\langle x_{\alpha}(t)\rangle dt + \sum_{\alpha,\beta}(\sum_{\gamma} C_{\alpha\gamma}(t)P_{i\gamma})\,(\phi_{\eta}^{-1})_{\alpha\beta}(dz_{\beta} - \sum_{\gamma} C_{\beta\gamma}(t)\langle x_{\gamma}\rangle dt))$$

$$\times(\sum_{\alpha} A_{j\alpha}(t)\langle x_{\alpha}(t)\rangle dt + \sum_{\alpha,\beta}(\sum_{\gamma} C_{\alpha\gamma}(t)P_{j\gamma})\,(\phi_{\eta}^{-1})_{\alpha\beta}(dz_{\beta} - \sum_{\gamma} C_{\beta\gamma}(t)\langle x_{\gamma}\rangle dt))$$

$$= (\sum_{\alpha,\beta}(\sum_{\gamma} C_{\alpha\gamma}(t)P_{i\gamma})\,(\phi_{\eta}^{-1})_{\alpha\beta}(dz_{\beta} - \sum_{\gamma} C_{\beta\gamma}(t)\langle x_{\gamma}\rangle dt))$$

$$\times(\sum_{r,s}(\sum_{\gamma} C_{s\gamma}(t)P_{j\gamma})\,(\phi_{\eta}^{-1})_{rs}(dz_r - \sum_{\gamma} C_{r\gamma}(t)\langle x_{\gamma}\rangle dt))$$

$$= \sum_{(\alpha,\beta)}(\sum_{\gamma} C_{\alpha\gamma}(t)P_{i\gamma})\,(\sum_{\gamma} C_{\beta\gamma}(t)P_{j\gamma})\,(\phi_{\eta}^{-1})_{\alpha\beta}(\phi_{\eta}^{-1})_{\alpha\beta}(\phi_{\eta})_{\alpha\beta}$$

$$= \sum_{(\alpha,\beta)}(\sum_{\gamma} C_{\alpha\gamma}(t)P_{i\gamma})\,(\sum_{\gamma} C_{\beta\gamma}(t)P_{j\gamma})\,(\phi_{\eta}^{-1})_{\alpha\beta}. \tag{16}$$

After combining (13)-(16), we have

$$dP_{ij} = d\langle x_i x_j\rangle - \langle x_i\rangle d\langle x_j\rangle - \langle x_j\rangle d\langle x_i\rangle - d\langle x_i\rangle d\langle x_j\rangle$$

$$= (\sum_{\alpha} A_{i\alpha}(t)\langle x_{\alpha}(t)x_j(t)\rangle + \sum_{\alpha} A_{j\alpha}(t)\langle x_{\alpha}(t)x_i(t)\rangle + (GG^T)_{ij}(t))dt + \sum_{\alpha,\beta}\sum_{\gamma}(P_{i\gamma}\langle x_j\rangle + P_{j\gamma}\langle x_i\rangle)C_{\alpha\gamma}(t)$$

$$\times(\phi_{\eta}^{-1})_{\alpha\beta}(dz_{\beta}-\sum_{\gamma}C_{\beta\gamma}(t)\langle x_{\gamma}\rangle dt)-(\sum_{\alpha}A_{j\alpha}(t)\langle x_{i}(t)\rangle\langle x_{\alpha}(t)\rangle)dt$$

$$-\sum_{\alpha,\beta}(\sum_{\gamma}C_{\alpha\gamma}(t)P_{j\gamma}\langle x_{i}(t)\rangle)(\phi_{\eta}^{-1})_{\alpha\beta}(dz_{\beta}-\sum_{\gamma}C_{\beta\gamma}(t)\langle x_{\gamma}\rangle dt)$$

$$-\sum_{\alpha}A_{i\alpha}(t)\langle x_{j}(t)\rangle\langle x_{\alpha}(t)\rangle dt-\sum_{\alpha,\beta}(\sum_{\gamma}C_{\alpha\gamma}(t)P_{i\gamma})\langle x_{j}(t)\rangle(\phi_{\eta}^{-1})_{\alpha\beta}(dz_{\beta}-\sum_{\gamma}C_{\beta\gamma}(t)\langle x_{\gamma}\rangle dt)$$

$$-\sum_{(\alpha,\beta)}(\sum_{\gamma}C_{\alpha\gamma}(t)P_{i\gamma})(\sum_{\gamma}C_{\beta\gamma}(t)P_{j\gamma})(\phi_{\eta}^{-1})_{\alpha\beta}dt$$

$$=(\sum_{\alpha}A_{i\alpha}(t)(\langle x_{\alpha}(t)x_{j}(t)\rangle-\langle x_{j}(t)\rangle\langle x_{\alpha}(t)\rangle)+\sum_{\alpha}A_{j\alpha}(t)(\langle x_{\alpha}(t)x_{i}(t)\rangle-\langle x_{i}(t)\rangle\langle x_{\alpha}(t)\rangle)+(GG^{T})(t))dt$$

$$+\sum_{\alpha,\beta}\sum_{\gamma}(P_{i\gamma}\langle x_{j}\rangle+P_{j\gamma}\langle x_{i}\rangle)C_{\alpha\gamma}(t)(\phi_{\eta}^{-1})_{\alpha\beta}(dz_{\beta}-\sum_{\gamma}C_{\beta\gamma}(t)\langle x_{\gamma}\rangle dt)$$

$$-\sum_{\alpha,\beta}(\sum_{\gamma}C_{\alpha\gamma}(t)P_{j\gamma}\langle x_{i}(t)\rangle)(\phi_{\eta}^{-1})_{\alpha\beta}(dz_{\beta}-\sum_{\gamma}C_{\beta\gamma}(t)\langle x_{\gamma}\rangle dt)$$

$$-\sum_{\alpha,\beta}(\sum_{\gamma}C_{\alpha\gamma}(t)P_{i\gamma})\langle x_{j}(t)\rangle)(\phi_{\eta}^{-1})_{\alpha\beta}(dz_{\beta}-\sum_{\gamma}C_{\beta\gamma}(t)\langle x_{\gamma}\rangle dt)$$

$$-\sum_{\alpha,\beta}(\sum_{\gamma}C_{\alpha\gamma}(t)P_{i\gamma})(\sum_{\gamma}C_{\beta\gamma}(t)P_{j\gamma})(\phi_{\eta}^{-1})_{\alpha\beta}dt$$

$$=(\sum_{\alpha}A_{i\alpha}(t)P_{j\alpha}(t)+\sum_{\alpha}P_{i\alpha}(t)A_{j\alpha}(t)+(GG^{T})_{ij}(t)-\sum_{\alpha,\beta}(\sum_{\gamma}C_{\alpha\gamma}(t)P_{i\gamma})(\sum_{\gamma}C_{\beta\gamma}(t)P_{j\gamma})(\phi_{\eta}^{-1})_{\alpha\beta})dt.$$

Alternatively, the standard Kalman filtering equations, the above equation coupled with (12), are recast as

$$d\langle x_{t}\rangle=A_{t}\langle x_{t}\rangle dt+P_{t}C_{t}^{T}\phi_{\eta}^{-1}(dz_{t}-C_{t}\langle x_{t}\rangle dt),$$

$$dP_{t}=(A_{t}P_{t}+P_{t}A_{t}^{T}+(GG^{T})(t)-P_{t}C_{t}^{T}\phi_{\eta}^{-1}C_{t}P_{t})dt.$$

The Kalman filtering for the continuous state-continuous state measurement system is given by the following set of two equations by adopting the vec function setting, i.e.

$$dvec(\langle x(t)\rangle) = (\langle x_t\rangle^T \otimes I)vec(A_t)dt + ((dz_t^T - \langle x_t\rangle^T C_t^T dt)\phi_\eta^{-1}C_t \otimes I)vec(P_t), \tag{17a}$$

$$dvec(P_t) = ((I \otimes A_t + A_t \otimes I)vec(P_t) + vec((GG^T)(t)) - (P_t C_t^T \phi_\eta^{-1} C_t \otimes I)vec(P_t). \tag{17b}$$

Equations (17a)-(178b) describe the Kalman filtering for the continuous state-continuous measurement system using the vec function interpretation.

Remark 2. The stationary Kalman filtering for the continuous state-continuous measurement has an algebraic variance equation and constant coefficients. Alternatively, the conditional variance equation is

$$0 = (I \otimes A + A \otimes I)vec(P_t) + vec(GG^T) - (PC^T\phi_\eta^{-1}C \otimes I)vec(P_t).$$

Theorem 3. Suppose the filtering model of a vector time-varying *bilinear* Stratonovich stochastic differential system is the following:

$$dx_t = (A_0(t) + A_t x_t)dt + (G(t) + x_t B_t^T) \circ dW_t,\ dz_t = C_t x_t dt + d\eta_t,$$

where the solution $x_t \in U$, the phase space $U \subset R^n$. Note that the vector Brownian motion process $W = \{(W_1(t), W_2(t), ..., W_d(t)) \in R^d, 0 \le t < \infty\}$. The former part of the filtering model denotes the vector Brownian motion-driven vector time-varying Stratonovich bilinear stochastic differential equation. The latter denotes the noisy observation equation.

In the component-wise description, the coupled bilinear filtering equations for the filtering model are the following:

$$\begin{aligned} d\hat{x}_i(t) &= (A_0^i(t) + \sum_\alpha A_{i\alpha}(t)\hat{x}_\alpha(t) + \frac{1}{2}\sum_\varphi (G_{i\varphi}(t)B_\varphi(t) + B_\varphi^2(t)\hat{x}_i(t)))dt \\ &+ \sum_{\alpha,\beta}(\sum_p P_{ip}(t)C_{\alpha p}(t))(\phi_\eta^{-1})_{\alpha\beta}(dz_\beta \\ &- \sum C_{\beta\gamma}(t)\hat{x}_\beta(t)dt), \end{aligned} \tag{18a}$$

$$\begin{aligned} dP_{ij} &= (\sum_p P_{ip}A_{jp}(t) + \sum_p P_{jp}A_{ip}(t) + \sum_\varphi G_{i\varphi}(t)G_{j\varphi}(t) + \hat{x}_i\sum_\varphi G_{j\varphi}(t)B_\varphi(t) \\ &+ \sum_\varphi G_{i\varphi}(t)B_\varphi(t)\hat{x}_j(t) + \hat{x}_i\hat{x}_j\sum_\varphi B_\varphi^2(t) + P_{ij}\sum_\varphi B_\varphi^2(t) \end{aligned} \tag{18b}$$

The Kalman filtering becomes a special case of the coupled bilinear filtering equations, i.e. (18a)-(18b),

Table 1. Summary of Properties of Kalman filtering

Properties	Kalman filtering for the continuous state-discrete measurement system	Kalman filtering for the continuous state-continuous measurement system
Evolution of conditional characteristic function (Lipster and Shreve, 1977)	Ordinary differential equation	Stochastic integro-differential equation
Riccati equation	Does not hold	Matrix ordinary differential equation
Lyapunov equation	Matrix ordinary differential equation	Does not hold
The Itô vs. Stratonovich (Manella and McClintok, 2012)	Invariant	Invariant
Vec function (Brewer, 1978)		

Proof. The proof of the coupled bilinear filtering equations involves the following steps: First, construct the stochastic evolution of conditional characteristic function for the bilinear Stratonovich stochastic filtering model of Theorem 3 of the chapter. Then, consider and to compute the component-wise bilinear stochastic filtering equations, equations (18a)–(18b). Since a procedure to construct the proof of the bilinear stochastic filtering equations is similar to the proof of Theorem 2, we omit the detail. The following is worth to mention: (i) after ignoring the term,

from the conditional mean (18a), we get the conditional mean equation of the Kalman filter. Secondly, after ignoring five correction terms from the conditional variance equation of the bilinear filter, see (18b),

we arrive at the conditional variance evolution equation of the Kalman filter.

Theorem 3 is about the filtering equations for the bilinear Stratonovich stochastic differential equation. An alternative interpretation is the bilinear Itô stochastic differential equation. The structure of the bilinear filtering equations hinges on stochastic interpretations. On the other hand, the Kalman filtering equations do not hinge on Itô and Stratonovich stochastic interpretations. The Itô-Stratonovich dilemma does not influence the Kalman filtering since the Itô and Stratonovich stochastic integrals coincide for linear stochastic differential equations. A good source for the Stratonovich perspective on the bilinear filtering can be found in (Rathore et al., 2020). Table 1 sum up five systems-theoretic properties of the Kalman filtering. Some of them are scattered in the literature, and some of them are relatively very less known.

CONCLUDING REMARKS

In this chapter, we have developed three Kalman filtering Theorems and their formal proof for the continuous state-discrete measurement system and continuous state-continuous measurement system. The chapter exploits the conditional characteristic function evolution equation and the Itô calculus to achieve that. The results of this chapter reveals how Kalman filtering equations become a special case of the stochastic evolution of conditional characteristic function for linear stochastic differential systems and linear measurement systems. Three influential results of stochastic processes, i.e., Kalman filtering, conditional characteristic function, and the Itô calculus are unified by adopting formal and systematic frameworks that were not available previously in the literature. This chapter differentiates between the Kalman filtering for the continuous state-continuous measurement system as well as the continuous state-discrete measurement system. Table 1 displays five significant ingredients of the Kalman filter, conditional characteristic function evolution, the Riccati equation, the Lyapunov matrix equation, the vec function, the Itô-Stratonovich dilemma.

REFERENCES

Anderson, B. D. O., & Moore, J. B. (1991). Kalman Filtering: Whence, What and Whither? In A. C. Antoulas (Ed.), *Mathematical System Theory* (pp. 41–54). doi:10.1007/978-3-662-08546-2_4

Brewer, J. W. (1978). Kronecker products and matrix calculus in system theory. *IEEE Transactions on Circuits and Systems*, *25*(9), 772–781. doi:10.1109/TCS.1978.1084534

Fujisaki, M., Kallianpur, G., & Kunita, H. (1972). Stochastic differential equations for the nonlinear filtering problem. *Osaka Journal of Mathematics*, *9*(1), 19–40.

Gattami, A. (2014). Kalman meets Shannon. *19th IFAC World Congress, IFAC Proceedings*, *47*(3), 2376–2381.

Holbrook, J., & Bhatia, R. (2006). An Old Question Asked in a New Context Presents Strange Aspects. *The Mathematical Intelligencer*, *28*(1), 32–39. doi:10.1007/BF02987000

Jazwinski, A. H. (1970). *Stochastic Processes and Filtering Theory*. Academic Press.

Kalman, R. E. (1960). A new approach to linear filtering and prediction problems. *The ASME Transactions. Journal of Basic Engineering*, *82*(1), 34–45. doi:10.1115/1.3662552

Karatzas, I., & Shreve, S. E. (1991). *Brownian Motion and Stochastic Calculus*. Springer-Verlag.

Kunita, H. (2010). Itô's stochastic calculus: Its surprising power for applications. *Stochastic Processes and Their Applications*, *120*(5), 622–652. doi:10.1016/j.spa.2010.01.013

Liptser, R. S., & Shiryaev, A. N. (1977). *Statistics of Random Processes* (Vol. 1). Springer. doi:10.1007/978-1-4757-1665-8

Mannella, R., & McClintock, P. V. E. (2012). Itô versus Stratonovich: 30 Years Later. *Fluctuation and Noise Letters*, *11*(1), 1240010–1240019. doi:10.1142/S021947751240010X

Patel, H. G., & Sharma, S. N. (2014). Third-order continuous-discrete filtering equations for a non-linear dynamical system, *The ASME Transactions. Journal of Computational and Nonlinear Dynamics*, *9*(3), 034502–034509. doi:10.1115/1.4026064

Rathore, S., & Sharma, S. N. (2018). Consensus on the Itô vs Stratonovich dilemma revisited. *IFAC-PapersOnLine*, *51*(1), 719–724. doi:10.1016/j.ifacol.2018.05.121

Rathore, S., Sharma, S. N., Bhatt, D. S., & Nagarsheth, S. H. (2020). Non-linear filering for bilinear stochastic differential systems: A Stratonovich perspective. *Transactions of the Institute of Measurement and Control*, *42*(10), 1–14. doi:10.1177/0142331219895711

Sharma, S. N., & Gawalwad, B. G. (2016). Wiener meets Kolmogorov. In *IEEE Conference on Norbert Wiener in the 21st Century (Thinking Machines in the Physical World)*. University of Melbourne.

APPENDIX

Calculation of the term

$$\left\langle \exp(s^T x_t)\sum_\gamma C_{\alpha\gamma}(t)x_\gamma(t)\right\rangle - \left\langle \exp(s^T x_t)\right\rangle\left\langle \sum_\gamma C_{\alpha\gamma}(t)x_\gamma(t)\right\rangle.$$

Here, we prove that

$$\left\langle \exp(s^T x_t)\sum_\gamma C_{\alpha\gamma}(t)x_\gamma(t)\right\rangle - \left\langle \exp(s^T x_t)\right\rangle\left\langle \sum_\gamma C_{\alpha\gamma}(t)x_\gamma(t)\right\rangle = \sum_\gamma C_{\alpha\gamma}(t)(P_{i\gamma}\left\langle x_j\right\rangle + P_{j\gamma}\left\langle x_i\right\rangle),$$

where the terms of the above relation are associated with the filtering model of the Theorem 2 of the chapter. The term

$$\left\langle \exp(s^T x_t)\sum_\gamma C_{\alpha\gamma}(t)x_\gamma(t)\right\rangle - \left\langle \exp(s^T x_t)\right\rangle\left\langle \sum_\gamma C_{\alpha\gamma}(t)x_\gamma(t)\right\rangle$$

arises in the stochastic evolution of conditional characteristic function for the linear stochastic differential system coupled with the continuous linear measurement system. We wish to compute the term, where $\exp(s^T x_t) = x_i x_j$. Then, the resulting expression is embedded in the conditional variance evolution of the standard Kalman filter. Thus, we have

$$\left\langle \exp(s^T x_t)\sum_\gamma C_{\alpha\gamma}(t)x_\gamma(t)\right\rangle - \left\langle \exp(s^T x_t)\right\rangle\left\langle \sum_\gamma C_{\alpha\gamma}(t)x_\gamma(t)\right\rangle$$

$$= \left\langle x_i x_j\sum_\gamma C_{\alpha\gamma}(t)x_\gamma(t)\right\rangle - \left\langle x_i x_j\right\rangle\left\langle \sum_\gamma C_{\alpha\gamma}(t)x_\gamma(t)\right\rangle.$$

Since the conditional expectation operator is linear, the right-hand side of the above becomes

$$\left\langle \exp(s^T x_t)\sum_\gamma C_{\alpha\gamma}(t)x_\gamma(t)\right\rangle - \left\langle \exp(s^T x_t)\right\rangle\left\langle \sum_\gamma C_{\alpha\gamma}(t)x_\gamma(t)\right\rangle$$

$$= \left\langle x_i x_j\sum_\gamma C_{\alpha\gamma}(t)x_\gamma(t)\right\rangle - \left\langle x_i x_j\right\rangle\left\langle \sum_\gamma C_{\alpha\gamma}(t)x_\gamma(t)\right\rangle.$$

$$= \sum_\gamma C_{\alpha\gamma}(t)\left\langle x_i x_j x_\gamma\right\rangle - \tag{A.1}$$

After considering the Gaussian assumption, even powers will contribute, and odd power contributions will vanish. Thus, the right-hand side of (A.1) becomes

$$\sum_{\gamma} C_{\alpha\gamma}(t)\left\langle x_i x_j x_\gamma \right\rangle - \sum_{\gamma} C_{\alpha\gamma}(t)\left\langle x_i x_j \right\rangle\left\langle x_\gamma \right\rangle$$

$$= \sum_{\gamma} C_{\alpha\gamma}(t)\left(\frac{1}{2}\sum_{p,q} P_{pq}\frac{\partial^2 \left\langle x_i \right\rangle\left\langle x_j \right\rangle\left\langle x_\gamma \right\rangle}{\partial\left\langle x_p \right\rangle\partial\left\langle x_q \right\rangle} - \frac{1}{2}\sum_{p,q} P_{pq}\frac{\partial^2 \left\langle x_i \right\rangle\left\langle x_j \right\rangle}{\partial\left\langle x_p \right\rangle\partial\left\langle x_q \right\rangle}\left\langle x_\gamma \right\rangle\right)$$

$$= \frac{1}{2}\sum_{\gamma} C_{\alpha\gamma}(t)\left(\sum_{p,q} P_{pq}\frac{\partial^2 \left\langle x_i \right\rangle\left\langle x_j \right\rangle\left\langle x_\gamma \right\rangle}{\partial\left\langle x_p \right\rangle\partial\left\langle x_q \right\rangle} - \sum_{p,q} P_{pq}\frac{\partial^2 \left\langle x_i \right\rangle\left\langle x_j \right\rangle}{\partial\left\langle x_p \right\rangle\partial\left\langle x_q \right\rangle}\left\langle x_\gamma \right\rangle\right)$$

$$= \frac{1}{2}\sum_{\gamma} C_{\alpha\gamma}(t)\sum_{p,q} P_{pq}\left(\frac{\partial}{\partial\left\langle x_p \right\rangle}(\delta_{iq}\left\langle x_j \right\rangle\left\langle x_\gamma \right\rangle + \delta_{jq}\left\langle x_i \right\rangle\left\langle x_\gamma \right\rangle\right.$$

$$\left. + \delta_{q\gamma}\left\langle x_i \right\rangle\left\langle x_j \right\rangle) - \delta_{ip}\delta_{jq}\left\langle x_\gamma \right\rangle - \delta_{jp}\delta_{iq}\left\langle x_\gamma \right\rangle\right).$$

The right-hand side of the above expression is further simplified to

$$\frac{1}{2}\sum_{\gamma} C_{\alpha\gamma}(t)\sum_{p,q} P_{pq}(\delta_{iq}\delta_{jp}\left\langle x_\gamma \right\rangle + \delta_{iq}\delta_{\gamma p}\left\langle x_j \right\rangle + \delta_{jq}\delta_{ip}\left\langle x_\gamma \right\rangle$$

$$+ \delta_{jq}\delta_{\gamma p}\left\langle x_i \right\rangle + \delta_{q\gamma}\delta_{ip}\left\langle x_j \right\rangle + \delta_{q\gamma}\delta_{jp}\left\langle x_i \right\rangle - \delta_{ip}\delta_{jq}\left\langle x_\gamma \right\rangle - \delta_{jp}\delta_{iq})\left\langle x_\gamma \right\rangle)$$

$$= \frac{1}{2}\sum_{\gamma} C_{\alpha\gamma}(t)\sum_{p,q} P_{pq}(\delta_{iq}\delta_{jp}\left\langle x_\gamma \right\rangle + \delta_{iq}\delta_{\gamma p}\left\langle x_j \right\rangle + \delta_{jq}\delta_{ip}\left\langle x_\gamma \right\rangle + \delta_{jq}\delta_{\gamma p}\left\langle x_i \right\rangle + \delta_{q\gamma}\delta_{ip}\left\langle x_j \right\rangle + \delta_{q\gamma}\delta_{jp}\left\langle x_i \right\rangle$$

$$- \delta_{ip}\delta_{jq}\left\langle x_\gamma \right\rangle - \delta_{jp}\delta_{iq})\left\langle x_\gamma \right\rangle)$$

$$= \sum_{\gamma} C_{\alpha\gamma}(t)\sum_{p,q} P_{pq}(\delta_{iq}\delta_{\gamma p}\left\langle x_j \right\rangle + \delta_{jq}\delta_{\gamma p}\left\langle x_i \right\rangle) = \sum_{\gamma} C_{\alpha\gamma}(t)(P_{i\gamma}\left\langle x_j \right\rangle + P_{j\gamma}\left\langle x_i \right\rangle).$$

Thus,

$$\left\langle \exp(s^T x_t)\sum_{\gamma} C_{\alpha\gamma}(t)x_\gamma(t) \right\rangle - \left\langle \exp(s^T x_t) \right\rangle\left\langle \sum_{\gamma} C_{\alpha\gamma}(t)x_\gamma(t) \right\rangle$$

$$= \left\langle x_i x_j \sum_\gamma C_{\alpha\gamma}(t) x_\gamma(t) \right\rangle - \left\langle x_i x_j \right\rangle \left\langle \sum_\gamma C_{\alpha\gamma}(t) x_\gamma(t) \right\rangle = \sum_\gamma C_{\alpha\gamma}(t) (P_{i\gamma} \left\langle x_j \right\rangle + P_{j\gamma} \left\langle x_i \right\rangle). \tag{A.2}$$

QED

Chapter 12
Project Control:
A Bayesian Model

Franco Caron
Politecnico Milano, Italy

ABSTRACT

The capability to elaborate a reliable estimate at completion for a project since the early stage of project execution is the prerequisite in order to provide an effective project control. The non-repetitive and uncertain nature of projects and the involvement of multiple stakeholders increase project complexity and raise the need to exploit all the available knowledge sources in order to improve the forecasting process. Therefore, drawing on a set of case studies, this chapter proposes a Bayesian approach to support the elaboration of the estimate at completion in those industrial fields where projects are denoted by a high level of uncertainty and complexity. The Bayesian approach allows the authors to integrate experts' opinions, data records related to past projects, and data related to the current performance of the ongoing project. Data from past projects are selected through a similarity analysis. The proposed approach shows a higher accuracy in comparison with the traditional formulas typical of the earned value management (EVM) methodology.

INTRODUCTION

Forecasting is a critical activity in project management: relying upon sound estimates to complete, the project manager can steer the ongoing project in order to meet specific time and cost objectives (Dvir and Lechler, 2004). Moreover, foresight is needed to avoid constantly being forced to manage emergencies, since emergency is often a lack of foresight. Without anticipation there can be no rationale in making a decision and we'll have to be at least adaptable to changing circumstances.

Planning and forecasting are strictly intertwined both in the early stage when the project baseline must be determined and throughout the entire project life cycle when project objectives have to be pursued (Hogarth and Makridakis, 1981). In the project control process the role of the Estimate To Complete (ETC) is critical, since the information drawn from the ETC, in comparison with the project baseline, may highlight the need for and the type of corrective action that may change the project plan. In fact,

DOI: 10.4018/978-1-7998-4706-9.ch012

ETC is the base for any effective corrective action. This approach to project control corresponds to a *feed-forward* type control loop (Anbari, 2003; Christensen, 1996), since analysis of the future informs present-day decisions.

From a recent survey (Merrow, 2011), analyzing the data of more than 300 global mega-projects, it appeared that in 2010 65% of the industrial projects with a minimum budget of 1 billion US dollars did not succeed in meeting the objectives of cost, duration and quality. This means that the forecasting accuracy is a critical problem for the project control process and, in particular, the methodologies commonly applied for forecasting purposes require an improvement.

To explain this kind of poor performance of the forecasting process, some considerations must be developed about the knowledge sources feeding the process, the forecasting techniques to be applied and the mitigating measures taken in order to avoid possible biases affecting the forecasting process.

As shown in Fig. 1, at a given time of the project duration, i.e. the time now (TN), a certain amount of the work will be already completed (Work Completed, WC), while the rest of the work will be ahead, corresponding to the Work Remaining (WR). The cost and time performance related to the Work Completed will be known, while a forecast will have to be developed for the WR.

Figure 1. Estimation at Completion at Time Now (internal view)

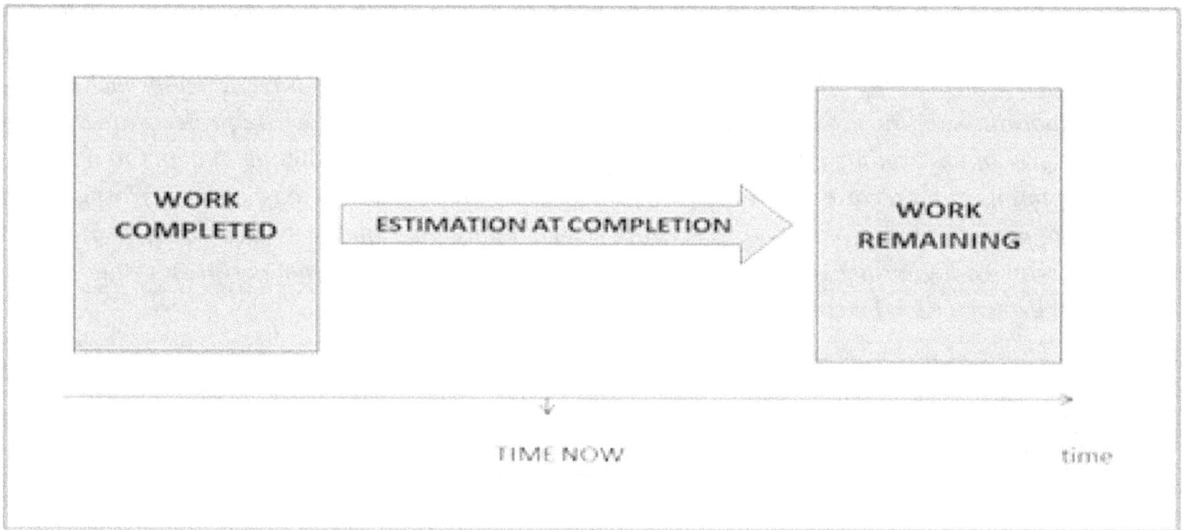

It should be noted that both the *accuracy of the forecast about WR* and the *impact of the corrective actions* that may be implemented based on the forecast depend on the progress of the project at the time now. The effectiveness of the corrective actions is greater in the early stages of the project execution and progressively diminishes while progress increases: in fact, as progress increases, the degrees of freedom available to steer the project tend to reduce progressively. On the other hand, the capability to forecast the project final duration and the final cost follows an opposite trend. In fact, at an early time in the execution phase, the knowledge available to the decision maker is scant and rapidly evolving; therefore, the capability to provide a *reliable forecast* is jeopardized, particularly if the forecast is only based upon the analysis of the performance of the ongoing project until the time now.

Drawing on a set of case studies (Caron et al., 2013a; Caron et al. 2013b), this paper will propose a Bayesian approach to determine the estimate to complete for a project. The paper has a twofold objective:

- A general objective related to the exploitation of all the available knowledge in order to improve the forecasting process;
- A specific objective related to the use of a Bayesian approach in order to integrate in a formal and rigorous way the diverse knowledge sources.

The second section analyzes the different techniques available to foresee the future; the third section identifies the different knowledge sources available; the fourth section addresses the mitigation of possible systematic biases during the forecasting process; the fifth section introduces the Earned Value Management system frequently used to determine the estimate at completion for a project, both in terms of cost and time; the sixth section introduces the general structure of a Bayesian model and eventually some results are given stemming from the application of the model to some projects in the oil & gas industry.

FORECASTING METHODOLOGIES

The future is something having its seeds in the present, in particular in the interests and behavior of project stakeholders (Kuosa, 2012). This is the basis for the forecasting process.

In general, the basic approaches available in order to improve the forecasting process during project control can be summarized as follows:

- *Trend analysis*; based on the extrapolation into the future, i.e. into the WR, of the actual trend at Time Now of a performance index, e.g. the Schedule Performance Index. This is a typical technique applied in the Earned Value Management.
- *Network analysis*; a network model of the project may indicate the activities making up the project and the precedence relationships between them; making assumptions about the activities the overall behavior of the system may be obtained.
- *Pattern analysis*; for instance based on S curves that have been identified as an invariable law describing the grand pattern of the progress of a class of similar projects. Exploiting the identification of typical patterns, e.g. described in terms of S-curves characterizing the progress of a class of similar projects, the effort required by the current project to achieve a certain milestone can be estimated.
- *Monte Carlo Simulation* allows to perform a what if analysis of the future development of the project; provided a mathematical/ logical model of the project and assuming suitable values for the input variables simulation allows for the analysis of possible future scenarios.
- *Stakeholder analysis* allows for the identification, assessment and management of the main stakeholders of the project. Much of the future is already in the stakeholders' interest, attitude, behavior, actions, influence on the project. Stakeholders are basically the main sources of risk – both positive and negative – for the project. On the other hand, since stakeholders are the main sources of knowledge about the project, their early engagement may increase significantly the effective-

ness of the planning and control process. The forecasting process becomes a participatory process involving a wide variety of stakeholders, increasing legitimacy of forecasting results and a shared sense of commitment.

Independently of the methodologies applied, an effective process of forecasting/planning depends on utilizing all the different types of available knowledge, in particular when facing a high level of project complexity.

KNOWLEDGE SOURCES

As mentioned above, all the knowledge available should be used in order to address the planning and control process for a complex project (Caron, 2014; Reich et al., 2014; Schindler and Eppler, 2003).

In general, the knowledge available to the project team may be classified in two ways: explicit/ tacit and internal/external. Explicit external knowledge corresponds to data records about projects completed in the past, including measures of forecasting capability in terms of the difference between estimated and actual overall cost and duration. Taking into consideration past experience should mitigate possible "optimistic" bias in estimating future project performance (Lovallo and Kahneman, 2003). Explicit internal knowledge corresponds to data records concerning the work completed WC, allowing for an evaluation of project performance at Time Now. Tacit external knowledge concerns the identification of similarities between the current project and some past projects in order to allow for transferring past data to the current project. Tacit internal project is about possible events/situations affecting the project's work remaining.

Depending on the types of knowledge used, three alternatives may be envisioned:

- Utilizing only data records related to WC, by extrapolating the current performance trend into the future;
- Adjusting the trend stemming from data records through experts' judgment about the expected performance during WR;
- Integrating the internal view of the project, i.e. data records related to WC and experts' judgment related to WR, with data records deriving from similar projects completed in the past.

The first alternative corresponds to the basic approach used in EVM based on the linear extrapolation of data records related to the WC of the current project.

In the second alternative, besides the explicit knowledge, also the tacit knowledge of the project team may be used to give a significant contribution to forecasting capability, particularly at the project outset when available explicit information is scant and critical decisions are to be taken. In order to elicit properly the experts knowledge, a set of requirements must be met: probabilistic reasoning, team work, keeping track of forecasting error and open mind (not seeing just what we know already) (Kuosa, 2014).

The specific contribution given by tacit knowledge i.e. by experts about the future development of the project, may concern:

- making sense of the weak signals indicating emerging situations possibly affecting project performance (e.g. weak signals such a series of scope changes, permits timeliness, engineering sequence not aligned with construction, high rate of rework in construction, missing data in design process, etc.) (Merrow, 2011; Williams et al., 2012; Haji-Kazemi et al., 2013).

- monitoring drivers explaining the project development during WC, and presumably affecting also WR, i.e. answering the question what kind of plausible drivers may have generated the actual development of the project till Time Now and how they will also influence the future? (e.g. drivers such as schedule aggressiveness, engineering completeness at execution start up, owner involvement, turnover in project leadership, anomalous low bid from subcontractors, unsatisfied stakeholders, new technology, project team integration, project team staffing, etc.) (Merrow, 2011);
- anticipating certain/uncertain events or conditions affecting project performance during WR which may originate both internally and externally to the project. Certain events may include planned corrective actions or contractual constraints, while uncertain events, i.e. risks, may arise both in terms of threats (i.e. adverse weather conditions) and opportunities (i.e. more efficient solutions deriving from suppliers collaboration);
- anticipating possible behaviors of the stakeholders involved in the project, e.g. opportunistic behavior; it should be noted that in this case the focus moves from risk events to risk sources, i.e. to the stakeholders as risk sources; the more a system is susceptible to human influence, the lower its predictability becomes.

As for the third approach, beside the use of internal knowledge, both explicit and tacit, external knowledge related to similar projects completed in the past may also be used. The use of data records related to similar projects completed in the past has been introduced both with reference to the project outset in order to improve the estimate of the project budget, in particular when a proposal has to be prepared, and with reference to the project control process at a generic Time Now, in order to implement effective corrective measures. Moreover, data related to past projects may help to avoid possible biases in the forecasting process.

FORECASTING BIAS

Even though project management systems based on EVM have been extensively implemented in the recent years, failures in meeting planned objectives are common, in particular in large engineering and construction projects such as in the oil & gas industry (Merrow, 2011). However, it remains an open question whether these failures are due to a lack of project efficiency during execution or to a lack of forecasting accuracy during the planning phase. In the former case, both positive and negative deviations from the baseline should be expected, depending on the specific evolution of the project. On the contrary, a systematic overrun in terms of cost and time may be easier explained as a weakness of the forecasting process since the project's outset.

Kahneman's and Tversky's studies (1979) show that a major source of forecasting failure, which influences the accuracy of final cost and duration estimates, is linked to an exclusively *"internal"* view approach, i.e. based only on knowledge deriving from inside the current project. Subsequently, the focus has moved to the psychological and political factors affecting the project planning process (Lovallo & Kahneman, 2003), and, in particular, two main sources of planning failure have been identified (Flyvbjerg, 2006; Flyvbjerg, 2009).

Firstly, the *cognitive illusions*, entailing two major aspects: over-optimism, i.e. the common attitude to assess future projects with greater optimism than justified by the actual previous experience, and anchoring, i.e. the attitude to deal with complex decisions selecting an initial reference point in the past experience and anchoring the estimate onto it.

Secondly, the *strategic and political pressures*, that may typically emerge during proposal preparation. Indeed, the approval of a project pre-supposes a competition involving different proposals, which often causes a voluntary underestimation of cost and duration by the project proposers in order to make their own proposal as attractive as possible.

Even more so, the need emerges to exploit all the available knowledge during the planning process, in order to minimize any bias effect. In fact, as shown in figure 1, the traditional control process often focuses only on data related to the current project, corresponding to an exclusively *"internal"* view (Flyvbjerg, 2006). An integration is needed between the *"internal"* and the *"external"* view, the latter is based on knowledge related to projects completed in the past (see figure 2) (Flyvbjerg, 2006).

Figure 2. Internal and external view

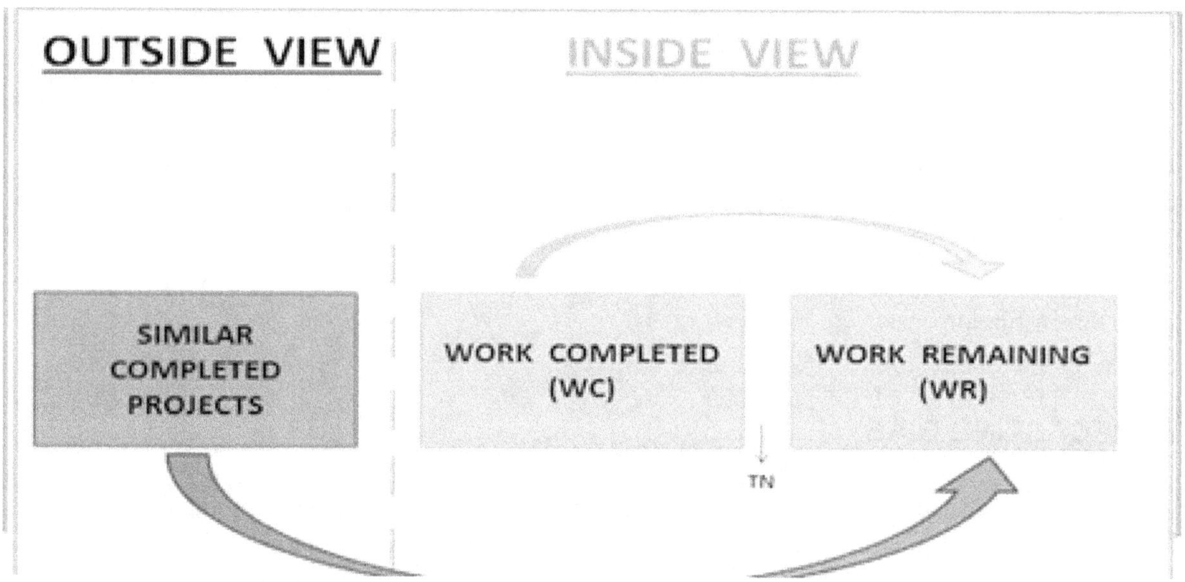

In fact, it may be assumed that the current project can be viewed as belonging to a cluster of similar projects completed in the past. Note that the selection of the cluster of similar projects is basically subjective since it depends on the similarity criteria adopted (Savio & Nikoloupolos, 2011; Green & Armstrong, 2007). Some cases, in fact, may express strong ambiguity. For example, if a company has to estimate the costs of an investment in a new technology and in an unfamiliar technological domain, should it take into account the set of highly innovative projects developed in different technological domains or the set of barely innovative projects but belonging to the same technological domain? Neither the former nor the latter option may be the best solution but both should be considered (Kahneman & Tversky, 1979).

Besides similarity criteria, the subjective assessment should also consider the trade-off between using a large number of past projects, leading to the risk of including projects substantially different from the current one, and a small number of projects, leading to the risk of losing statistic significance. Also the transferability of past data to the current project and the possible presence of outliers in the distribution should be critically evaluated.

EARNED VALUE MANAGEMENT SYSTEM

The most popular methodology used to develop an estimate at completion is the EVM. It is based on compact indices, easy to be evaluated, making it a synthetic and efficient approach (Fleming and Koppelman, 2006).

It is based on linear extrapolation since the performance related to WC is linearly extrapolated to WR. The extrapolation normally concerns the parameters Cost Performance Index (CPI) and Schedule Performance Index (SPI). These indexes may take into account not only data records related to WC but also subjective expectations of the decision maker about the WR.

As Project Management Institute (2013) stated, the main processes involved in project management are: initiating, planning, executing, monitoring, controlling and closing. In particular, Earned Value Management (EVM) represents an effective way of addressing the project control process. EVM is an efficient performance measurement and reporting technique for estimating cost and time at completion (Project Management Institute, 2013; Marshall et al., 2008). The following basic parameters are used in EVM, where TN indicates Time Now, i.e. the time along the project life cycle at which the control process is implemented:

- Planned Value (PV), the budget cost of work scheduled at TN;
- Earned Value (EV), the budget cost of work completed at TN;
- Actual Cost (AC), the actual cost of work completed at TN.

EVM was improved by Lipke (2002_a ; 2000_b ; 2003), who introduced the concept of Earned Schedule (ES) for obtaining a measure of the schedule performance index based on time units and overcoming the flaws associated with a Schedule Performance Index SPI defined as the ratio between EV and PV, both of them expressed in monetary terms. Earned Schedule is the time at which the EV value achieved at TN should have been obtained according to the project baseline. The new Schedule Performance Index SPI(t) at TN, defined as the ratio between ES and TN, represents a more effective approach, since it avoids the problem of the convergence of the EV and PV values toward BAC, i.e. Budget At Completion, as the project reaches completion (Lipke, 2006_a;Lipke, 2006_b).

The above three parameters and the ES, all of them evaluated at TN, allow for the calculation of a set of indices and variances at TN. The most important of these are:

- Cost Performance Index

$$CPI = EV / AC$$

- Cost Variance

$$CV = EV - AC$$

- Schedule Performance Index

$SPI_{(t)} = ES / TN$;

- Schedule Variance

$SV_{(t)} = ES - TN$

Variances CV and SV(t) summarize the project's past performance during WC, while indexes CPI and SPI(t) may be used in order to extrapolate the current trend and estimate the future performance during WR (Anbari, 2003).

Many Estimate at Completion formulas have been proposed during almost 50 years of EVM application but none of them has proved to be always more accurate than any other one (Christensen, 1993). In the basic approach, the estimations of final cost (i.e. EAC) and final duration (i.e. TAC, Time at Completion) are based on the following equations:

$$EAC = AC + (BAC - EV) / CPI_f \quad (1)$$

where:

AC = Actual Cost at TN

BAC = Budget at Completion

EV = Earned Value at TN

CPI_f = Cost Performance Index estimated for the work remaining (WR)

$TAC = TN + (PAC - ES) / SPI_f$ (2)where:

TN = Time Now

PAC = Planned at Completion, i.e. the planned duration of the project

ES = Earned Schedule at Time Now

SPI_f = Schedule Performance Index estimated for the work remaining (WR)

It should be noted that future performance values may significantly differ from past performance. The new performance indices CPI_f and $SPI(t)_f$ have been introduced in equations 1 and 2 with reference to the Work Remaining and may consider a possible evolution of the project different from the expected, in particular differing from past performance. While the generic indices CPI and SPI(t) are related to the overall WC, CPI_f and $SPI_{(t)f}$ will be related to the overall WR. In fact, relying only on past performance while developing a forecast could be misleading, since considering only past values of CPI and $SPI_{(t)}$ is similar to driving a car whilst looking just in the rear view mirror, so making it impossible to dodge the obstacles that may lie on the route ahead.

Both equations 1 and 2 indicate that the values assigned to the performance indexes CPI_f and SPI_f play a critical role in order to obtain an accurate estimate of the final cost and duration. As a consequence, forecasting capability can be improved by utilizing all the available knowledge about the performance indexes CPIf and SPIf (Liu & Zu, 2007; Goodwin, 2005). The Bayesian approach allow for integrating in a rigorous way the knowledge content stemming from diverse knowledge sources.

BAYESIAN APPROACH

The Bayes Theorem represents a rigorous and formal approach allowing for an update of a prior estimate, which expresses the experts' preliminary opinion, by means of the data gathered in the field. For instance, the project team may assume a prior estimate of the final budget overrun, based on subjective expectations about the development of the current project, and this prior estimate may be updated based on the actual performance of the current project at Time Now (Caron et al., 2013). It should be noted that the two knowledge sources should be considered independent, so avoiding any anchoring bias. In this way knowledge stemming from two different sources, data records and expert's judgment, can be integrated. In a Bayesian framework, the experts' preliminary opinions, are an example of use of subjective probability, the only probability concept applicable to non-repetitive processes such as the projects. In fact the essentially subjective dimension inherent in any project, must be recognized. Subjective probability is defined as the degree of belief in the occurrence of an event, by a given person at a given time and with a given set of information. It should be noted that increasing the level of knowledge available may modify the value of the subjective probability assigned to a future event (De Finetti, 1937; Caron and al., 2013). While the metaphor of 'frequency based' probability is the dice throwing (i.e. a repetitive process), the metaphor of the subjective probability is the bet (i.e. a unique process). In general, we can assume that any proactive action involving the future is a bet. In fact, De Finetti defined probability as a price. Let's consider a lottery where if the event E occurs the better wins 1, if it doesn't happen the better wins 0. How much is the better prepared to pay for accepting the lottery? That price is the subjective probability value associated to the event E. The probability value corresponds to the degree of belief about event E occurring.

The essence of Bayesian inference is using probability to describe our state of knowledge about some event or parameter of interest. A prior distribution (based on expert's tacit knowledge) is updated by means of experimental observations (data records) collected during the execution process, in order to obtain a posterior distribution, integrating both knowledge types.

For instance, the project team may assume a prior distribution of the final budget overrun, based on subjective expectations about the development of the current project, and this prior distribution may be updated based on the actual performance of the current project until Time Now (Caron et al., 2013).

Hence, Bayesian approaches to formulate valid forecasts have been proposed in literature (Palomo et al., 2006; Gardoni et al., 2007; Kim and Reinschmidt, 2009) . The adoption of a Bayesian approach enables the elaboration of probabilistic time and cost estimates at completion based on multiple knowledge sources. This statement is easier to understand if the formulation of the Bayesian theorem is considered:

$$f\left(\mu \mid y_1, y_2, \ldots, y_n\right) \propto L\left(\mu; y_1, y_2, \ldots, y_n\right) \cdot f\left(\mu\right) \tag{3}$$

Eq. (3) consists of three main components:

- $f(\mu)$ is the prior distribution of the parameter of inference, μ : it summarizes the initial subjective opinion detained by the decision-maker about the probability density function of the parameter μ ;

- $L(\mu; y_1, y_2, \ldots, y_n)$ is the likelihood function, obtained after the collection of the *n* experimental data $y_1, y_2, \ldots, y_n$;
- $f(\mu | y_1, y_2, \ldots, y_n)$ is the posterior distribution, i.e. the distribution that expresses the knowledge acquired about the parameter μ after updating the initial subjective judgments with the experimental data.

APPLICATION

Firstly, the application of the above model requires the definition of the inference parameter, for instance the performance indices CPI_f and $SPI(t)_f$ in order to forecast the final variance between budget cost and actual cost. Secondly, it requires the definition of the prior and likelihood distributions in order to obtain the posterior distribution. The prior distribution is based on the elicitation of the experts' tacit knowledge. It may be integrated by the data records related to projects completed in the past. The likelihood distribution is based on data records related to the current project. The Bayesian approach (see fig. 2) allows for the integration of both knowledge sources. The prior distribution updated by the likelihood function gives the posterior distribution. It should be noted that the Bayesian approach doesn't offer a point estimate, such as in the traditional EVM approach, but a distribution estimate, so increasing the knowledge available for the project team. Moving from the prior distribution to the posterior distribution the degree of belief in the estimate increases, since a larger knowledge content has been used. Assuming that standard distributions, e.g. the normal distribution, have been used, this improvement corresponds to a reduction of the standard deviation of the posterior distribution compared with the prior one.

The Bayesian approach has been applied to a set of large engineering projects in the oil and gas industry (Caron et al. 2013a; Caron et al. 2013b) in order to forecast the actual final duration and the actual final budget. These projects were characterized by a high level of uncertainty and complexity. After the projects reached completion the forecasting error related to the Bayesian model were compared with the one that would have been obtained if traditional EVM forecasting formulas had been applied. The result was that the Bayesian approach allowed for a better forecasting accuracy independently from the type of project (on shore, off shore, subsea) and in particular in the early stage of a project when the information is scant and the potential impact of decisions is very high.

CONCLUSION

Drawing on a set of case studies, this paper proposes a Bayesian approach to determine the estimate to complete, both in terms of cost and time, for a project. The paper focuses on the knowledge sources necessary to feed the forecasting process and the mitigating measures taken in order to avoid possible estimating biases affecting the process.

The Bayesian approach allows for increasing the forecasting accuracy, mainly in those fields where projects are denoted by a high degree of complexity and uncertainty and in particular in the early phase of the project when knowledge available is scant.

The future research may develop a Bayesian model allowing for integrating a multiple set of knowledge sources. For instance, the project team may assume a prior estimate of the final budget overrun, based both on subjective expectations about the current project and data records related to similar past projects, and this prior estimate may be updated based on the actual performance of the current project at Time Now.

REFERENCES

Anbari, F. (2003). Earned Value: Project management method and extension. *Project Management Journal*, *34*(4), 12–23. doi:10.1177/875697280303400403

Ansoff, H. I. (1975). Managing strategic surprise by response to weak signals. *California Management Review*, *17*(2), 21–23. doi:10.2307/41164635

Caron, F. (2014). *Managing the continuum: certainty, uncertainty, unpredictability in large engineering projects*. Springer.

Caron, F., Ruggeri, F., & Borgarucci, C. (2013b). Bayesian integration of internal and external views in forecasting project performance. *Journal of Modern Project Management*, *1*(2), 112–121.

Caron, F., Ruggeri, F., & Merli, A. (2013a). A Bayesian approach to improve estimate at completion in Earned Value Management. *Project Management Journal*, *44*(1), 1, 3–16. doi:10.1002/pmj.21303

Christensen, D. (1996). Project Advocacy and the estimate at completion problem. *Journal of Cost Analysis and Management*, 35-60.

De Finetti, B. (1937). La prévison: Ses lois logiques, ses sources subjectives. *Annales de l'Institut Henri Poincaré*, *7*(1), 1–68.

Dvir, D., & Lechler, T. (2004). Plans are nothing, changing plans is everything: The impact of changes on project success. *Research Policy*, *33*(1), 1–15. doi:10.1016/j.respol.2003.04.001

Fleming, Q. W., & Koppelman, J. (2006). *Earned value project management* (3rd ed.). Project Management Institute.

Flyvbjerg, B. (2006). From Nobel prize to project management: Getting risk right. *Project Management Journal*, *37*(3), 5–15. doi:10.1177/875697280603700302

Flyvbjerg, B. (2009). Survival of the un-fittest: Why the worst infrastructure gets built – and what we can do about it. *Oxford Review of Economic Policy*, *25*(3), 344–367. doi:10.1093/oxrep/grp024

Flyvbjerg, B., Holm, M.S. & Buhl, S. (2002), Underestimating costs in public works projects: error or lie? *Journal of the American Planning Association, 68*(3), 279-295.

Gardoni, P., Reinschmidt, K. F., & Kumar, R. (2007). A probabilistic framework for Bayesian adaptive forecast. *Computer-Aided Civil and Infrastructure Engineering*, *22*(3), 182–196. doi:10.1111/j.1467-8667.2007.00478.x

Goodwin, P. (2005). How to integrate management judgment with statistical forecasts. *Foresight*, *1*, 8–12.

Green, K. C., & Armstrong, J. S. (2007). Structured analogies for forecasting. *International Journal of Forecasting, 23*(3), 365–367. doi:10.1016/j.ijforecast.2007.05.005

Haji-Kazemi, S., Andersen, B., & Krane, H. P. (2013). A review on possible approaches for detecting early warning signs in projects. *Project Management Journal, 44*(5), 55–69. doi:10.1002/pmj.21360

Hogarth, R. M., & Makridakis, S. (1981). Forecasting and planning: An evaluation. *Management Science, 27*(2), 115–138. doi:10.1287/mnsc.27.2.115

Kahneman, D., & Tversky, A. (1979). Intuitive prediction: Biases and corrective procedures. *TIMS Studies in Management Science, 12*, 313–327.

Kim, B., & Reinschmidt, K. F. (2009). Probabilistic forecasting of project duration using Bayesian inference and the beta distribution. *Journal of Construction Engineering and Management, 135*(3), 178–186. doi:10.1061/(ASCE)0733-9364(2009)135:3(178)

Kolltveit, B.J., Karlsen, J.T. & Gronhaug, K. (2004, Oct.). Exploiting opportunities in uncertainty during the early project phase. *Journal of Management in Engineering*.

Kuosa, T. (2012). *The evolution of strategic foresight*. Gower.

Lipke, W. (2002a). A study of the normality of Earned Value Management indicators. *The Measurable News*, (December), 1–16.

Lipke, W. (2002b). Statistical process control of project performance. *Crosstalk, 13*(March), 16–20.

Lipke, W. (2003). Schedule is different. *The Measurable News*, 31-34.

Lipke, W. (2006a). Earned Schedule Leads to Improved Forecasting. *Proceedings of the 3rd international conference on project management (PROMAC 2006).*

Lipke, W. (2006b). Statistical methods applied to EVM ...the next frontier. *The Measurable News*, (Winter), 18–30.

Liu, L. & Zhu, K. (2007). Improving cost estimates of construction projects using phased cost factors. *Journal of Construction Engineering and Management, 133*(1).

Lovallo, D., & Kahneman, D. (2003). Delusion of Success: How optimism undermines executives' decisions. *Harvard Business Review, 81*, 56–63. PMID:12858711

Merrow, E. W. (2011). Oil industry megaprojects: our recent track record. *Offshore Technology Conference*. 10.4043/21858-MS

Palomo, J., Ruggeri, F., Rios Insua, D., Cagno, E., Caron, F., & Mancini, M. (2006). On Bayesian forecasting of procurement delays: A case study. *Applied Stochastic Models in Business and Industry, 22*(2), 181–192. doi:10.1002/asmb.627

Project Management Institute. (2013). *A Guide to the Project Management Body of Knowledge* (5th ed.). Author.

Reich, B. H., Gemino, A., & Sauer, C. (2014). How knowledge management impacts performance in projects: An empirical study. *International Journal of Project Management, 32*(4), 590–602. doi:10.1016/j.ijproman.2013.09.004

Schindler, M., & Eppler, M. J. (2003). Harvesting project knowledge: A review of project learning methods and success factors. *International Journal of Project Management, 21*(3), 219–228. doi:10.1016/S0263-7863(02)00096-0

Weick, K. E. (1995). *Sense-making in organizations*. SAGE.

Williams, T., Klakegg, O. J., Walker, D. H. T., Andersen, B., & Magnussen, O. M. (2012). Identifying and Acting on Early Warning Signs in Complex Projects. *Project Management Journal, 43*(2), 37–53. doi:10.1002/pmj.21259

Williams, T., Samset, K., & Sunnevag, K. J. (Eds.). (2009). *Making essential choices with scant information*. Palgrave Macmillan. doi:10.1057/9780230236837

KEY TERMS AND DEFINITIONS

Bayes Theorem: Represents a rigorous approach to update a prior distribution, which expresses the experts' preliminary opinion, through the experimental data gathered on the field.

Earned Value Management System (EVMS): Management system aiming at an integrated control of project cost and schedule based on the concept of earned value i.e. budget cost of work performed at Time Now.

Estimate to Complete (ETC): Effort required to complete the project, in terms of cost and time.

Monte Carlo Simulation: "What if" analysis of the future project scenarios, provided a mathematical/ logical model of the project implemented on a computer.

Project Control: Project management process aiming at identifying and implementing possible corrective actions based on the expected performance of the project.

Subjective Probability: The degree of belief in the occurrence of an event, by a given person at a given time and with a given set of information.

Trend Analysis: Linear extrapolation to the work remaining of the data records related to indices of project performance, e.g. productivity, during the execution of the work completed in order to estimate cost and time to complete the project.

Compilation of References

Wang, R. (2017). Research on data security technology based on cloud storage. *Procedia Engineering*, *174*, 1340–1355. doi:10.1016/j.proeng.2017.01.286

Prerna, P., & Agarwal, P. (2017). Cryptography Based Security for Cloud Computing System. *International Journal of Advanced Research in Computer Science*, *8*(5).

Kumar, G. K., & Gobi, D. M. (2017). Role of Cryptography & its Related Techniques in Cloud Computing Security. *International Journal for Research in Applied Science and Engineering Technology, 5.*

Jaber, A. N., & Zolkipli, M. F. B. (2013, November). Use of cryptography in cloud computing. In *2013 IEEE International conference on control system, computing and Engineering* (pp. 179-184). IEEE. 10.1109/ICCSCE.2013.6719955

Brindha, K., & Jeyanthi, N. (2015, February). Securing cloud data using visual cryptography. In *International Confernce on Innovation Information in Computing Technologies* (pp. 1–5). IEEE. doi:10.1109/ICIICT.2015.7396073

Chakraborty, M., Jana, B., & Mandal, T. (2018, July). A Secure Cloud Computing Authentication Using Cryptography. In *2018 International Conference on Emerging Trends and Innovations In Engineering And Technological Research (ICETIETR)* (pp. 1-4). IEEE. 10.1109/ICETIETR.2018.8529100

Harmon, E. (2018). *Strategies used by cloud security managers to implement secure access methods*. Academic Press.

Sujitha, K., & Praveen, K. (2015, January). Analysing cloud simulation results using big data analytics model. In *2015 International Conference on Computer Communication and Informatics (ICCCI)* (pp. 1-6). IEEE.

Deshmukh, S., & Sumeet, S. (2015, December). Big Data Analytics Using Public Cloud Infrastructure: Use Cases and Cost Economics. In *2015 International Conference on Computational Intelligence and Communication Networks (CICN)* (pp. 782-784). IEEE. 10.1109/CICN.2015.159

Khan, Z., Anjum, A., & Kiani, S. L. (2013, December). Cloud based big data analytics for smart future cities. In *2013 IEEE/ACM 6th International Conference on Utility and Cloud Computing* (pp. 381-386). IEEE. 10.1109/UCC.2013.77

Manekar, A. K., & Pradeepini, G. (2015, December). Cloud based big data analytics a review. In *2015 International Conference on Computational Intelligence and Communication Networks (CICN)* (pp. 785-788). IEEE. 10.1109/CICN.2015.160

Mayilvaganan, M., & Sabitha, M. (2013, December). A cloud-based architecture for Big-Data analytics in smart grid: A proposal. In *2013 IEEE International Conference on Computational Intelligence and Computing Research* (pp. 1-4). IEEE. 10.1109/ICCIC.2013.6724168

Ramamoorthy, S., & Rajalakshmi, S. (2013, July). Optimized data analysis in cloud using BigData analytics techniques. In *2013 Fourth International Conference on Computing, Communications and Networking Technologies (ICCCNT)* (pp. 1-5). IEEE. 10.1109/ICCCNT.2013.6726631

Yadav, S. L., & Sohal, A. (2017). Review Paper on Big Data Analytics in Cloud Computing. *International Journal of Computer Trends and Technology*, *49*, 156–160.

Sayeed, Z., Liao, Q., Faucher, D., Grinshpun, E., & Sharma, S. (2015, June). Cloud analytics for wireless metric prediction-framework and performance. In *2015 IEEE 8th International Conference on Cloud Computing* (pp. 995-998). IEEE. 10.1109/CLOUD.2015.135

Vuppala, S. K., Dinesh, M. S., Viswanathan, S., Ramachandran, G., Bussa, N., & Geetha, M. (2017, November). Cloud based big data platform for image analytics. In *2017 IEEE International Conference on Cloud Computing in Emerging Markets (CCEM)* (pp. 11-18). IEEE. 10.1109/CCEM.2017.11

Abdelhamid, N., Thabtah, F., & Abdel-jaber, H. (2017, July). Phishing detection: A recent intelligent machine learning comparison based on models content and features. In 2017 IEEE international conference on intelligence and security informatics (ISI) (pp. 72-77). IEEE. doi:10.1109/ISI.2017.8004877

Abrar Ul Haq, M., Ayub, K., & Ullah, M. I. (2015). Micro-level Determinants of Rural Poverty in Pakistan. *International Journal of Scientific and Research Publications*, *5*(2), 1–4.

Afroz, S., Garg, V., McCoy, D., & Greenstadt, R. (2013, September). *Honor among thieves: A common's analysis of cybercrime economies. In 2013 APWG eCrime Researchers Summit*. IEEE.

Akram, F., Metwally, M. K., Han, H. S., Jeon, H. J., & Kim, T. S. (2013, February). A novel P300-based BCI system for words typing. In *2013 International Winter Workshop on Brain-Computer Interface (BCI)* (pp. 24-25). IEEE.

Alkhatib, M., Talaei-Khoei, A., & Ghapanchi, A. (2016). *Analysis of research in healthcare data analytics*. arXiv preprint arXiv:1606.01354

Al-Suwaidi, N., Nobanee, H., & Jabeen, F. (2018). Estimating Causes of Cyber Crime: Evidence from Panel Data FGLS Estimator. *International Journal of Cyber Criminology*, *12*(2), 392–407.

Anand, S. (1983). *Inequality and Poverty in Malaysia: Measurement and Decomposition*. Oxford University Press.

Anbari, F. (2003). Earned Value: Project management method and extension. *Project Management Journal*, *34*(4), 12–23. doi:10.1177/875697280303400403

Anderson, B. D. O., & Moore, J. B. (1991). Kalman Filtering: Whence, What and Whither? In A. C. Antoulas (Ed.), *Mathematical System Theory* (pp. 41–54). doi:10.1007/978-3-662-08546-2_4

An, J., & Kim, H. W. (2018). A data analytics approach to the cybercrime underground economy. *IEEE Access: Practical Innovations, Open Solutions*, *6*, 26636–26652. doi:10.1109/ACCESS.2018.2831667

Ansoff, H. I. (1975). Managing strategic surprise by response to weak signals. *California Management Review*, *17*(2), 21–23. doi:10.2307/41164635

Armin, J., Thompson, B., Ariu, D., Giacinto, G., Roli, F., & Kijewski, P. (2015, August). 2020 cybercrime economic costs: No measure no solution. In *2015 10th International Conference on Availability, Reliability and Security* (pp. 701-710). IEEE.

Aven, T., & Eidesen, K. (2007). A predictive Bayesian approach to risk analysis in health care. *BMC Medical Research Methodology*, *7*(1), 38. doi:10.1186/1471-2288-7-38 PMID:17714597

Ballard, D. H. (1982). *Computer Vision*. Prentice Hall.

Balov, N. (2016, August). *Bayesian hierarchical models in Stata. In 2016 Stata Conference (No. 30)*. Stata Users Group.

Basu, A. (2014). Social network analysis: A methodology for studying terrorism. In *Social networking* (pp. 215–242). Springer. doi:10.1007/978-3-319-05164-2_9

Baulch, B. (2002). P*overty monitoring and targeting using ROC curves: Examples from Vietnam. IDS Working Paper, 161*(August).

Belavkin, V. (1980). Quantum filtering of Markov signals with white quantum noise. *Radiotechnika I Electronica, 25*, 1445-1453.

Belavkin, V. (1983). Theory of control of observable quantum systems. *Automation and Remote Control*, *44*, 178–188.

Benneyan, J. C., Lloyd, R. C., & Plsek, P. E. (2003). Statistical process control as a tool for research and healthcare improvement. *BMJ Quality & Safety*, *12*(6), 458–464. doi:10.1136/qhc.12.6.458 PMID:14645763

Bhatt, A. B., & Karandikar, R. L. (1999). *On filtering with Ornstein-Uhlenbeck process as observation noise*. Retrieved from www.isid.ac.in/~statmath/eprints/2002/isid200232.ps

Bhatt, U., Iyyani, D., Jani, K., & Mali, S. (2018, April). Troll-Detection Systems Limitations of Troll Detection Systems and AI/ML Anti-Trolling Solution. In *2018 3rd International Conference for Convergence in Technology (I2CT)* (pp. 1-6). IEEE.

Bhunia, A. K., Shaikh, A. A., Sharma, G., & Pareek, S. (2015). A two storage inventory model for deteriorating items with variable demand and partial backlogging. *J. Ind. Prod. Eng.*, *32*(4), 263–272. doi:10.1080/21681015.2015.1046508

Birbaumer, N. (2006). Breaking the silence: Brain–computer interfaces (BCI) for communication and motor control. *Psychophysiology*, *43*(6), 517–532. doi:10.1111/j.1469-8986.2006.00456.x PMID:17076808

Bircan, H. (2004). Lojistik Regresyon Analizi : T ı p Verileri Üzerine Bir Uygulama. *Kocaeli Ünv Sosyal Bilimler Ens Derg.*, *2*, 185–208.

Biyase, M., & Zwane, T. (2018). An empirical analysis of the determinants of poverty and household welfare in South Africa. *Journal of Developing Areas*, *52*(1), 115–130. doi:10.1353/jda.2018.0008

Botrel, L., & Kübler, A. (2018, January). Reliable predictors of SMR BCI performance—Do they exist? In *2018 6th International Conference on Brain-Computer Interface (BCI)* (pp. 1-3). IEEE.

Boulkedid, R., Abdoul, H., Loustau, M., Sibony, O., & Alberti, C. (2011). Using and reporting the Delphi method for selecting healthcare quality indicators: A systematic review. *PLoS One*, *6*(6), e20476. doi:10.1371/journal.pone.0020476 PMID:21694759

Bourguignon, F., & Chakravarty, S. (2003). The measurement of multidimensional poverty. *The Journal of Economic Inequality*, *1*(1), 25–49. doi:10.1023/A:1023913831342

Bouten, L. (2004). *Filtering and control in quantum optics* (PhD Thesis Dissertation). Nottingham University, UK.

Bouvrie, J. (2006). *Notes on convolutional neural networks*. Academic Press.

Brachman, R. J., & Levesque, H. J. (2004). *Knowledge representation and reasoning*. Morgan Kaufman.

Bradley, A. P. (1997). The use of the area under the ROC curve in the evaluation of machine learning algorithms. *Pattern Recognition*, *30*(7), 1145–1159. doi:10.1016/S0031-3203(96)00142-2

Breiman, L. (2001). Statistical modeling: The two cultures (with comments and a rejoinder by the author). *Statistical Science*, *16*(3), 199–231. doi:10.1214s/1009213726

Brewer, J. W. (1978). Kronecker products and matrix calculus in system theory. *IEEE Transactions on Circuits and Systems*, *25*(9), 772–781. doi:10.1109/TCS.1978.1084534

Butler, J., Foot, C., Bomb, M., Hiom, S., Coleman, M., Bryant, H., Vedsted, P., Hanson, J., & Richards, M.ICBP Working Group. (2013). The international cancer benchmarking partnership: An international collaboration to inform cancer policy in Australia, Canada, Denmark, Norway, Sweden and the United Kingdom. *Health Policy (Amsterdam)*, *112*(1-2), 148–155. doi:10.1016/j.healthpol.2013.03.021 PMID:23693117

Cacace, M., Ettelt, S., Mays, N., & Nolte, E. (2013). Assessing quality in cross-country comparisons of health systems and policies: Towards a set of generic quality criteria. *Health Policy (Amsterdam)*, *112*(1-2), 156–162. doi:10.1016/j.healthpol.2013.03.020 PMID:23628482

Canbay, T., & Selim, S. (2010). Türkiye'de Hanehalkı Yoksulluğu. *Ege Akademik Bakış / Ege Academic Review*, *10*(2), 627–649.

Carminati, M., Caron, R., Maggi, F., Epifani, I., & Zanero, S. (2014, June). BankSealer: An online banking fraud analysis and decision support system. In *IFIP International Information Security Conference* (pp. 380-394). Springer. 10.1007/978-3-642-55415-5_32

Caron, F. (2014). *Managing the continuum: certainty, uncertainty, unpredictability in large engineering projects*. Springer.

Caron, F., Ruggeri, F., & Borgarucci, C. (2013b). Bayesian integration of internal and external views in forecasting project performance. *Journal of Modern Project Management*, *1*(2), 112–121.

Caron, F., Ruggeri, F., & Merli, A. (2013a). A Bayesian approach to improve estimate at completion in Earned Value Management. *Project Management Journal*, *44*(1), 1, 3–16. doi:10.1002/pmj.21303

Chapra, S. C. (2012). *Applied Numerical Methods*. McGraw-Hill.

Chatterjee, A., Roy, S., & Shrivastava, R. A Machine Learning Approach to Prevent Cancer. In *Handbook of Research on Disease Prediction Through Data Analytics and Machine Learning* (pp. 112–141). IGI Global.

Chatterjee, A., & Sinha, T. (2019). Correlation between Absence, Interest in the Field and Grades in an Organization using Regression Model. *International Journal of Engineering and Advanced Technology*, *8*(6), 1436–1441. doi:10.35940/ijeat.F8118.088619

Chaudhari, P., Agrawal, H., & Kotecha, K. (2019). Data augmentation using MG-GAN for improved cancer classification on gene expression data. *Soft Computing*, 1–11.

Chauhan, A., & Singh, A. P. (2015). A note on the inventory models for deteriorating items with Verhulst's model type demand rate. *Int. J. Oper. Res.*, *22*(2), 243–261. doi:10.1504/IJOR.2015.067340

Chen, T., & Chefd'Hotel, C. (2014, September). Deep learning based automatic immune cell detection for immunohistochemistry images. In *International workshop on machine learning in medical imaging* (pp. 17-24). Springer. 10.1007/978-3-319-10581-9_3

Choudhary, P., Singh, U., Dalal, S., & Bisen, D. (2019, February). Social Network Analysis (SNA): A Vision for Counter-Terrorism Approach in Modern Indian Defence Sector. *Proceedings of 2nd International Conference on Advanced Computing and Software Engineering (ICACSE)*. 10.2139srn.3349028

Christensen, D. (1996). Project Advocacy and the estimate at completion problem. *Journal of Cost Analysis and Management*, 35-60.

Chun, J., Bae, B., & Jo, S. (2016, February). BCI based hybrid interface for 3D object control in virtual reality. In *2016 4th International Winter Conference on Brain-Computer Interface (BCI)* (pp. 1-4). IEEE.

Cierniak, R. (2005). *Computed tomography*. Exit.

Citko, W., Jakóbczak, D., & Sieńko, W. (2005, September). *On Hurwitz - Radon matrices based signal processing*. Paper presented at the workshop Signal Processing at Poznan University of Technology, Poznań, Poland.

Clerc, M., Bougrain, L., & Lotte, F. (Eds.). (2016). Brain-computer interfaces. Academic Press.

Cocozza-Thivent, C., Eymard, R., Mercier, S., & Roussignol, M. (2006). Characterization of the Marginal Distributions of Markov Processes Used in Dynamic Reliability. *Journal of Applied Mathematics and Stochastic Analysis*, 1–18.

Collins, G. W. II. (2003). *Fundamental Numerical Methods and Data Analysis*. Case Western Reserve University.

Dahlquist, G., & Bjoerck, A. (1974). *Numerical methods*. Prentice Hall.

Dahlquist, G., & Bjoerck, A. (1974). *Numerical Methods*. Prentice Hall.

Dal, S. (2013). *Türkiye'de Hanehalkı Yoksulluğunu Belirleyen Etmenler*. Gazi Üniversitesi, Sosyal Bilimler Enstitüsü.

Dansuk, E. (1997), *Türkiye'de Yoksulluğun Ölçülmesi ve Sosyo-Ekonomik Yapılarla Ölçülmesi,* (Published Specialization Thesis). DPT, Sosyal Sektörler ve Koordinasyon Genel Müdürlüğü Ücretler ve Gelirler Dairesi Başkanlığı, Yayın No: DPT: 2472, Ankara.

Dari, S., & Sani, B. (2017). An EPQ model for delayed deteriorating items with quadratic demand and shortages. Asian Journal of Mathematics and Computer Research, 22(2), 87–103.

Das, P., & Das, A. K. (2017, January). Behavioural analysis of crime against women using a graph based clustering approach. In *2017 International Conference on Computer Communication and Informatics (ICCCI)* (pp. 1-6). IEEE. 10.1109/ICCCI.2017.8117714

De Finetti, B. (1937). La prévison: Ses lois logiques, ses sources subjectives. *Annales de l'Institut Henri Poincaré*, *7*(1), 1–68.

Deaton, A. (2003). Regional poverty estimates for India, 1999-2000. Research Program in Development Studies, Princeton University.

Dejdumrong, N. (2007). A Shape Preserving Verification Techniques for Parametric Curves. *Computer Graphics, Imaging and Visualization. CGIV, 2007*, 163–168.

De, P. K., & Rawat, A. (2013). Optimal order quantity of an EOQ model using expected value of a fuzzy function. *IEEE Int. Conf. Fuzzy Syst.* 10.1109/FUZZ-IEEE.2013.6622534

Dirican, A. (2001). Tani Testí Performanslarının Değerlendírílmesí Ve Kıyaslanması. *Cerrahpaşa Tıp Dergisi*, *32*(1), 25–30.

Dumanlı, R. (1996). *Yoksulluk ve Türkiye'deki Boyutları*. DPT-Sosyal Sektörler ve Koordinasyon Genel Müdürlüğü.

Dvir, D., & Lechler, T. (2004). Plans are nothing, changing plans is everything: The impact of changes on project success. *Research Policy*, *33*(1), 1–15. doi:10.1016/j.respol.2003.04.001

Dyn, N., Levin, D., & Gregory, J. A. (1987). A 4-Point Interpolatory Subdivision Scheme for Curve Design. *Computer Aided Geometric Design*, *4*, 257–268.

Eckmann, B. (1999). Topology, algebra, analysis- relations and missing links. *Notices of the American Mathematical Society*, *5*(46), 520–527.

Einstein, A. (1956). *Investigations on the theory of the Brownian movement*. Dover Publications, Inc.

Epo, B. N. (2010). *Determinants of Poverty in Cameroon: A Binomial and Polychotomous Logit Analysis*. SSRN.

Fagin, R., Halpern, J. Y., Moses, Y., & Vardi, M. Y. (1995). *Reasoning about knowledge*. MIT Press.

Feller, W. (2000). *An Introduction to Probability Theory and its Applications* (Vol. 2). John Wiley and Sons.

Flach, P., Blockeel, H., Ferri, C., Hernández-Orallo, J., & Struyf, J. (2003). Decision support for data mining. In Data Mining and Decision Support. Springer US. doi:10.1007/978-1-4615-0286-9_7

Fleming, Q. W., & Koppelman, J. (2006). *Earned value project management* (3rd ed.). Project Management Institute.

Flyvbjerg, B., Holm, M.S. & Buhl, S. (2002), Underestimating costs in public works projects: error or lie? *Journal of the American Planning Association, 68*(3), 279-295.

Flyvbjerg, B. (2006). From Nobel prize to project management: Getting risk right. *Project Management Journal, 37*(3), 5–15. doi:10.1177/875697280603700302

Flyvbjerg, B. (2009). Survival of the un-fittest: Why the worst infrastructure gets built – and what we can do about it. *Oxford Review of Economic Policy, 25*(3), 344–367. doi:10.1093/oxrep/grp024

Forde, I., Morgan, D., & Klazinga, N. S. (2013). Resolving the challenges in the international comparison of health systems: The must do's and the trade-offs. *Health Policy (Amsterdam), 112*(1-2), 4–8. doi:10.1016/j.healthpol.2013.01.018 PMID:23434265

Fox, R. F. (1986). Uniform convergence to an effective Fokker-Planck equation for weakly colored noise. *Physical Review A, 34*(5), 4525–4527. doi:10.1103/PhysRevA.34.4525 PMID:9897829

Franceschetti, M., Dousse, O., David, N. C., & Thiran, P. (2007). Closing the gap in the capacity of wireless networks via percolation theory. *IEEE Transactions on Information Theory, 53*(3), 1009–1018. doi:10.1109/TIT.2006.890791

Fujisaki, M., Kallianpur, G., & Kunita, H. (1972). Stochastic differential equations for the nonlinear filtering problem. *Osaka Journal of Mathematics, 9*(1), 19–40.

Gajewski, B. J., Mahnken, J. D., & Dunton, N. (2008). Improving quality indicator report cards through Bayesian modeling. *BMC Medical Research Methodology, 8*(1), 77. doi:10.1186/1471-2288-8-77 PMID:19017399

Gao, S., Wang, Y., Gao, X., & Hong, B. (2014). Visual and auditory brain–computer interfaces. *IEEE Transactions on Biomedical Engineering, 61*(5), 1436–1447. doi:10.1109/TBME.2014.2300164 PMID:24759277

Gardoni, P., Reinschmidt, K. F., & Kumar, R. (2007). A probabilistic framework for Bayesian adaptive forecast. *Computer-Aided Civil and Infrastructure Engineering, 22*(3), 182–196. doi:10.1111/j.1467-8667.2007.00478.x

Garza-Rodríguez, J. (2015). The determinants of Poverty in the Mexican States of the US-Mexico Border. *Estudios Fronterizos (México), 17*(33), 141–167. doi:10.21670/ref.2016.33.a06

Gattami, A. (2014). Kalman meets Shannon. *19th IFAC World Congress, IFAC Proceedings, 47*(3), 2376–2381.

Geweke, J., Gowrisankaran, G., & Town, R. J. (2003). Bayesian inference for hospital quality in a selection model. *Econometrica, 71*(4), 1215–1238. doi:10.1111/1468-0262.00444

Goodwin, P. (2005). How to integrate management judgment with statistical forecasts. *Foresight, 1*, 8–12.

Green, K. C., & Armstrong, J. S. (2007). Structured analogies for forecasting. *International Journal of Forecasting, 23*(3), 365–367. doi:10.1016/j.ijforecast.2007.05.005

Grosse-Wentrup, M. (2019, February). The elusive goal of BCI-based communication with CLIS-ALS patients. In *2019 7th International Winter Conference on Brain-Computer Interface (BCI)* (pp. 1-2). IEEE.

Guan, C., Robinson, N., Handiru, V. S., & Prasad, V. A. (2017, January). Detecting and tracking multiple directional movements in EEG based BCI. In *2017 5th International Winter Conference on Brain-Computer Interface (BCI)* (pp. 44-45). IEEE. 10.1109/IWW-BCI.2017.7858154

Guo, Z. (2005). a Verhulst Model on Time Series Error Corrected for Port Throughput Forecasting. *Journal of the Eastern Asia Society for Transportation Studies*, *6*, 881–891.

Gwak, K., Leeb, R., Millán, J. D. R., & Kim, D. S. (2013, February). A novel tactile stimulation system for BCI feedback. In *2013 International Winter Workshop on Brain-Computer Interface (BCI)* (pp. 29-31). IEEE. 10.1109/IWW-BCI.2013.6506619

Gwanda, Y. I., Bari, A. N., & Singh, V. V. (2019). Optimal Production Model for Inventory Items with Verhulst's Demand and Time Dependent Amelioration Rate. *Palest. J. Math.*, *8*(2), 413–425.

Haji-Kazemi, S., Andersen, B., & Krane, H. P. (2013). A review on possible approaches for detecting early warning signs in projects. *Project Management Journal*, *44*(5), 55–69. doi:10.1002/pmj.21360

Häkkinen, U., Iversen, T., Peltola, M., Seppälä, T. T., Malmivaara, A., Belicza, É., Fattore, G., Numerato, D., Heijink, R., Medin, E., & Rehnberg, C. (2013). Health care performance comparison using a disease-based approach: The Euro-HOPE project. *Health Policy (Amsterdam)*, *112*(1-2), 100–109. doi:10.1016/j.healthpol.2013.04.013 PMID:23680074

Han, C., Hayashi, H., Rundo, L., Araki, R., Shimoda, W., Muramatsu, S., . . . Nakayama, H. (2018, April). GAN-based synthetic brain MR image generation. In *2018 IEEE 15th International Symposium on Biomedical Imaging (ISBI 2018)* (pp. 734-738). IEEE. 10.1109/ISBI.2018.8363678

Han, X., Zhang, S., & Gao, X. (2019, February). A study on reducing training time of BCI system based on an SSVEP dynamic model. In *2019 7th International Winter Conference on Brain-Computer Interface (BCI)* (pp. 1-2). IEEE. 10.1109/IWW-BCI.2019.8737318

Hänggi, P. (1995). The functional derivative and its use in the description of noisy dynamical systems. In L. Pesquera & M. Rodriguez (Eds.), Stochastic Processes Applied to Physics (pp. 69-95). World Scientific.

Hänggi, P., & Jung, P. (1995). Colored noise in dynamical systems. In I. Prigogine & S. A. Rice (Eds.), *Advances in Chemical Physics* (pp. 239–323). John Wiley and Sons.

Hanley, A. J., & McNeil, J. B. (1982). The Meaning and Use of the Area under a Receiver Operating Characteristic (ROC) Curve. *Radiology*, *143*(1), 29–36. doi:10.1148/radiology.143.1.7063747 PMID:7063747

Hati, K. K., & Majumder, R. (2011). *Health for development: a district level study in West Bengal*. Academic Press.

Hida, T., Kuo, H.-H., Pontho, J., & Streit, L. (1993). *White Noise: An Infinite Dimensional Calculus*. Kluwer, Academic Publishers. doi:10.1007/978-94-017-3680-0

Hogarth, R. M., & Makridakis, S. (1981). Forecasting and planning: An evaluation. *Management Science*, *27*(2), 115–138. doi:10.1287/mnsc.27.2.115

Holbrook, J., & Bhatia, R. (2006). An Old Question Asked in a New Context Presents Strange Aspects. *The Mathematical Intelligencer*, *28*(1), 32–39. doi:10.1007/BF02987000

Hossein, Z., & Gerard, A. (2013). Trends in cost sharing among selected high income countries—2000–2010. *Health Policy (Amsterdam)*, *112*(1-2), 35–44. doi:10.1016/j.healthpol.2013.05.020 PMID:23809913

Hudson, R. L., & Parthasarathy, K. R. (1984). Quantum Itô formula and stochastic evolutions. *Communications in Mathematical Physics*, *93*(3), 301–323. doi:10.1007/BF01258530

Hu, H., Guan, Q., Chen, S., Ji, Z., & Yao, L. (2017). Detection and recognition for life state of cell cancer using two-stage cascade CNNs. *IEEE/ACM Transactions on Computational Biology and Bioinformatics*. PMID:29990223

Indolia, S., Goswami, A. K., Mishra, S. P., & Asopa, P. (2018). Conceptual understanding of convolutional neural network- a deep learning approach. *Procedia Computer Science*, *132*, 679–688. doi:10.1016/j.procs.2018.05.069

Jakóbczak, D. (2006). *Application of discrete, orthogonal operator of Hurwitz - Radon in compression and reconstruction of monochromatic images' contours* (Unpublished doctoral dissertation). Polish - Japanese Institute of Information Technology, Warsaw, Poland.

Jakóbczak, D. (2011). Curve Extrapolation and Data Analysis using the Method of Hurwitz-Radon Matrices. *Folia Oeconomica Stetinensia*, *9*(17), 121-138.

Jakóbczak, D. (2007). 2D and 3D image modeling using Hurwitz - Radon matrices. *Polish Journal of Environmental Studies*, *4A*(16), 104–107.

Jakóbczak, D. (2007). 2D and 3D Image Modeling Using Hurwitz-Radon Matrices. *Polish Journal of Environmental Studies*, *4A*(16), 104–107.

Jakóbczak, D. (2009). Curve Interpolation Using Hurwitz-Radon Matrices. *Polish Journal of Environmental Studies*, *3B*(18), 126–130.

Jakóbczak, D. (2010). Object Modeling Using Method of Hurwitz-Radon Matrices of Rank k. In W. Wolski & M. Borawski (Eds.), *Computer Graphics: Selected Issues* (pp. 79–90). University of Szczecin Press.

Jakóbczak, D. (2010). Shape Representation and Shape Coefficients via Method of Hurwitz-Radon Matrices. *Lecture Notes in Computer Science*, *6374*, 411–419.

Jakóbczak, D. (2011). Curve Parameterization and Curvature via Method of Hurwitz-Radon Matrices. *Image Processing & Communications-. International Journal (Toronto, Ont.)*, *1-2*(16), 49–56.

Jakóbczak, D. (2011). Data Extrapolation and Decision Making via Method of Hurwitz-Radon Matrices. *Lecture Notes in Computer Science*, *6922*, 173–182.

Jakóbczak, D. (2013). Probabilistic Modeling of Signature using the Method of Hurwitz-Radon Matrices. *Global Perspectives on Artificial Intelligence*, *1*(1), 1–7.

Jakóbczak, D., & Kosiński, W. (2007). Application of Hurwitz - Radon matrices in monochromatic medical images decompression. In Z. Kowalczuk & B. Wiszniewski (Eds.), *Intelligent data mining in diagnostic purposes: Automatics and informatics* (pp. 389–398). PWNT.

Jakóbczak, D., & Kosiński, W. (2007). Hurwitz - Radon operator in monochromatic medical image reconstruction. *Journal of Medical Informatics & Technologies*, *11*, 69–78.

Jana, D. K., Das, B., & Roy, T. K. (2013). *A partial backlogging inventory model for deteriorating item under fuzzy inflation and discounting over random planning horizon: A fuzzy genetic algorithm approach* (Vol. 2013). Adv. Oper. Res.

Jana, D. K., Maity, K., & Roy, T. K. (2013). A Bi-fuzzy Approach to a Production-Recycling-Disposal Inventory Problem with Environment Pollution Cost via Genetic Algorithm. *International Journal of Computers and Applications*, *61*(22), 975–8887.

Jani, S. (2018). *The Growth of Crypto currency in India: Its Challenges& Potential Impacts on Legislation.* Research Gate Publication.

Jazwinski, A. H. (1970). *Stochastic Processes and Filtering Theory.* Academic Press.

Kahneman, D., & Tversky, A. (1979). Intuitive prediction: Biases and corrective procedures. *TIMS Studies in Management Science, 12*, 313–327.

Kakkad, V., Shah, H., Patel, R., & Doshi, N. (2019). A Comparative study of applications of Game Theory in Cyber Security and Cloud Computing. *Procedia Computer Science, 155*, 680–685. doi:10.1016/j.procs.2019.08.097

Kalman, R. E. (1960). A new approach to linear filtering and prediction problems. *The ASME Transactions. Journal of Basic Engineering, 82*(1), 34–45. doi:10.1115/1.3662552

Kang, E. S., Kim, B. C., & Suk, H. I. (2016, February). An empirical suggestion for collaborative learning in motor imagery-based BCIs. In *2016 4th International Winter Conference on Brain-Computer Interface (BCI)* (pp. 1-3). IEEE. 10.1109/IWW-BCI.2016.7457450

Kaongoen, N., & Jo, S. (2017, January). The effect of selective attention on multiple ASSRs for future BCI application. In *2017 5th International Winter Conference on Brain-Computer Interface (BCI)* (pp. 9-12). IEEE. 10.1109/IWW-BCI.2017.7858144

Karatzas, I., & Shreve, S. E. (1991). Brownian Motion and Stochastic Calculus (graduate text in mathematics). New York: Springer.

Karatzas, I., & Shreve, S. E. (1991). *Brownian Motion and Stochastic Calculus.* Springer-Verlag.

Kim, M., & Kim, S. P. (2018, January). A comparsion of artifact rejection methods for a BCI using event related potentials. In *2018 6th International Conference on Brain-Computer Interface (BCI)* (pp. 1-4). IEEE. 10.1109/IWW-BCI.2018.8311530

Kim, B., & Reinschmidt, K. F. (2009). Probabilistic forecasting of project duration using Bayesian inference and the beta distribution. *Journal of Construction Engineering and Management, 135*(3), 178–186. doi:10.1061/(ASCE)0733-9364(2009)135:3(178)

Kitrungrotsakul, T., Iwamoto, Y., Han, X. H., Takemoto, S., Yokota, H., Ipponjima, S., . . . Chen, Y. W. (2019, May). A cascade of CNN and LSTM network with 3D anchors for mitotic cell detection in 4D microscopic image. In *ICASSP 2019-2019 IEEE International Conference on Acoustics, Speech and Signal Processing (ICASSP)* (pp. 1239-1243). IEEE.

Kızılgöl, Ö. (2008). Yoksulluğun Belirleyicileri: Türkiye Örneği. *Finans Politik & Ekonomik Yorumlar, 45*(526).

Klazinga, N., Fischer, C., & Ten Asbroek, A. (2011). Health services research related to performance indicators and benchmarking in Europe. *Journal of Health Services Research & Policy, 16*(2_suppl), 38-47.

Kobayashi, N., & Nakagawa, M. (2018). BCI-based control of electric wheelchair using fractal characteristics of EEG. *IEEJ Transactions on Electrical and Electronic Engineering, 13*(12), 1795–1803. doi:10.1002/tee.22742

Kolltveit, B.J., Karlsen, J.T. & Gronhaug, K. (2004, Oct.). Exploiting opportunities in uncertainty during the early project phase. *Journal of Management in Engineering.*

Kozera, R. (2004). *Curve modeling via interpolation based on multidimensional reduced data.* Silesian University of Technology Press.

Kozera, R. (2004). *Curve Modeling via Interpolation Based on Multidimensional Reduced Data.* Silesian University of Technology Press.

Kriegman, D. J., & Ponce, J. (1990). On recognizing and positioning curved 3-D objects from image contours. *IEEE Transactions on Pattern Analysis and Machine Intelligence*, *12*(12), 1127–1137. doi:10.1109/34.62602

Kshetri, N. (2018). Introducing the IT Economics Department. *IT Professional*, *20*(1), 83–87. doi:10.1109/MITP.2018.011311501

Kumar, N., Singh, S. R., & Kumari, R. (2012). *An inventory model with time-dependent demand and limited storage facility under inflation* (Vol. 2012). Adv. Oper. Res.

Kumar, P. V. (2016, March). Growing cyber crimes in India: A survey. In *2016 International Conference on Data Mining and Advanced Computing (SAPIENCE)* (pp. 246-251). IEEE. 10.1109/SAPIENCE.2016.7684146

Kumar, S., & Agarwal, D. (2018). Hacking attacks, methods, techniques and their protection measures. *International Journal of Advance Research in Computer Science and Management*, *4*(4), 2253–2257.

Kumar, S., Koley, S., & Kuamr, U. (2015). Present Scenrio of Cyber Crime in INDIA and its Preventions. *IJSER*, *6*(4), 1972–1976.

Kunita, H. (2010). Itô's stochastic calculus: Its surprising power for applications. *Stochastic Processes and Their Applications*, *120*(5), 622–652. doi:10.1016/j.spa.2010.01.013

Kuo, H.-H. (2009). White noise stochastic integration. Stochastic Analysis: Classical and Quantum, eProceedings, 57-71.

Kuo, H.-H., & Russek, A. (1988). White noise approach to stochastic integration. *Journal of Multivariate Analysis*, *24*(2), 218–236. doi:10.1016/0047-259X(88)90037-1

Kuosa, T. (2012). *The evolution of strategic foresight*. Gower.

Kuppulakshmi & Sugapriya. (2020). Effective Economic Production Quantity model for Penalty Maintenance cost with Rework Allowing Price Discount and Shortage. *Test Engineering and Management*, *83*, 16267-16286.

Lahariya, C. (2018). 'Ayushman Bharat' program and universal health coverage in India. *Indian Pediatrics*, *55*(6), 495–506. doi:10.100713312-018-1341-1 PMID:29978817

Lamdan, Y., Schwartz, J. T., & Wolfson, H. J. (1990). Affine invariant model-based object recognition. *IEEE Transactions on Robotics and Automation*, *5*(6), 578–589. doi:10.1109/70.62047

Larsen, E. A. (2011). *Classification of EEG signals in a brain-computer interface system* (Master's thesis). Institutt for datateknikkoginformasjonsvitenskap.

Lasko, T. A., Bhagwat, J. G., Zou, K. H., & Ohno-Machado, L. (2005). The use of receiver operating characteristic curves in biomedical informatics. *Journal of Biomedical Informatics*, *38*(5), 404–415. doi:10.1016/j.jbi.2005.02.008 PMID:16198999

Latecki, L. J., & Lakaemper, R. (1999). Convexity rule for shape decomposition based on Discrete Contour Evolution. *Computer Vision and Image Understanding*, *3*(73), 441–454. doi:10.1006/cviu.1998.0738

Lau, R. Y., Xia, Y., & Ye, Y. (2014). A probabilistic generative model for mining cybercriminal networks from online social media. *IEEE Computational Intelligence Magazine*, *9*(1), 31–43. doi:10.1109/MCI.2013.2291689

Lee, D. H., Seo, T. S., & Ko, Y. T. (2001). Spiral CT of the gastric carcinoma: Staging and enhancement pattern. *Clinical Imaging*, *25*(1), 32–37. doi:10.1016/S0899-7071(01)00245-5 PMID:11435037

Li, L., Kabadi, S. N., & Nair, K. P. K. (2002). Fuzzy models for single-period inventory problem. *Fuzzy Sets and Systems*, *132*(3), 273–289. doi:10.1016/S0165-0114(02)00104-5

Lim, H., & Ku, J. (2019, February). High engagement in BCI action observation game by relevant character's movement. In *2019 7th International Winter Conference on Brain-Computer Interface (BCI)* (pp. 1-3). IEEE. 10.1109/IWW-BCI.2019.8737252

Lim, J. H., Hwang, H. J., & Im, C. H. (2013, February). "Eyes-closed" SSVEP-based BCI for binary communication of individuals with impaired oculomotor function. In *2013 International Winter Workshop on Brain-Computer Interface (BCI)* (pp. 79-80). IEEE. 10.1109/IWW-BCI.2013.6506637

Lipke, W. (2003). Schedule is different. *The Measurable News*, 31-34.

Lipke, W. (2002a). A study of the normality of Earned Value Management indicators. *The Measurable News*, (December), 1–16.

Lipke, W. (2002b). Statistical process control of project performance. *Crosstalk*, *13*(March), 16–20.

Lipke, W. (2006a). Earned Schedule Leads to Improved Forecasting. *Proceedings of the 3rd international conference on project management (PROMAC 2006).*

Lipke, W. (2006b). Statistical methods applied to EVM ...the next frontier. *The Measurable News*, (Winter), 18–30.

Lipton, M., & Ravallion, M. (1995). Poverty and policy. Poverty and Policy. Handbook of Development Economics, 3, 2551–2657.

Liptser, R. S., & Shiryayev, A. N. (1977). *Statistics of Random Processes* (Vol. 1). Springer. doi:10.1007/978-1-4757-1665-8

Liu, C., Xie, S., Xie, X., Duan, X., Wang, W., & Obermayer, K. (2018, January). Design of a video feedback SSVEP-BCI system for car control based on improved MUSIC method. In *2018 6th International Conference on Brain-Computer Interface (BCI)* (pp. 1-4). IEEE. 10.1109/IWW-BCI.2018.8311499

Liu, L. & Zhu, K. (2007). Improving cost estimates of construction projects using phased cost factors. *Journal of Construction Engineering and Management, 133*(1).

Liu, T., & Geiger, D. (1999). Approximate tree matching and shape similarity. *Int. Conf. Computer Vision.*

Lorton, A., Fouladirad, M., & Grall, A. (2013). A Methodology for Probabilistic Model-based Prognosis. *European Journal of Operational Research*, *225*, 443–454.

Lovallo, D., & Kahneman, D. (2003). Delusion of Success: How optimism undermines executives' decisions. *Harvard Business Review*, *81*, 56–63. PMID:12858711

Lowe, D. G. (1999, September). *Object recognition from local scale-invariant features.* Paper presented at the International Conference on Computer Vision, Corfu, Greece.

Lowe, D. G. (2001). *Local feature view clustering for 3D object recognition.* Paper presented at the IEEE Conference on Computer Vision and Pattern Recognition, Kauai, HI. 10.1109/CVPR.2001.990541

Lowe, D. G. (1991). Fitting parameterized three-dimensional models to images. *IEEE Transactions on Pattern Analysis and Machine Intelligence*, *5*(13), 441–450. doi:10.1109/34.134043

Mahapatra, G. S., Adak, S., & Kaladhar, K. (2019). A fuzzy inventory model with three parameter Weibull deterioration with reliant holding cost and demand incorporating reliability. *Journal of Intelligent & Fuzzy Systems*, *36*(6), 5731–5744. doi:10.3233/JIFS-181595

Mani, N. B. S., Suri, S., Gupta, S., & Wig, J. D. (2001). Two-phase dynamic contrast-enhanced computed tomography with water-filling method for staging of gastric carcinoma. *Clinical Imaging*, *25*(1), 38–43. doi:10.1016/S0899-7071(99)00134-5 PMID:11435038

Mannella, R., & McClintock, P. V. E. (2012). Itô versus Stratonovich: 30 Years Later. *Fluctuation and Noise Letters*, *11*(1), 1240010–1240019. doi:10.1142/S021947751240010X

Marchal, B., Dedzo, M., & Kegels, G. (2010). Turning around an ailing district hospital: A realist evaluation of strategic changes at Ho Municipal Hospital (Ghana). *BMC Public Health*, *10*(1), 1–16. doi:10.1186/1471-2458-10-787 PMID:21184678

Markman, A. B. (1998). *Knowledge representation*. Lawrence Erlbaum Associates.

Mattia, D., Astolfi, L., Toppi, J., Petti, M., Pichiorri, F., & Cincotti, F. (2016, February). Interfacing brain and computer in neurorehabilitation. In *2016 4th International Winter Conference on Brain-Computer Interface (BCI)* (pp. 1-2). IEEE. 10.1109/IWW-BCI.2016.7457446

Merrow, E. W. (2011). Oil industry megaprojects: our recent track record. *Offshore Technology Conference*. 10.4043/21858-MS

Metz, C. E. (1978). Basic principles of ROC analysis. *Seminars in Nuclear Medicine*, *8*(4), 283–298. doi:10.1016/S0001-2998(78)80014-2 PMID:112681

Metz, C. E. (2008). ROC analysis in medical imaging: A tutorial review of the literature. *Radiological Physics and Technology*, *1*(1), 2–12. doi:10.100712194-007-0002-1 PMID:20821157

Minot, N., & Baulch, B. (2002). *The spatial distribution of poverty in Vietnam and the potential for targeting*. Policy Research Working Paper; 2829.

Mohapatra, S. (n.d.). *Public Health Expenditure and its Effect on Health Outcomes: A New Methodological approach in the Indian Context*. Academic Press.

Mukherjee, S. (2017). Anatomy and Significance of Public Healthcare Expenditure and Economic Growth Nexus in India: Its Implications for Public Health Infrastructure Thereof. In Social, Health, and Environmental Infrastructures for Economic Growth (pp. 120-144). IGI Global.

Müller, K. R. (2016, February). Machine learning for BCI: towards analysing cognition. In *2016 4th International Winter Conference on Brain-Computer Interface (BCI)* (pp. 1-2). IEEE. 10.1109/IWW-BCI.2016.7457453

Müller, K. R. (2014, February). Multimodal imaging, non-stationarity and BCI. In *2014 International Winter Workshop on Brain-Computer Interface (BCI)* (pp. 1-1). IEEE.

Müller-Putz, G. R., Ofner, P., Schwarz, A., Pereira, J., Pinegger, A., Dias, C. L., . . . Sburlea, A. I. (2017, January). Towards non-invasive EEG-based arm/hand-control in users with spinal cord injury. In *2017 5th International Winter Conference on Brain-Computer Interface (BCI)* (pp. 63-65). IEEE.

Nallala, S., Swain, S., Das, S., Kasam, S. K., & Pati, S. (2015). Why medical students do not like to join rural health service? An exploratory study in India. *Journal of Family & Community Medicine*, *22*(2), 111. doi:10.4103/2230-8229.155390 PMID:25983608

Nallaperumal, K. (2018, December). CyberSecurity Analytics to Combat Cyber Crimes. In *2018 IEEE International Conference on Computational Intelligence and Computing Research (ICCIC)* (pp. 1-4). IEEE.

Nam, S., Kim, K. H., & Kim, D. S. (2013, February). Motor trajectory decoding based on fMRI-based BCI—A simulation study. In *2013 International Winter Workshop on Brain-Computer Interface (BCI)* (pp. 89-91). IEEE. 10.1109/IWW-BCI.2013.6506641

Niëns, L. M., & Brouwer, W. B. F. (2013). Measuring the affordability of medicines: Importance and challenges. *Health Policy (Amsterdam)*, *112*(1-2), 45–52. doi:10.1016/j.healthpol.2013.05.018 PMID:23827263

Nurse, E. S., Karoly, P. J., Grayden, D. B., & Freestone, D. R. (2015). A generalizable brain-computer interface (BCI) using machine learning for feature discovery. *PLoS One*, *10*(6), e0131328. doi:10.1371/journal.pone.0131328 PMID:26114954

Obata, N. (1994). White Noise Calculus and Fock Space. *Lecture Notes in Mathematics*, 1577.

Oluwatayo, I. B., & Babalola, M. A. (2020). Asset Ownership and Income as Drivers of Household Poverty In South Africa. *Journal of Developing Areas*, *54*(3). Advance online publication. doi:10.1353/jda.2020.0028

Ornstein, L. S., & Uhlenbeck, G. E. (1930). On the theory of the Brownian motion. *Physical Review*, *36*, 823–841.

Orphanides, C., Akhgar, B., & Bayerl, P. S. (2016, August). Discovering knowledge in online drug transactions using conceptual graphs and formal concept analysis. In *2016 European Intelligence and Security Informatics Conference (EISIC)* (pp. 100-103). IEEE. 10.1109/EISIC.2016.026

Paddock, S. M. (2014). Statistical benchmarks for health care provider performance assessment: A comparison of standard approaches to a hierarchical Bayesian histogram-based method. *Health Services Research*, *49*(3), 1056–1073. doi:10.1111/1475-6773.12149 PMID:24461071

Palomo, J., Ruggeri, F., Rios Insua, D., Cagno, E., Caron, F., & Mancini, M. (2006). On Bayesian forecasting of procurement delays: A case study. *Applied Stochastic Models in Business and Industry*, *22*(2), 181–192. doi:10.1002/asmb.627

Pando, V., San-José, L. A., & Sicilia, J. (2019). Profitability ratio maximization in an inventory model with stock-dependent demand rate and non-linear holding cost. *Applied Mathematical Modelling*, *66*, 643–661. doi:10.1016/j.apm.2018.10.007

Pan, H., Xu, Z., & Huang, J. (2015, October). An effective approach for robust lung cancer cell detection. In *International Workshop on Patch-based Techniques in Medical Imaging* (pp. 87-94). Springer. 10.1007/978-3-319-28194-0_11

Paoli, F., Schmidt, I., Wigzell, O., & Ryś, A. (2019). An EU approach to health system performance assessment: Building trust and learning from each other. *Health Policy (Amsterdam)*, *123*(4), 403–407. doi:10.1016/j.healthpol.2019.02.004 PMID:30777300

Parthasarathy, K. R. (1992). *An Introduction to Quantum Stochastic Calculus*. Birkhäuser.

Patel, H. G., & Sharma, S. N. (2014). Third-order continuous-discrete filtering equations for a non-linear dynamical system, *The ASME Transactions. Journal of Computational and Nonlinear Dynamics*, *9*(3), 034502–034509. doi:10.1115/1.4026064

Perez, L., & Wang, J. (2017). *The effectiveness of data augmentation in image classification using deep learning*. arXiv preprint arXiv:1712.04621

Pergler, M., & Freeman, A. (2008). Probabilistic Modeling as an Exploratory Decision-Making Tool. *McKinsey Working Papers on Risk, 6*, 1-18.

Pillai N, V. (2016). *Panel data analysis with stata part 1 fixed effects and random effects models*. Academic Press.

Pitt, M. K., & Shephard, N. (1999). Filtering via simulations: Auxiliary particle filtering. *Journal of the American Statistical Association*, *94*(446), 590–599. doi:10.1080/01621459.1999.10474153

Pope, A. R., & Lowe, D. G. (2004). Probabilistic models of appearance for 3-D object recognition. *International Journal of Computer Vision*, *2*(40), 149–167.

Prathap, B. R., & Ramesha, K. (2018, February). Twitter sentiment for analysing different types of crimes. In *2018 International Conference on Communication, Computing and Internet of Things (IC3IoT)* (pp. 483-488). IEEE. 10.1109/IC3IoT.2018.8668140

Project Management Institute. (2013). *A Guide to the Project Management Body of Knowledge* (5th ed.). Author.

Protter, P. E. (1991). *Stochastic Integration and Differential Equations*. Springer, Berlin.

Punsawad, Y., & Wongsawat, Y. (2017). Multi-command SSAEP-based BCI system with training sessions for SSVEP during an eye fatigue state. *IEEJ Transactions on Electrical and Electronic Engineering*, *12*, S72–S78. doi:10.1002/tee.22441

Raghupathi, V., & Raghupathi, W. (2015). *Benchmarking hospital performance using health analytics*. Academic Press.

Ralston, A. (1965). *A first course in numerical analysis*. McGraw-Hill Book Company.

Ralston, A., & Rabinowitz, P. (2001). *A First Course in Numerical Analysis* (2nd ed.). Dover Publications.

Ramsey, N. F. (2014, February). Exploration of the brain for optimal placement of BCI implants in paralyzed people. In *2014 International Winter Workshop on Brain-Computer Interface (BCI)* (pp. 1-3). IEEE. 10.1109/iww-BCI.2014.6782543

Rao, Y. S., Saini, H., & Panda, T. C. (2014). *Effect of Cyber Crime Indian Economy*. Academic Press.

Rashid, H., Tanveer, M. A., & Khan, H. A. (2019, July). Skin Lesion Classification Using GAN based Data Augmentation. In *2019 41st Annual International Conference of the IEEE Engineering in Medicine and Biology Society (EMBC)* (pp. 916-919). IEEE. 10.1109/EMBC.2019.8857905

Rathore, S., & Sharma, S. N. (2018). Consensus on the Itô vs Stratonovich dilemma revisited. *IFAC-PapersOnLine*, *51*(1), 719–724. doi:10.1016/j.ifacol.2018.05.121

Rathore, S., Sharma, S. N., Bhatt, D. S., & Nagarsheth, S. H. (2020). Non-linear filering for bilinear stochastic differential systems: A Stratonovich perspective. *Transactions of the Institute of Measurement and Control*, *42*(10), 1–14. doi:10.1177/0142331219895711

Reich, B. H., Gemino, A., & Sauer, C. (2014). How knowledge management impacts performance in projects: An empirical study. *International Journal of Project Management*, *32*(4), 590–602. doi:10.1016/j.ijproman.2013.09.004

Revathi, S., & Suriakala, M. (2018, February). An Intelligent and Novel Algorithm for Securing Vulnerable Users of Online Social Network. In *2018 Second International Conference on Computing Methodologies and Communication (ICCMC)* (pp. 214-219). IEEE. 10.1109/ICCMC.2018.8487760

Revuz, D., & Yor, M. (1991). *Continuous Martingales and Brownian Motion*. Springer-Verlag. doi:10.1007/978-3-662-21726-9

Ribeiro, F. L. (2017). An attempt to unify some population growth models from first principles. *Revista Brasileira de Ensino de Física*, *39*(1), 1–11. doi:10.1590/S0102-47442004000100001

Risken, H. (1984). *The Fokker-Planck Equation: Method of Solutions and Applications*. Springer-Verlag. doi:10.1007/978-3-642-96807-5

Rogers, D. F. (2001). *An Introduction to NURBS with Historical Perspective*. Morgan Kaufmann Publishers.

Rohleder, T. R., Lewkonia, P., Bischak, D. P., Duffy, P., & Hendijani, R. (2011). Using simulation modeling to improve patient flow at an outpatient orthopedic clinic. *Health Care Management Science*, *14*(2), 135–145. doi:10.100710729-010-9145-4 PMID:21152989

Ruangnapakul, N., Salam, Y. D., & Shawkat, A. R. (2019). A Systematic Analysis of Cyber bullying in Southeast Asia Countries. *International Journal of Innovative Technology and Exploring Engineering*, *8*(8S), 104–111.

Rubin-Delanchy, P., Lawson, D. J., & Heard, N. A. (2016). Anomaly detection for cyber security applications. In Dynamic Networks and Cyber-Security (pp. 137-156). doi:10.1142/9781786340757_0006

Rubin, M., Stein, O., Turko, N. A., Nygate, Y., Roitshtain, D., Karako, L., Barnea, I., Giryes, R., & Shaked, N. T. (2019). TOP-GAN: Stain-free cancer cell classification using deep learning with a small training set. *Medical Image Analysis*, *57*, 176–185. doi:10.1016/j.media.2019.06.014 PMID:31325721

Saber, E., Yaowu, X., & Murat Tekalp, A. (2005). Partial shape recognition by sub-matrix matching for partial matching guided image labeling. *Pattern Recognition*, *38*, 1560–1573.

Saduanov, B., Alizadeh, T., An, J., & Abibullaev, B. (2018, January). Trained by demonstration humanoid robot controlled via a BCI system for telepresence. In *2018 6th International Conference on Brain-Computer Interface (BCI)* (pp. 1-4). IEEE. 10.1109/IWW-BCI.2018.8311508

Salehinejad, H., Valaee, S., Dowdell, T., Colak, E., & Barfett, J. (2018, April). Generalization of deep neural networks for chest pathology classification in x-rays using generative adversarial networks. In *2018 IEEE International Conference on Acoustics, Speech and Signal Processing (ICASSP)* (pp. 990-994). IEEE. 10.1109/ICASSP.2018.8461430

Sattar, Z., Riaz, S., & Mian, A. U. (2018, November). Challenges of Cybercrimes to Implementation of Legal Framework. In *2018 14th International Conference on Emerging Technologies (ICET)* (pp. 1-5). IEEE. 10.1109/ICET.2018.8603645

Schindler, M., & Eppler, M. J. (2003). Harvesting project knowledge: A review of project learning methods and success factors. *International Journal of Project Management*, *21*(3), 219–228. doi:10.1016/S0263-7863(02)00096-0

Schumaker, L. L. (2007). *Spline Functions: Basic Theory*. Cambridge Mathematical Library.

Sebastian, T. B., & Klein, P. N. (2003). On aligning curves. *IEEE Transactions on Pattern Analysis and Machine Intelligence*, *25*(1), 116–124.

Sen, A. (1976). Poverty: An Ordinal Approach to Measurement. *Econometrica*, *44*(2), 219–231. doi:10.2307/1912718

Sengupta, R., & Mukherjee, S. (2018). Crime, Deprivation and Social Sustainability—Evidence across States in India. *Indian Journal of Human Development*, *12*(3), 354–377. doi:10.1177/0973703018811173

Shah, N. H., & Naik, M. K. (2018). Inventory model for non – instantaneous deterioration and price- sensitive trended demand with learning effects. *Int. J. Inventory Research.*, *5*(1), 60–77. doi:10.1504/IJIR.2018.092356

Sharma, S. N. (2009). A Kushner approach for small random perturbations of a stochastic Duffing-van der Pol system. *Automatica*, *45*(4), 1097–1099. doi:10.1016/j.automatica.2008.12.010

Sharma, S. N., & Gawalwad, B. G. (2016). Wiener meets Kolmogorov. In *IEEE Conference on Norbert Wiener in the 21st Century (Thinking Machines in the Physical World)*. University of Melbourne.

Sieńko, W., & Citko, W. (2002). *Hamiltonian Neural Net based signal processing*. Paper presented at the International Conference on Signal and Electronic System ICSES, Wrocław – Świeradów Zdrój, Poland.

Sieńko, W., Citko, W., & Wilamowski, B. (2002). *Hamiltonian Neural Nets as a universal signal processor.* Paper presented at the 28th Annual Conference of the IEEE Industrial Electronics Society IECON, Sevilla, Spain. 10.1109/IECON.2002.1182910

Sieńko, W., Citko, W., & Jakóbczak, D. (2004). Learning and system modeling via Hamiltonian Neural Networks. In L. Rutkowski, J. Siekmann, R. Tadeusiewicz, & A. Zadeh (Eds.), *Lecture notes on artificial intelligence: Artificial intelligence and soft computing - ICAISC 2004* (pp. 266–271). Springer - Verlag. doi:10.1007/978-3-540-24844-6_36

Silas, N. (2010). Onyanga, Naftali Omollo- Ongati, Nyakinda Joseph Otula, "On the walraisian – Samuelson price adjustment model. *International Journal of Pure and Applied Mathematics*, *61*(2), 211–218.

Singh, J., Singh, G., & Singh, R. (2017). Optimization of sentiment analysis using machine learning classifiers. *Human-centric Computing and information. The Sciences*, *7*(1), 32.

Singh, P., Hashmi, G., & Swain, P. K. (2018). High prevalence of cesarean section births in private sector health facilities-analysis of district level household survey-4 (DLHS-4) of India. *BMC Public Health*, *18*(1), 613. doi:10.118612889-018-5533-3 PMID:29747609

Singh, S., Shekhar, C., Acharya, R., Moore, A. M., Stillman, M., Pradhan, M. R., ... Sundaram, A. (2018). The incidence of abortion and unintended pregnancy in India, 2015. *The Lancet. Global Health*, *6*(1), e111–e120. doi:10.1016/S2214-109X(17)30453-9 PMID:29241602

Song, M. R., & Lee, S. W. (2018, January). Meta BCI: Hippocampus-striatum network inspired architecture towards flexible BCI. In *2018 6th International Conference on Brain-Computer Interface (BCI)* (pp. 1-3). IEEE.

Song, Y., & Sepulveda, F. (2017, January). An online self-paced brain-computer interface onset detection based on sound-production imagery applied to real-life scenarios. In *2017 5th International Winter Conference on Brain-Computer Interface (BCI)* (pp. 46-49). IEEE. 10.1109/IWW-BCI.2017.7858155

Soussen, C., & Mohammad-Djafari, A. (2004). Polygonal and polyhedral contour reconstruction in computed tomography. *IEEE Transactions on Image Processing*, *11*(13), 1507–1523. doi:10.1109/TIP.2004.836159 PMID:15540458

Sowa, J. F. (2000). *Knowledge representation: logical, philosophical and computational foundations.* Brooks/Cole.

Staggs, V. S., & Gajewski, B. J. (2017). Bayesian and frequentist approaches to assessing reliability and precision of health-care provider quality measures. *Statistical Methods in Medical Research*, *26*(3), 1341–1349. doi:10.1177/0962280215577410 PMID:25788482

Straffin, P. D. (1993). *Game theory and strategy.* Mathematical Association of America.

Stratonovich, R. L. (1963). *Topics in the Theory of Random Noise* (Vol. 1). Gordan and Breach.

Sumner, A. (2007). Meanin versus Measurement: Why Do Economic Indicators of Poverty Still Predominate? *Development in Practice*, *17*(1), 4–13. doi:10.1080/09614520601092485

Syed, J. K., & Aziz, L. A. (2016). Fuzzy Inventory Model without Shortages Using Triangular Fuzzy Numbers and Signed Distance Method. *International Journal of Scientific Research*, *5*(7), 1179–1182.

Tadeusiewicz, R., & Flasiński, M. (1991). *Image Recognition.* PWN.

Tang, K. (2005). Geometric optimization algorithms in manufacturing. *Computer-Aided Design and Applications*, *2*(6), 747–757. doi:10.1080/16864360.2005.10738338

Tarokh, V., Jafarkhani, H., & Calderbank, R. (1999). Space-Time Block Codes from orthogonal designs. *IEEE Transactions on Information Theory*, *5*(45), 1456–1467. doi:10.1109/18.771146

Tatlıdil, H., & Demirağ, İ. (2014). Türkiye'de Yoksulluğun Sosyo-Ekonomik ve Demografik Değişkenlerle İlişkilerinin Lojistik Regresyon ve Mars Yöntemleri Kullanılarak İncelenmesi. *TISK Academy/TISK Akademi*, *9*(17).

Terdik, G. Y. (1990). Stationary solutions for bilinear systems with constant coefficients. In E. Cinlar, K. L. Chug, & R. K. Getoor (Eds.), *Seminar on Stochastic Processes* (pp. 196–206). Birkhäuser. doi:10.1007/978-1-4612-3458-6_12

Thuy, M. B. H., & Hoang, V. T. (2019, December). Fusing of deep learning, transfer learning and gan for breast cancer histopathological image classification. In *International Conference on Computer Science, Applied Mathematics and Applications* (pp. 255-266). Springer.

Tomak, L., & Bek, Y. (2009). Journal of Experimental and Clinical Medicine Deneysel ve Klinik Tıp Dergisi İşlem karakteristik eğrisi analizi ve eğri altında kalan alanların karşılaştırılması. *On Dokuz Mayıs Üniversitesi. Deneysel ve Klinik Tıp Dergisi*, *27*(2), 58–65.

TÜİK. (2008). *Tüketim Harcamaları, Yoksulluk ve Gelir Dağılımı: Sorularla Resmi İstatistikler Dizisi – 6. Türkiye İstatistik Kurumu Matbaası.*

TÜİK. (2011). *Gelir ve Yaşam Koşulları Araştırması (Kesit) 2011 Mikro Veri Seti.* Author.

Udayakumar, R., & Geetha, K. V. (2018). An EOQ model for non-instantaneous deteriorating items with two levels of storage under trade credit policy. *J. Ind. Eng. Int.*, *14*(2), 343–365. doi:10.100740092-017-0228-4

Ullman, S., & Basri, R. (1991). Recognition by linear combinations of models. *IEEE Transactions on Pattern Analysis and Machine Intelligence*, *10*(13), 992–1006. doi:10.1109/34.99234

Wang, M. C., & Uhlenbeck, L. S. (1945). On the theory of the Brownian motion II. *Reviews of Modern Physics*, *17*(2-3), 323–342. doi:10.1103/RevModPhys.17.323

Watson, J. (2002). *Strategy – an introduction to game theory*. University of California Press.

Wax, N. (Ed.). (1954). *Selected Papers on Noise and Stochastic Processes*. Dover Publications, Inc.

Weick, K. E. (1995). *Sense-making in organizations*. SAGE.

Williams, T., Klakegg, O. J., Walker, D. H. T., Andersen, B., & Magnussen, O. M. (2012). Identifying and Acting on Early Warning Signs in Complex Projects. *Project Management Journal*, *43*(2), 37–53. doi:10.1002/pmj.21259

Williams, T., Samset, K., & Sunnevag, K. J. (Eds.). (2009). *Making essential choices with scant information*. Palgrave Macmillan. doi:10.1057/9780230236837

Wodon, Q. T. (1997). Targeting the poor using ROC curves. *World Development*, *25*(12), 2083–2092. doi:10.1016/S0305-750X(97)00108-3

Wolpaw, J. R., Birbaumer, N., McFarland, D. J., Pfurtscheller, G., & Vaughan, T. M. (2002). Brain–computer interfaces for communication and control. *Clinical Neurophysiology*, *113*(6), 767–791. doi:10.1016/S1388-2457(02)00057-3 PMID:12048038

Xie, Y., Xing, F., Kong, X., Su, H., & Yang, L. (2015, October). Beyond classification: Structured regression for robust cell detection using convolutional neural network. In *International conference on medical image computing and computer-assisted intervention* (pp. 358-365). Springer. 10.1007/978-3-319-24574-4_43

Yadav, D., Singh, S. R., & Kumari, R. (2015). Retailer's optimal policy under inflation in fuzzy environment with trade credit. *International Journal of Systems Science*, *46*(4), 754–762. doi:10.1080/00207721.2013.801094

Yadav, S., Timbadia, M., Yadav, A., Vishwakarma, R., & Yadav, N. (2017, April). Crime pattern detection, analysis & prediction. In *2017 International conference of Electronics, Communication and Aerospace Technology (ICECA)* (Vol. 1, pp. 225-230). IEEE. 10.1109/ICECA.2017.8203676

Yang, J., Zhao, J. X., Cao, Q., Hao, L., Zhou, D., Gan, Z., Ji, L.-N., & Mao, Z. W. (2017). Simultaneously inducing and tracking cancer cell metabolism repression by mitochondria-immobilized rhenium (I) complex. *ACS Applied Materials & Interfaces*, *9*(16), 13900–13912. doi:10.1021/acsami.7b01764 PMID:28368110

You, P. S. (2003). Dynamic pricing of inventory with cancellation demand. *The Journal of the Operational Research Society*, *54*(10), 1093–1101. doi:10.1057/palgrave.jors.2601619

Zhang, J., Hu, H., Chen, S., Huang, Y., & Guan, Q. (2016, December). Cancer cells detection in phase-contrast microscopy images based on Faster R-CNN. In *2016 9th International Symposium on Computational Intelligence and Design (ISCID)* (Vol. 1, pp. 363-367). IEEE. 10.1109/ISCID.2016.1090

Zhang, X., Yao, L., Wang, X., Monaghan, J., Mcalpine, D., & Zhang, Y. (2019). *A survey on deep learning based brain computer interface: Recent advances and new frontiers*. arXiv preprint arXiv:1905.04149

Zhang, D., & Lu, G. (2004). Review of Shape Representation and Description Techniques. *Pattern Recognition*, *1*(37), 1–19.

Zhou, D. (2012). Grey Verhulst Model Based on BP Neural Network Optimization for Oil Production Forecasting. *Int. J. Energy Sci. IJES IJES*, *2*(3), 115–118.

Zweig, M. H., & Campbell, G. (1993). Receiver-operating characteristic (ROC) plots: A fundamental evaluation tool in clinical medicine. *Clinical Chemistry*, *39*(4), 561–577. doi:10.1093/clinchem/39.4.561 PMID:8472349

About the Contributors

Dariusz Jacek Jakóbczak was born in Koszalin, Poland, on December 30, 1965. He graduated in mathematics (numerical methods and programming) from the University of Gdansk, Poland in 1990. He received the Ph.D. degree in 2007 in computer science from the Polish – Japanese Institute of Information Technology, Warsaw, Poland. From 1991 to 1994 he was a civilian programmer in the High Military School in Koszalin. He was a teacher of mathematics and computer science in the Private Economic School in Koszalin from 1995 to 1999. Since March 1998 he has worked in the Department of Electronics and Computer Science, Technical University of Koszalin, Poland and since October 2007 he has been an Assistant Professor in the Chair of Computer Science and Management in this department. His research interests tie mathematics with computer science and include computer vision, shape representation, curve interpolation, contour reconstruction and geometric modeling, probabilistic methods and discrete mathematics.

* * *

Sugapriya C. obtained her B.Sc(maths) M.Sc (maths) M.phil(maths) and Ph.d from leading universities in Tamilnadu, India. Presently she is working as Assistant Professor in Queen Mary's College Chennai, Tamilnadu,India. She published more than twenty Research paper in international conferences and journals. Her area of interest includes Optimization techniques, Inventory control, Fuzzy logic and Neural network. She is active reviewer for many reputed journals and conferences.

Franco Caron, Professor with the Management, Economics and Industrial Engineering Department at Politecnico di Milano, he is in charge of the course of "Management of Large Engineering Projects" both in the Systems Engineering and Industrial Engineering Programs. He is also in charge at MIP-Politecnico of the course of Project Risk Analysis and Management both in the Master in Project Management (MPM) and the Master in Strategic Project Management - developed jointly by MIP-Politecnico di Milano, Heriot Watt University Edimburgh and UMEA University (Sweden). Member of PMI Northern Italy Chapter.

Ahan Chatterjee is an Engineering Student, in the Department of Computer Science and Engineering, The Neotia University, Sarisha. His research interests are in the areas of Machine Learning, Deep Learning, Applied Data Analytics in Healthcare, and Financial Market fields. Besides, he is a member of the International Association of Engineers (IAENG), Hong Kong. He has worked in several organizations namely, CSIR-CDRI, GoOffer Hyperlocal Pvt. Ltd., Research Guruji, and Daten & Wissen Pvt. Ltd as mainly Research Intern and Data Science Intern. He has contributed research papers in reputed journals at home and abroad, and also in edited books in different domains of computer science and data analytics published by Springer, IGI Global. Moreover, he has presented papers in 8 International Conferences and holds 1 Best Paper Award in an International Conference.

Nagarajan Deivanayagampillai did his doctorate from Manonmaniyam Sundaranar University. He has published more than 70 research papers in reputed journals. He got one patent grant. He is working as a professor in the Department of Mathematics, Hindustan Institute of Technology and science, India.

Hakan Demirgil is an Associate Professor Doctor on Econometrics at the Süleyman Demirel University Faculty of Economics specializing in applied econometrics. Since 2009, he has taken the position of the head of the Department of Econometrics. He attended Faculty of Economics at the İstanbul University. He obtained both his Master and PhD in Economics from the Süleyman Demirel University where he focused on firm's survival and growth performance due to their innovation capabilities. His research interest include: the firm performance, binary choice models, sample selection bias, volatility spillovers and social network analysis. Demirgil is an author of three books and, and is the author and co-author of over 20 articles in refereed journals, book chapters and proceedings; he supervised more than 10 graduate students.

Yash Gupta is a computer engineering student in The Neotia University. He has intermediate skills in programing in java and c programming languages. He has completed his 10 and 10+2 studies and is ambitious about his future in software development.

Jeganathan Kathirvel is currently Assistant Professor of Ramanujan Institute for Advanced Study in Mathematics at University of Madras, Chepauk, Chennai, India. He received the Ph.D degree in Mathematics for Stochastic Queueing-Inventory Modelling at Alagappa University, Karaikudi, India. He holds his M. Phil. in Mathematics from Madurai Kamaraj University, Madurai, India and M. Sc. in Mathematics from Cardamom Planters Association College, Bodinayakanur, India. He has published many papers in reputed journals like OPEARCH, International journal of Applied and Computational Mathematics, Mathematics and Computers in Simulations, etc. His research interests include Stochastic Modelling, Optimization Techniques, Inventory and Queueing systems.

Aniruddha Mandal is a diligent and competent student at The Neotia University, pursuing his career as a B.Tech student in the field of Computer Science and Data Analytics. He is a highly creative and dynamic student throughout his early education life and currently a freshman in his undergraduate course. He has programming experience in Java, C and Python language, along with a knowledge in HTML5 and has skills in working with Arduino based project. He has attended several workshops in premier institutes like IIT Kharagpur and Jadavpur University on Cyber Security and Machine Learning. His field of interest is Machine Learning, Data Science, Deep Learning and Network Security.

Shaival H. Nagarsheth received the B.E. degree in Instrumentation & Control from Sarvajanik College of Engineering & Technology, Surat, Gujarat, India in 2014, with a gold medal. In 2016 he received his M. Tech degree in Instrumentation & Control with specialisation in Control & Automation from Nirma University, Ahmedabad, Gujarat, India, with a gold medal. He is currently pursuing his doctoral degree in control theory from the Electrical Engineering Department at Sardar Vallabhbhai National

Institute of Technology, Surat, Gujarat, India. His research includes matrix calculus, controller design for multivariable systems, sensitivity integrals, fractional control design. He was a recipient of EECI (European Embedded Control Institute) Overseas Grant amongst the ten research scholars all around the world. He was awarded an amount of 500 Euros for attending 2019-IGSC (International Graduate School on Control) held at Automatic Control Laboratory, ETH Zürich, Zürich, Switzerland. He is a member of the IEEE Young Professionals, IEEE student member, IEEE Control System Society, IEAE student member, ISA student member, IFAC Affiliate, IFAC-ACDOS (NMO) member.

Aditi Priya completed her schooling from kendriya Vidyalaya and higher secondary schooling from jawahar vidya mandir and later she joined The Neotia University to pursue her career in computer science engineering with specialization In cyber security. She has performed fairly well all through her school life and is in the 1st year of Btech programme. She has a strong grasp over c++ language and she is also interested in the field of deep learning and machine learning.

Sandhya Rathore completed her Phd program from SVNIT,Surat, India. Her specialization in in application of stochastic theory to power electronics circuit. Currently she is teaching in Sarvajanik College of Engg and technology since 1998. Her other interests are in Electrical machine their design performance analysis and control.

Swagatam Roy is currently pursuing the third year of engineering at The Neotia University in the Department of Computer Science Engineering with a specialization in Data Analytics. His performance is good in college and the CGPA is increasing in the consecutive semesters. He has good knowledge in C, C++ with good application skills in python. Also, 2 papers of him had been published in a renounced journal. He is a member of IAENG, Hong Kong, along with that the best paper awardee in an International Conference. His area of interest lies in Data Analytics, machine learning, deep learning, artificial intelligence.

Shambhu N. Sharma received the B.E. degree from Government College of Engineering, Rewa (M.P.), India in 1994, the M Tech degree from Banaras Hindu University (Now IIT BHU), UP, India in 2000, and the PhD degree from Delhi University in 2007. Currently, he is working as a Professor in the Electrical Engineering Department of the National Institute of Technology, Surat, India. He specialises in the areas of stochastic systems, control theory, stochastic filtering, and stochastic differential equations with applications to electrical and electronic networks. One of his works is known as a pioneering work in stochastic systems. In addition to these, he is also interested in multivariable system theory as well as non-linear system theory. He had a visiting academic appointment at the Department of Systems and Control of Jožef Stefan Institute, Ljubljana, the Republic of Slovenia under the joint program of the Indian National Science Academy (INSA, New Delhi) and the Slovenian Academy of Sciences and Arts (SASA, Ljubljana).

Shruti Sinha is currently a student in The Neotia University, pursuing Computer Science Engineering with a specialization of Cyber Security. She is a member of International Association of Engineers (IAENG). She is a talented Engineering student with industrious and systematic approach to learning information. She is an open and clear worker with disciplined execution and methodical nature having extensive knowledge of Java and Python and Data Handling skills. She is interested in the field of Cyber Security, Artificial Intelligence and Open Source.

Trisha Sinha completed her initial studies from a convent school and later joined The Neotia University to pursue her career in Robotics Engineering. She has performed fairly well all through her school life and is in the penultimate year of her B.Tech Programme. She has a strong grasp over C and C++ language, along with some handy skill in Arduino programming and working with various sensors. She has interest in the fields of Automation, Deep Learning, Machine Learning. She has co-authored a paper which has been published in IJEAT(a Scopus Indexed Journal).

Kuppulakshmi V. is a PhD Scholar (Full Time) PG and Research Department of mathematics Queen Mary's College, Chennai. Attended the national and international conferences and published journals in the field inventory management and Fuzzy Environment.

Index

www.ingramcontent.com/pod-product-compliance
Lightning Source LLC
LaVergne TN
LVHW060419300926
849162LV00006B/1466